INTERNATIONAL SERIES OF MONOGRAPHS IN
SOLID STATE PHYSICS
GENERAL EDITORS: N. KURTI AND R. SMOLUCHOWSKI

VOLUME 7

STRUCTURE AND PROPERTIES OF INORGANIC SOLIDS

STRUCTURE AND PROPERTIES OF INORGANIC SOLIDS

BY

FRANCIS S. GALASSO
United Aircraft Research Laboratories

ILLUSTRATED BY

W. DARBY

PERGAMON PRESS
Oxford · New York · Toronto
Sydney · Braunschweig

Pergamon Press Ltd., Headington Hill Hall, Oxford
Pergamon Press Inc., Maxwell House, Fairview Park, Elmsford, New York 10523
Pergamon of Canada Ltd., 207 Queen's Quay West, Toronto 1
Pergamon Press (Aust.) Pty. Ltd., 19a Boundary Street, Rushcutters Bay, N.S.W. 2011, Australia
Vieweg & Sohn GmbH, Burgplatz 1, Braunschweig

First edition 1970

Library of Congress Catalog Card No. 70-104123

PRINTED IN GREAT BRITAIN BY PAGE BROS. (NORWICH) LTD.
08 006873 1

CONTENTS

PREFACE

THIS is an up-to-date reference book written to meet the needs of teachers of advanced undergraduate and graduate courses and of researchers in the various disciplines which make up the field of materials sciences. It describes the structure of metals, intermetallics, halides, hydrides, carbides, borides and other inorganic phases as well as some of their properties of current interest. This book was not intended to be as complete as such excellent and thorough texts as *Structural Inorganic Chemistry* by Alexander Wells and *Crystal Structures*, Vols. I, II and III, by Professor Ralph G. Wycoff, which deal primarily with structures. It is rather intended as mainly a survey of structures of those solids which have properties of considerable contemporary interest. Some of the structures and properties which fall into this category are Cu_2AlMn, spinel, magnetoplumbite and garnet structures (ferromagnetic properties), fluorite and scheelite structures (laser properties), perovskite structure (ferroelectric properties), the β-tungsten structure (superconducting properties) and the zinc blende structure (semiconducting properties).

In order to simplify and systematize the presentation of structures, I divided them into related groups, describing the simple ones first, and using these in describing the more complex structures. By placing similar structures together, it is possible to visualize a number of them as simple variations of a common one. Several figures of each structure, always using the convention described in the introduction, are included with the descriptions.

Because this book was written for readers with diverse backgrounds such as engineering, chemistry, metallurgy, physics, ceramics and mineralogy, I felt that background information on crystallography and properties should be presented in Chapter 1 to assist the readers who may need it to understand some of the data and topics covered in subsequent chapters. Each of these chapters begins with a brief description of the related structures. For each structure the name, a common compound with the structure, the structure type, the space group, and the symmetry are given. A simple description of the structure is presented with several figures, the atomic positions are listed for the reader with a knowledge of crystallography, unit cell data for phases with the structure are tabulated and a discussion of the properties of more important materials is given. For some structures physical constants of compounds are tabulated. The properties of elemental solids are treated very briefly because they are generally well known. On the other hand, properties of compounds are considered in greater detail. The structure and properties of intermetallics, borides, carbides, halides, hydrides and oxides are discussed in the appropriate chapters and they are summarized in the last chapter. The properties of a number of other phases not covered in Chapters 1–12 are listed in the Appendix.

I am grateful to those responsible for my writing this book and for their help in deciding what should be included. I became convinced that a book which emphasized structures and properties of materials of current interest and use would be of great value. Professor Martin Buerger influenced the decision to present these structures in a systematic manner and Professor Roman Smoluchowski suggested the emphasis on properties.

I am grateful to members of the United Aircraft Research Laboratories for helping me prepare basic discussions of the properties found in the introduction. These scientists are Drs. George Yntema and Fred Otter, Jr. (superconductivity), Mr. Frank Douglas (dielectric properties), Dr. Alexander Shuskus (magnetic properties), Dr. Michael Brienza (optical properties) and Drs. Earl Thompson, Robert Sierakowski and Kenneth Kreider (mechanical properties). I owe Dr. Gerry Peterson a special debt of gratitude for reviewing these discussions and I must thank Professor Lewis Katz of the University of Connecticut, Dr. Michael Kestigian of Sperry-Rand Research Laboratories and Dr. John Longo of Lincoln Laboratories for each reviewing several chapters.

I thank my daughter, Cynthia, and Wilda Darby for their assistance in collecting the data, United Aircraft Research Laboratories personnel, Kathy Donahue, Joyce Hurlburt and Bonnie True and my wife, Lois, for preparing the manuscript and Nancy Letendre, Carol Begansky and Willie Jenkins for helping Wilda Darby with the illustrations.

CHAPTER 1

INTRODUCTION

THIS chapter provides an introductory treatment of crystallography and of material properties as background for those readers who may not be familiar with these topics. This simplifies the presentation of data and discussions in the following chapters.

1.1. Crystallography

Lattice and Unit Cells

The materials considered in this book are crystalline which means that atoms in each crystal are arranged in a three-dimensional periodic pattern in space. Starting at any point in the crystal and moving in any direction, points with the same environment as the original point will be found. These points will be evenly spaced apart along each particular direction. A three-dimensional set of such points obtained by using three non-coplanar directions originating from one point is called a Bravais lattice. All crystals are based on one of the fourteen Bravais lattices shown in Fig. 1.1a.

The three directions and the repeat distance along each one of them determine a parallelepiped which reproduced indefinitely will fill out the space. These directions called crystalline axes are usually chosen to coincide with axes of high symmetry (see below) although this is not possible for the triclinic system and for two directions in the monoclinic system. The smallest parallelepiped thus defined is called the unit cell. In the face-centered, body-centered and rhombohedral systems the preferred unit cell is larger than the minimum Bravais cell because the latter is more difficult to visualize and use. The unit cell has dimensions a, b and c and angles α, β and γ representing the angles between b and c, c and a, and a and b respectively. Table 1.1a presents data which are used to describe unit cells of various types.

Symmetry

Symmetry operations are essential in describing crystalline structures. In many instances, for example, only half the structure need be identified because the other half is its mirror image. Thus, it is said that there is a mirror plane between these halves.

A unit cell is said to have a n-fold rotation axis if a rotation of $360/n$ degrees around this axis brings it to an identical place in the cell. A tetragonal unit cell is said to have a fourfold rotation axis perpendicular to one set of faces. A hexagonal unit cell has one sixfold rotation axis. Some of these symmetry operations are illustrated in Fig. 1.1b and the minimum symmetry for each system is given in Table 1.1a.

Since symmetry operations are used only sparingly in this book, no further description will be given.

Planes and Directions

The orientation of a plane is given by its Miller index which is defined as the

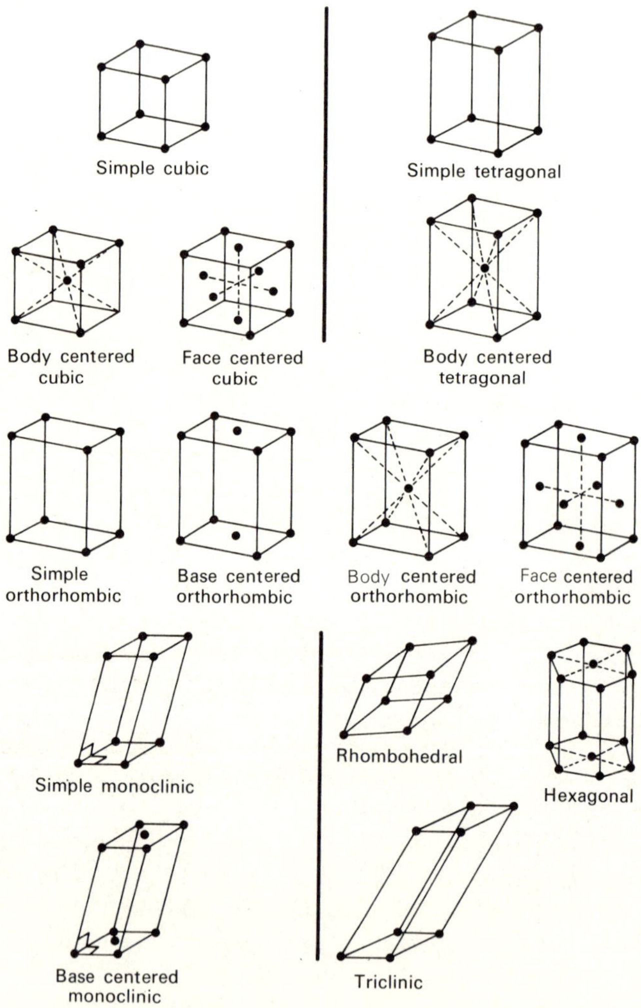

FIG. 1.1a The Bravais lattices

TABLE 1.1a. *Crystal Systems*

System	Unit cell dimensions	Lattice type	Minimum symmetry
Cubic = isometric	$a = b = c; \alpha = \beta = \gamma = 90°$	P, I, F	Four 3-fold rotation axes (four 3 or $\bar{3}$)
Tetragonal	$a = b \neq c; \alpha = \beta = \gamma = 90°$	P, I	One 4-fold rotation (or rotation inversion) axis (4 or $\bar{4}$)
Orthorhombic	$a \neq b \neq c; \alpha = \beta = \gamma = 90°$	P, I, F, C	Three perpendicular 2-fold rotation (or rotation inversion) axes (222 or $\overline{222}$)
Monoclinic	$a = b \neq c; \alpha = \gamma = 90° \neq \beta$	P, I	One 2-fold rotation (or rotation-inversion) axis (2 or $\bar{2}$)
Triclinic	$a \neq b \neq c; \alpha \neq \beta \neq \gamma$	P	None
Hexagonal	$a = b \neq c; \alpha = \beta = 90°, \gamma = 120°$	P	One 6-fold rotation (or rotation inversion) axis (6 or $\bar{6}$)
Rhombohedral	$a = b = c; \alpha = \beta = \gamma \neq 90°$	P	One 3-fold rotation (or rotation inversion) axis (3 or $\bar{3}$)

set of reciprocals of the fractional intercepts which the plane makes with the crystallographic axes. The intercepts have to be expressed in terms of the unit distances *a*, *b* and *c* along the three axes. In the cubic system the [*hkl*] direction is perpendicular to the (*hkl*) plane. Examples of some planes and directions are shown in Fig. 1.1c.

Crystal Structure

The atoms in a structure are present in space either on the points of a Bravais lattice or in some fixed relation to these points. The atomic positions are given in terms of fractions *x*, *y*, *z* of the unit cell edges *a*, *b*, *c* in directions parallel to these edges. In this book the origin of the unit cell is always taken at the bottom, back left-hand corner of cubic, tetragonal and orthorhombic cells and the back bottom corner of hexagonal cells. The values *x*, *y* and *z* are taken from the origin.

When the number of atoms in a particular unit cell is being considered, it must be remembered that there are unit cells around it which may share some of the atoms. For example, an atom at the origin is at 0,0,0, but because this atom is shared by eight other unit cells, only one-eighth of the atom belongs to a single unit cell. Thus an atom in the face of a unit cell is also in

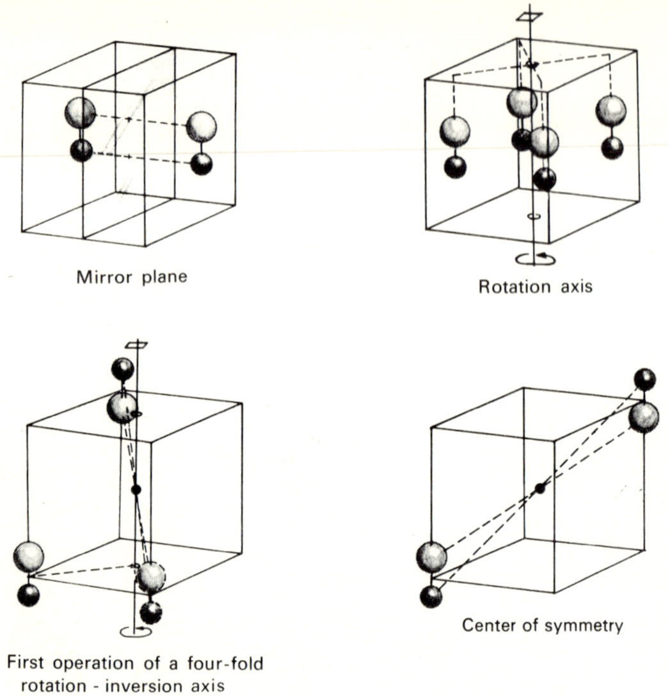

FIG. 1.1b Symmetry elements

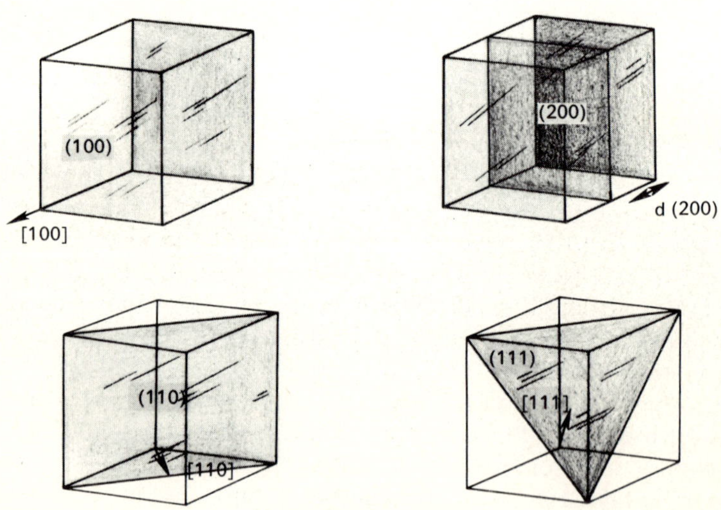

FIG. 1.1c Miller indices

the opposite face and an atom at an edge is also on three other edges. Examples are shown in Fig. 1.1d.

In this book the *x*, *y* and *z* values for each atom in the unit cell are given unless the number of atoms is very large. In trying to visualize structures from data for complex structures and for those presented in other books, there are several factors which should be noted. If a unit cell has a center of symmetry, then only half the atomic positions are sometimes given. The other atomic positions are obtained by changing the sign of each value of *x*, *y* and

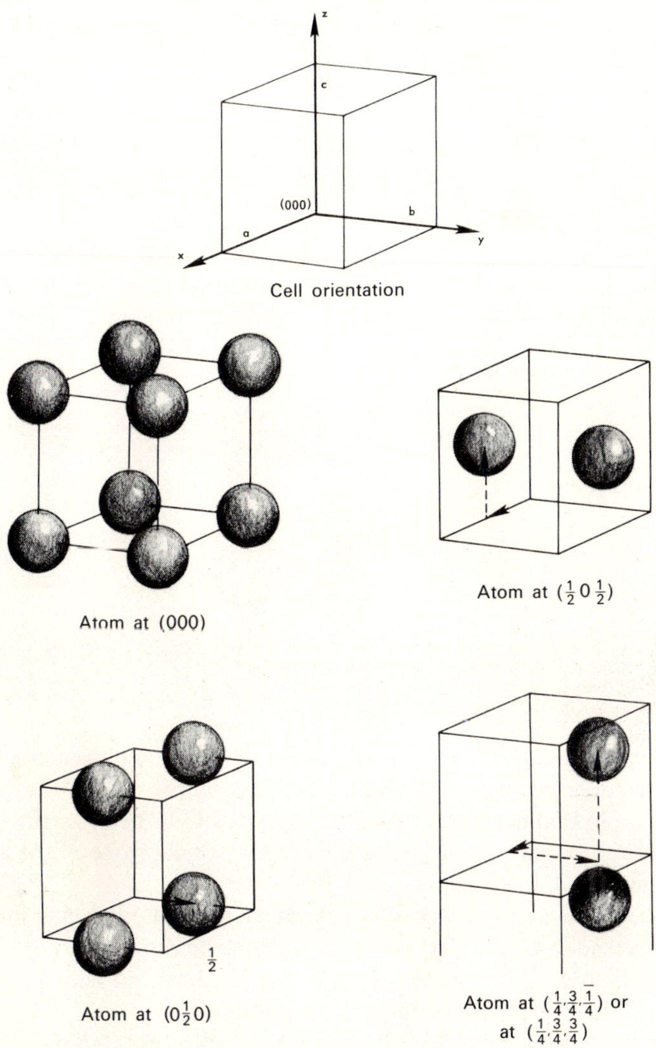

FIG. 1.1d Atomic positions

z. For example, if an atom is located at $\bar{x}$, y, $\bar{z}$, and another one at x, $\bar{y}$, z, then these two positions can be written as $\pm(x, \bar{y}, z)$.

Sometimes, x, y and z values will be given which are to be added to each x', y' and z' value listed below them. For example, if 0,0,0; $\frac{1}{2}, \frac{1}{2}, \frac{1}{2}$ is written above $\frac{1}{4}, \frac{1}{4}, \frac{1}{4}$, this means that there are two atoms, one at $\frac{1}{4}, \frac{1}{4}, \frac{1}{4}$ and the other at $\frac{3}{4}, \frac{3}{4}, \frac{3}{4}$. Since these positions (0, 0, 0; $\frac{1}{2}, \frac{1}{2}, \frac{1}{2}$) are positions in a body-centered lattice, B.C. is sometimes written instead of the two positions or F.C. instead of (0, 0, 0; $\frac{1}{2}\frac{1}{2}$, 0; $\frac{1}{2}$, 0, $\frac{1}{2}$; 0, $\frac{1}{2}, \frac{1}{2}$) for positions in a face-centered lattice.

Another way of presenting the atomic positions is by giving the space group symbol or number and the letter for each group of equivalent positions. The letter and positions are listed in the *International Tables for X-ray Crystallography*, Vol. 1, The Kynock Press, Birmingham, England, 1952.

Fundamental Configurations

Within many structures groups of atoms form distinct configurations. Two of the most common configurations or polyhedra are the tetrahedron and the octahedron. A tetrahedral configuration of atoms is formed by placing three atoms close packed together in a triangle and then setting another atom on top (see Fig. 1.1e). If these atoms are large, the hole in the center can be

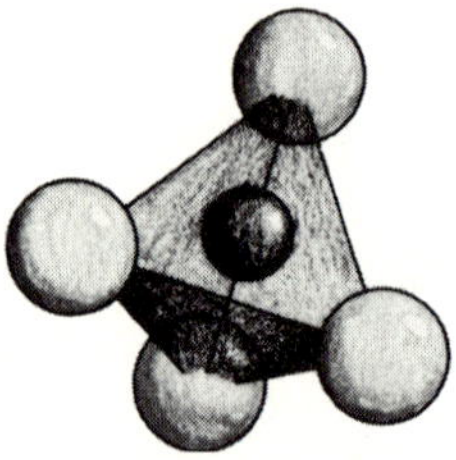

FIG. 1.1e Tetrahedral arrangement of atoms of one element about the atom of another element

occupied by a small atom. Such a tetrahedral hole is found in each corner of a face-centered cubic lattice. The hole is situated in the center of each octant of the cube. In this book this position is referred to as being *in* the corner of the cube as opposed to an atom *at* the corner, which is one of the apexes of the tetrahedra.

An octahedral arrangement of atoms can be formed by placing four atoms in a square and setting one atom above and one below it (Fig. 1.1f). The hole in this configuration is larger than the one in the tetrahedral arrangement of atoms and therefore must be filled by larger atoms.

The coordination number, i.e. the number of nearest neighbor atoms within a tetrahedron is four, and it is six for an octahedron.

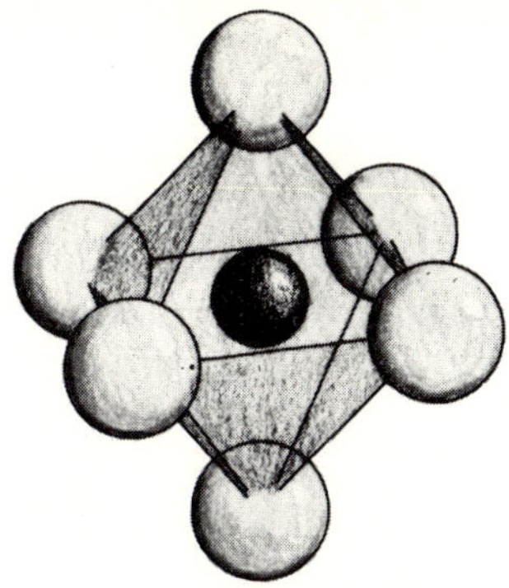

FIG. 1.1f Octahedral arrangement of atoms of one element about the atom of another element

Atomic and Ionic Radii and the Periodic Table

The structures of many compounds depend upon the radii and the ratio of radii of atoms and ions. Unfortunately there is no rigorous way of measuring or even defining these radii and thus several systems are in use. The radii are listed in Table 1.1b.

Certain trends in the radii can be noted by referring to the periodic table (Fig. 1.1g) which for convenience has been divided into A_1 the alkali and alkaline earth elements, A_2 the transition elements and B_1 and B_2 the post-transition metals. In general, the radii of the elements decrease from left to

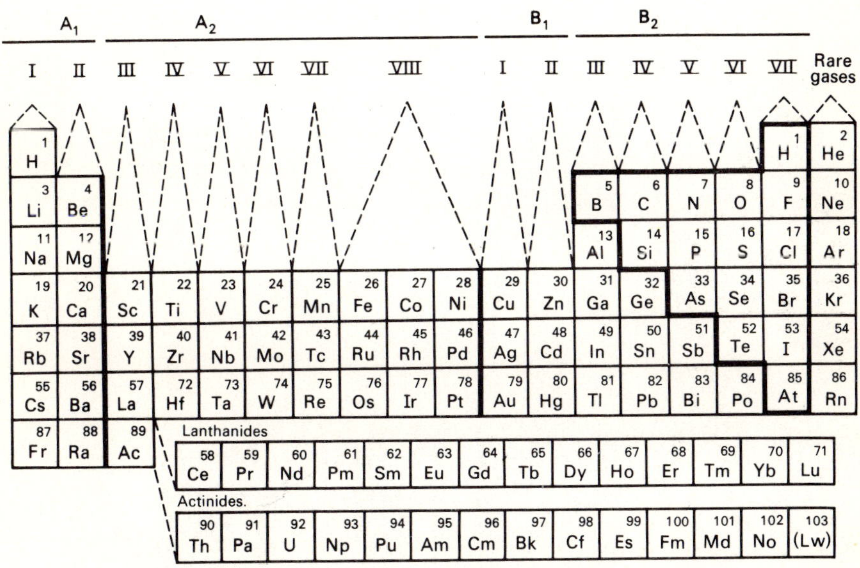

FIG. 1.1g Periodic chart of the elements

TABLE 1.1b. *Atomic and Ionic Radii* (Å)

Element	Atomic radius[1]	Metallic radius[2]	Valence	Ionic radius: Goldschmidt[3]	Ionic radius: Pauling[4]	Ionic radius: Ahrens[5]	Ionic radius: Other
Ac	1.95		+3			1.18	
Ag	1.60	1.44	+1	1.13	1.26	1.26	
			+2			0.89	
Al	1.25	1.43	+3	0.57	0.50	0.51	
Am	1.75		+3			1.07	
			+4			0.92	
Ar							
As	1.15	1.25–1.57	−3		2.22		
			+3			0.58	
			+5		0.47	0.46	
At			+7			0.62	
Au	1.35	1.44	+1		1.37	1.37	
			+3			0.85	
B	0.85	~1.0	+1			0.23	
			+3		0.20		
Ba	2.15	2.17	+2	1.43	1.35	1.34	
Be	1.05	1.1–1.14	+2	0.34	0.31	0.35	
Bi	1.60	1.55–1.74	+1		0.74		
			+3			0.96	
			+5			0.74	
Br	1.15		−1	1.96	1.95		1.80[6, 10]
			+5			0.47	
			+7		0.39	0.39	
C	0.70	0.71–0.77	−4		2.60		
			+4	0.20	0.15	0.16	
Ca	1.80	1.97	+2	1.06	0.99	0.99	
Cd	1.55	1.49–1.64	+2	1.03	0.97	0.97	
Ce	1.85	1.8248[8]	+1		1.01		
			+3	1.18		1.07	1.034[7]
			+4	1.02		0.94	
Cl	1.00		−1	1.81	1.81		1.64,[6] 1.65[10]
			+5			0.34	
			+7		0.26	0.27	
Co	1.35	1.25	+2	0.82		0.72	
			+3			0.63	
Cr	1.40	1.25	+3	0.65		0.63	
			+6	0.34–0.4	0.52	0.52	
Cs	2.60	2.63	+1	1.65	1.69	1.67	1.86,[6] [1.80[10]
Cu	1.35	1.28	+1		0.96	0.96	
			+2			0.72	

TABLE 1.1b (*cont.*)

Element	Atomic radius[1]	Metallic radius[2]	Valence	Ionic radius			
				Goldschmidt[3]	Pauling[4]	Ahrens[5]	Other
Dy	1.75	1.7952[8]	+3			0.92	0.908[7]
Er	1.75	1.7794[8]	+3	1.04		0.89	0.881[7]
Eu	1.85	1.994[8]	+2				1.09[9]
			+3			0.98	0.950[7]
F	0.50		−1	1.33	1.36		1.16,[6] 1.19[10]
			+7		0.07	0.08	
Fe	1.40	1.24	+2	0.83		0.74	
			+3	0.67		0.64	
Fr			+1			1.80	
Ga	1.30	1.22–1.40	+3	0.62	0.62	0.62	
Gd	1.80	1.810[8]	+3	1.11		0.97	0.938[7]
Ge	1.25	1.22	−4		2.72		
			+2			0.73	
			+4	0.53		0.53	
H	0.25		−1		2.08		
Hf	1.55	1.57–1.60	+4			0.78	
Hg	1.50	1.50	+2	1.12	1.10	1.10	
Ho	1.75	1.7887[8]	+3			0.91	0.894[7]
I	1.40		−1	2.20	2.16		2.05,[6] 2.01[10]
			+5			0.62	
			+7		0.50	0.50	
In	1.55	1.62–1.69	+3	0.92	0.81	0.81	
Ir	1.35	1.35	+4	0.66		0.68	
K	2.20	2.31	+1	1.33	1.33	1.33	1.49,[6] 1.51[10]
La	1.95	1.36–1.87, 1.8852[8]	+3	1.22	1.15	1.14	1.061[7]
Li	1.45	1.52	+1	0.78	0.60	0.68	0.94,[6] 0.90[10]
Lu	1.75	1.7516[8]	+3	0.99		0.85	0.848[7]
Mg	1.50	1.60	+2	0.78	0.65	0.66	
Mn	1.40	1.23–1.48	+2	0.91		0.80	
			+3	0.52		0.66	
			+4			0.60	
			+7		0.46	0.46	
Mo	1.45	1.36	+4	0.68		0.70	
			+6		0.62	0.62	
N	0.65		−3		1.71		
			+3			0.16	
			+5	0.1–0.2	0.11		

right in the table. In the 18-member periods, there are slight increases in the radii at the end of the A_2 elements. The decrease can be attributed to increasing positive charge in the nucleus pulling the electrons in closer, and the slight increase at the end of the A_2 metals can be explained as being due to screening of the nuclear charge by the d-electrons. Within a group, the atomic radii increase from the top to the bottom element.

The positive ions are smaller and the negative ions are larger than the corresponding neutral atoms. In the A_1 elements, it is easy to compare the ionic radii, since each group has one valence. In the transition metal series, on the other hand, many valence states are possible, and in the B_1 group there is more than one valence state possible for most elements. However, the radii of ions with isoelectronic structures follow the same trends as found for atomic radii.

Because of the decrease in the radii of elements which follow lanthanum (the lanthanide contraction), Zr^{4+} is approximately the same size as Hf^{4+} and Nb^{5+} and Ta^{5+} are about the same size. Normally, because of the decrease in size of ions for elements from left to right in the periodic table and from bottom to top in a group, ions of similar size are found on diagonals in the periodic table. These trends are useful in understanding the structure and properties of materials.

1.2. Properties

In this introductory treatment of the thermal, electrical, magnetic, dielectric, optical and mechanical properties emphasis is placed on properties of current interest, such as superconductivity, ferroelectricity, ferromagnetism and laser properties. Materials which exhibit these properties are given particular attention in the following chapters.

Thermal Properties

Thermal conductivity. The heat flux Q per unit area and unit of time caused by a temperature gradient, dT/dx is given by

$$Q = K\frac{dT}{dx}$$

where K is the thermal conductivity. Materials with larger K are therefore better thermal conductors. At room temperature metals conduct one or two orders of magnitude better than dielectrics while alloys do not conduct as well as pure metals. For example, the thermal conductivity of Cu is 0.94 cal/cm-sec-°C, NaCl is 0.017 cal/cm-sec-°C and Fe-2% Ni is ~0.1 cal/cm-sec-°C.

Kinetic theory of gases indicates that K is proportional to the number of heat carriers, their velocity and mean free path and to the specific heat. The carriers are either electrons or phonons (quantized lattice vibrations). In pure metals the thermal current is carried mostly by electrons, in dielectrics by phonons whereas in alloys both can make significant contributions.

In insulators at low temperatures ($T \ll \theta_D$, the Debye temperature) the phonon mean free path is limited by crystalline defects or sample boundaries. At higher temperatures the non-linearities (anharmonicity) in lattice vibrations become more important. Above $T \sim \theta_D/2$ the dominant resistance mechanisms are the Umklapp processes in which two phonons combine to make one propagating in an altogether different direction.

In the metals the mean free path of the electrons is limited primarily by scattering from lattice vibrations. Since the same carriers transport both heat and electrical current, there is a fixed ratio between the thermal and electrical conductivities (Wiedemann–Franz law),

$$K = L\sigma T$$

where σ is the electrical conductivity and L, the Lorenz number, $= 2.5 \times 10^{-8}$ watt/ohm/deg^2. In alloys disorder scattering often determines the mean free path for both electrons and phonons.

Electrical Properties

Electrical conductivity. Materials can be divided into conductors, semiconductors and insulators. The resisitivity of metals at room temperature is approximately 10^{-6} ohm-cm or higher; semiconductors have still higher resistivities; and insulators can have resistivities as high as 10^{18} ohm-cm. The resistivities of conductors increase with increasing temperature, while those of semiconductors and insulators decrease with temperature. The huge range of resistivities of solids is best understood in terms of their electronic energy diagrams. As shown in Fig. 1.2a, in metals there is a partly filled band

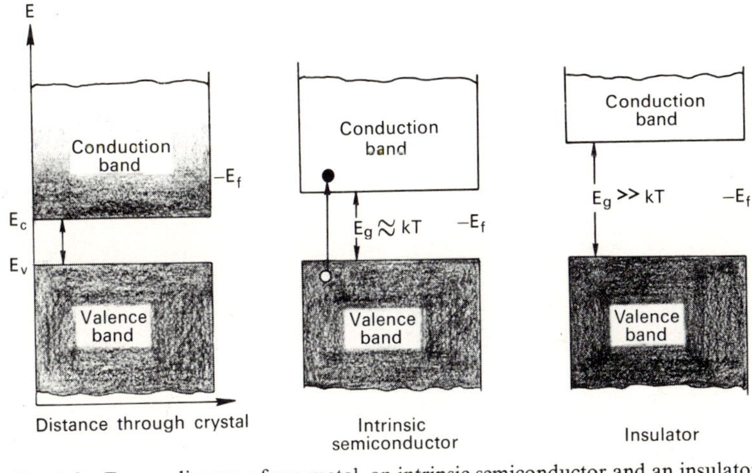

FIG. 1.2a Energy diagrams for a metal, an intrinsic semiconductor and an insulator

in which conduction electrons are free to move. A partly filled band may also arise from an overlap of a full and an empty band, as is found in the alkaline earth metals. E_f is the Fermi energy; at absolute zero all states with energies greater than E_f are vacant and those less than E_f are filled. E_f is defined as that level which has a probability of $\frac{1}{2}$ for being occupied.

In intrinsic semiconductors, at absolute zero all the electrons are in the valence band and thus there these solids are insulators. At higher temperatures, however, a few electrons can be excited into the conduction band through the energy gap E_g, which is of the order of KT, leaving holes in the valence band. This excitation permits intrinsic conductivity as given by the equation

$$\sigma = e(N_e\mu_e + N_h\mu_h)$$

where N_e is the equilibrium concentration of the electrons in the conduction band and is equal to N_h the concentration of holes in the valence band and μ_e and μ_h are the mobilities of the electrons and holes, respectively.

In impurity semiconductors, the current carrying electrons or holes are due primarily to foreign atoms or to other defects such as vacancies or non-stoichiometry. In Chapter 3 conductivity of oxides such as Cu_2O containing an excess of oxygen is discussed. These materials contain localized acceptor levels within the energy gap which may become occupied by electrons excited from the valence band (see Fig. 1.2b). The holes left in the valence band can conduct electrical current. These materials are called *p*-type conductors.

In Chapter 6 ZnO and other oxides which contain excess metal are covered. Here the donor levels are present in the energy gap close to the conduction band. When these electrons are excited to the conduction band, they can contribute to the conductivity (see Fig. 1.2c). These semiconductors

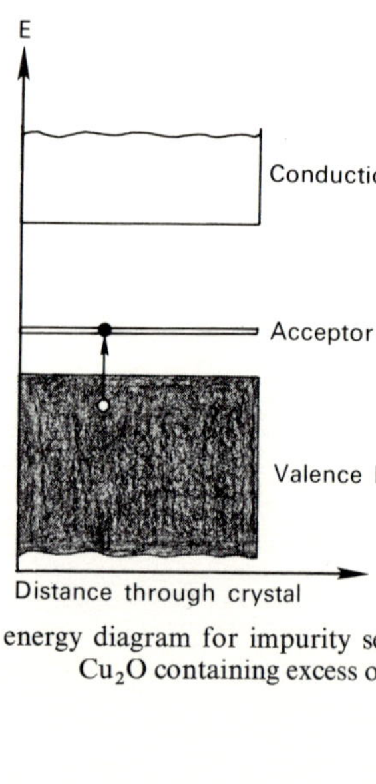

FIG. 1.2b Electronic energy diagram for impurity semiconductor, *p*-type, such as Cu_2O containing excess oxygen

are called *n*-type. The most important semiconductors, groups IV, III-V and II-VI, are discussed in Chapter 5.

Insulators differ from semiconductors essentially only by the magnitude of the energy gap E_g which is of the order of a few eV. Their weak electronic conductivity is usually caused by impurities or lattice defects. At high temperatures many of them show ionic conductivity in which the ions themselves carry the current across the lattice.

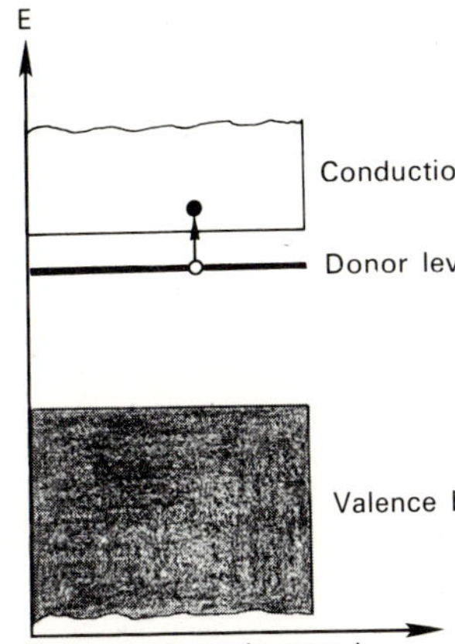

FIG. 1.2c Electronic energy diagram for impurity semiconductor, *n*-type such as ZnO containing excess zinc atoms

Thermoelectricity. There is a great deal of interest in materials which can convert thermal energy to electrical energy efficiently. These materials, called thermoelectrics, produce a relatively large voltage when a temperature gradient is maintained along the sample. The efficiency of such a material is given by the equation

$$Z = \frac{\alpha^2}{\rho K}$$

where α is the Seebeck coefficient, ρ is the electrical resistivity and K is the thermal conductivity. From this equation it can be seen that it is desirable to have a material with a low thermal conductivity, a low electrical resistivity and a large Seebeck coefficient.

The Seebeck coefficient and electrical and thermal conductivity of a number of materials are listed throughout the book. In Bi_2Te_3, which is one of the most efficient thermoelectrics, the Seebeck coefficient varies from 190 to 210 $\mu V/°C$; the electrical resistivity is approximately 10^{-3} ohm-cm; the thermal conductivity is 15 to 17 watts/cm²/°C; and thus $Z = 2.4 \times 10^{-3}$ $(°K)^{-1}$. There are dielectric materials with higher Seebeck coefficients, but with much higher electrical resistivities, and there are intermetallics with higher conductivities but with lower Seebeck coefficients.

Superconductivity. A superconductor is a material whose electrical resistivity vanishes below its transition temperature, T_c, known as the critical temperature. This critical temperature varies from a fraction of a degree Kelvin for some materials up to approximately 20.5°K for $Nb_3(Al_{0.8}Ge_{0.2})$. Superconductors also exhibit partial to complete shielding of magnetic fields from the interior regions of a bulk sample. When this shielding is complete ($B = 0$), the phenomenon is called the Meissner effect.

Superconductors are classified as type I or type II, according to their behavior in a magnetic field (see Fig. 1.2d). Superconductivity is destroyed in

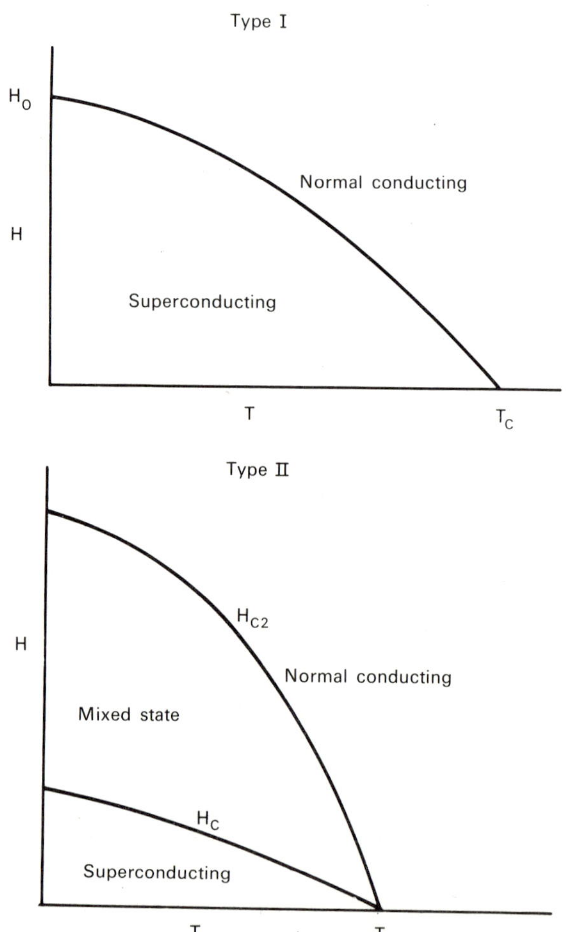

FIG. 1.2d. Magnetic field vs. temperature for superconductors

a type I material whenever the applied magnetic field is greater than the "thermodynamic critical field", H_c. The value of H_c is zero at $T = T_c$ and varies with temperature approximately as $H_c = H_0[1 - (T/T_c)^2]$, where H_0 depends on the material and is typically a few hundred oersteds. Superconductivity in a type II material exists in fields stronger than H_c up to a value known as the upper critical field H_{c2}, where it then also is destroyed and the ratio H_{c2}/H_c, which is roughly independent of T for a given material, ranges from just above 1 to about 50. Values of H_{c2} have been reported up to 200 kilo-oersteds. The mechanism that permits superconductivity to exist above the thermodynamic critical field in type II superconductors is the formation of a mixed state; i.e. a partial penetration of the magnetic field into the bulk of a sample by means of quantized magnetic flux lines called vortices.

Hard superconductors are type II superconductors which have been cold-worked or have had imperfections introduced by other means; they can carry large currents at high fields and exhibit irreversible behavior. Lattice imperfections in these materials serve to prevent motion of the vortices, the motion of which is dissipative. All elemental superconductors except niobium are of type I and most superconducting alloys and compounds are type II. Of the type II materials, most are hard superconductors unless special care is taken in their preparation.

At present, there is a microscopic theory which explains satisfactorily most of the phenomena which are associated with superconductivity. This is the Bardeen–Cooper–Schrieffer theory which is based on the assumption that superconductivity is a result of an attractive coupling between two electrons through a phonon interaction with the lattice. This coupling in turn leads to a cooperative state of the electrons which inhibits the scattering of individual carriers.

In addition to the BCS theory, there are a number of observations which have added to the knowledge of superconductivity. It has been noted, for example, that down to the lowest temperatures obtainable monovalent metals, good electrical conductors and ferromagnetic metals are not superconductors. Of the elements which are superconductors, the transition metals with three, five or seven outer electrons have the highest T_c values. For superconductivity in compounds, not only does the average electron per atom ratio appear to be important, but also the crystal structure. For example, many of the superconductors with the highest critical temperatures have either the sodium chloride, C15, or β-tungsten structure. While the reason for these relationships is not understood, their use has led to the discovery of many new superconductors with high critical temperatures.

Magnetic Properties

All substances subjected to a magnetic field, H, acquire a magnetic moment,

M, per unit volume; M is called the magnetization. For those materials whose magnetization is linearly dependent on field strength, a magnetic susceptibility, χ, is defined by

$$M = \chi H$$

Materials exhibiting linear magnetization characteristics generally have a susceptibility in the range $10^{-5} > \chi > -10^{-5}$. Those with negative susceptibilities are diamagnetic and those with positive susceptibilities are paramagnetic materials.

Diamagnetism. All atoms or ions produce a diamagnetic contribution to the total susceptibility although it may be masked by other types. This diamagnetism results from the perturbation of the electron orbits by the applied magnetic field. The motion of the electrons may be regarded as resistanceless current loops. The application of an external field H will induce a current in such a direction that the magnetic field from the induced current counteracts the applied field (Lenz's law). The induced current will persist as long as the external field is present. This diamagnetic contribution is temperature-independent. In metals there is also a diamagnetic contribution due to the electrons in the conduction band.

Paramagnetism. Paramagnetism requires the existence of a magnetic moment which arises from the uncompensated spin and angular moments of ions or molecules possessing an odd number of electrons. With no external magnetic field applied, the axes of the magnetic dipoles are randomly oriented. Application of the field causes the spins to assume a minimum energy configuration, i.e. parallel to the field. This, however, is opposed by thermal agitation, so that perfect alignment can be obtained only with an infinitely strong magnetic field or at absolute zero. Paramagnetic susceptibility is a temperature-dependent quantity except in metals where the so-called Pauli paramagnetism is temperature independent.

Ferromagnetism. Magnetic materials showing a highly nonlinear dependence of magnetization on the applied field exhibit a spontaneous magnetization below a certain critical temperature. This stems from an exchange interaction between the uncompensated electronic spins of neighboring atoms or ions of the solid. This spontaneous magnetization can be destroyed when the thermal energy is greater than the exchange energy, i.e. when the critical temperature is exceeded, and will result in a normal paramagnetic behavior.

A positive exchange energy results in a parallel alignment of neighboring spins which makes the material ferromagnetic. The ordering of spins extends over small regions called domains. In the absence of an external field, the directions of magnetization of different domains do not have to be the same. The critical temperature for a ferromagnetic material is called the ferromagnetic Curie point.

When the flux density B in a ferromagnetic material is plotted as a function of the applied magnetic field H, a hysteresis loop is obtained. At the top of the loop, shown in Fig. 1.2e, the directions of magnetization in all the domains are aligned and the B value is called the saturation magnetization, B_s.

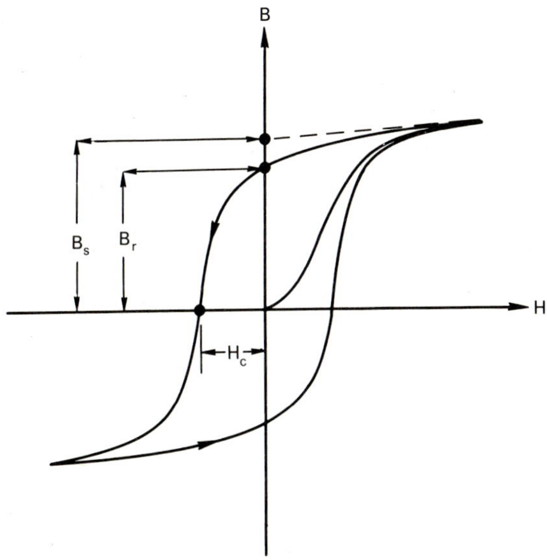

FIG. 1.2e Ferromagnetic hysteresis loop

When the applied field is removed, there is a domain readjustment and the final B_r is the residual magnetization or remanence. When the field is reversed, B goes to zero, at a field defined as the coercive force, H_c. When H_c of a material is large, it is hard to demagnetize, and it is considered to be a permanent magnetic material. Many of these materials are listed in Chapter 6.

When the exchange energy is negative the electron spins of adjacent atoms or ions are held in opposite orientations. If the magnetic moments of adjacent atoms are the same, this will result in a local cancellation of the moments and there will be no net observable magnetization. This condition is referred to as antiferromagnetism. If adjacent moments are not similar, there is a resultant net magnetization and the material is called ferrimagnetic. The transition temperature at which exchange forces in antiferromagnetic and in ferrimagnetic materials are overwhelmed by thermal excitation is called the Néel temperature.

Most ferrimagnetic materials (such as ferrites) are ionic compounds in which nonmagnetic ions play an important role as coupling agents between two neighboring magnetic ions which align their spins. This coupling is called super-exchange. Ferrimagnetic materials will be discussed further in the parts of this book dealing with the garnet, spinel, magnetoplumbite and perovskite structures.

Dielectric Properties and the Dielectric Constant

When a dielectric is placed between two parallel plates, the capacitance of this condenser is the ratio of q, the charge on either plate, to v, the applied voltage,

$$C = q/v$$

and the dielectric constant is the ratio of this capacitance to the capacitance of the condenser without the dielectric.

Ferroelectricity. In analogy to magnetization in a magnetic field there may occur an electric polarization P of a solid by an electric field E. The material acquires then a net electric moment. If P is not linear in E but exhibits a hysteresis analogous to the behavior of ferromagnetics then the material is ferroelectric. Ferroelectrics contain domains, within which the spontaneous polarization is in the same direction. When all the domains are aligned, the polarization is at the saturation value denoted by point P_s in Fig. 1.2f. Removal of the field allows the material to relax, the polarization drops to P_r,

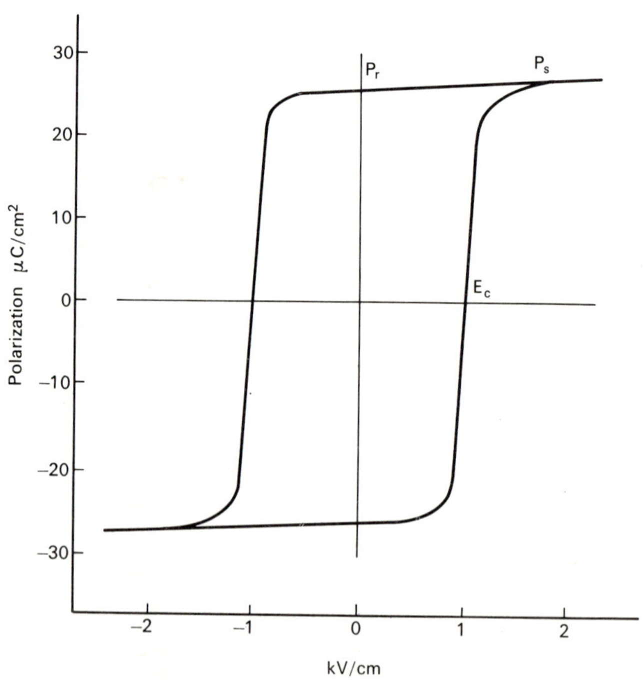

FIG. 1.2f Ferroelectric hysteresis loop

the remanent polarization, and a reversal of the applied field is required to decrease the polarization further. The strength of the electric field required to bring the polarization to zero, E_c, is called the coercive field.

Most of the ferroelectric materials which will be discussed in this book have slightly distorted perovskite structures or structures related to the perovskite structure. A necessary but not sufficient condition for a solid to be ferroelectric is the absence of a center of inversion symmetry. Of the twenty-one crystal classes (based on point group symmetry) which lack a center of symmetry, twenty are piezoelectric, i.e. the crystal can be polarized by application of an external stress. The piezoelectric property can be predicted on the basis of crystal structure. Ten out of the twenty piezoelectric classes exhibit pyroelectricity: in these materials, a spontaneous polarization exists, usually masked by compensating surface charges. However, when the temperature is changed, the polarization is changed, so that the polarization is no longer compensated, and this change is observable. Ferroelectric materials belong to the pyroelectric materials with the added property that the spontaneous polarization can be reversed by an applied electric field. This reversibility cannot be predicted on the basis of crystal structure alone.

Above a critical temperature, T_c, called the ferroelectric Curie point, the ferroelectric properties of a material disappear. Below this temperature, the spontaneous alignment of the electric dipoles results in a very high apparent dielectric "constant". Above the Curie temperature, the spontaneous polarization vanishes, with a corresponding drop in the dielectric constant and a linear relation is observed between polarization and applied electric field. For $BaTiO_3$, the Curie temperature is 120°C. In the subsequent chapters the Curie temperature is given for a number of other ferroelectric materials.

There exist also antiferroelectric crystals which show an anomaly in the dielectric constant versus temperature curve. They can best be described as having opposing lines of spontaneously polarized ions in their structures, and they do not exhibit hysteresis loops characteristic of ferroelectrics unless a ferroelectric state is induced by the application of high fields. Typical antiferroelectric materials are $PbZrO_3$ and $NaNbO_3$.

Optical Properties

The transmittance of electromagnetic radiation as a function of wavelength through several materials is given in the following chapters. Most of the data is for halides which transmit over a large range of wavelengths. In comparing these materials it is important to note that the intensity of the transmitted radiation decreases exponentially with the length of the light path. This information along with the indices of refraction at various wavelengths is useful for designing optical equipment, such as prisms and windows.

Fluorescence, lasers. When certain ions called activator ions are substituted in optical materials called hosts, they can absorb radiation and emit it as visible or near visible radiation. If the emission occurs within 10^{-8} sec of excitation the process is called fluorescence. If it takes a much longer time it is called phosphorescence. Many of the crystals which fluoresce have been used as lasers.

The laser (light amplification from the stimulated emission of radiation) is an optical oscillator. It consists of a resonant cavity containing an amplifying medium in which oscillations will occur at those frequencies for which the gain exceeds the loss. Typically, the cavity consists of two parallel opposing mirrors forming a Fabry–Perot resonator. In a solid state laser the amplifying medium is a crystal in which activator ions have been imbedded. The activator ions, situated in the host material, can absorb optical energy at one frequency and reradiate it at the lower frequency either spontaneously (fluorescence) or by an induced de-excitation (stimulated emission).

A particularly useful laser system utilizes a ruby crystal. The ruby rod is a single crystal of Al_2O_3 in which Cr^{3+} has been substituted for some of the aluminum ions. The three-level energy diagram for this system is shown in Fig. 1.2g. The optical "pump" energy is absorbed by an electronic transition

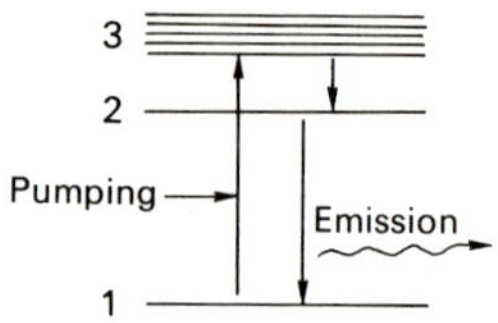

FIG. 1.2g Three-level energy diagram for a laser

from the ground state into a broad band above the upper laser level. The electron drops into the upper laser level by a nonradiative transition. The lifetime of this level is long ($\sim 10^{-3}$ sec) allowing electrons to accumulate. If pumping is severe enough, the population of the upper laser level can exceed the population of the lower laser level (the ground state in this case). Under such circumstances the population is said to be inverted and radiation at the laser wavelength will experience coherent gain by virtue of the stimulated transitions of electrons from the upper laser level to the ground state. Light directed along the axis of the cavity will retrace its path many times and be amplified.

It is not necessary that the ground state be the lower laser level. In many so-called four-level laser systems the lower laser level is a low-lying level just above the ground state. The relaxation from this terminal level to the ground state should be rapid in order to maintain a population inversion between it and the upper laser level. It is usually necessary that the energy separation of the lower laser level and the ground state be small enough so that the necessary inverted population can be created. This separation often imposes a low temperature restriction on the operation of some laser systems.

Most crystalline lasers are four-level systems and use trivalent rare earth ions as the active ions. Nd^{3+} is a particularly useful laser activator ion and like other ions has a spin-orbit component above the ground state acting as the terminal laser level. Other trivalent ions form the terminal level by splitting of the ground state of the ion by the crystalline field.

In some cases where the pumping level is too sharp or is not matched to a strong emission line of the pumping radiation, a second activator ion is added to the host. This "double doping" of the host is a way of sensitizing the system to the "pump" radiation. A $Y_3Al_5O_{12}$ crystal containing Cr^{3+} and Nd^{3+} is an example of such a system. The Cr^{3+} absorbs the "pump" radiation in a broad band from which the excited electrons relax to a metastable level. The excitation is then transferred to an excited level of the Nd^{3+} from which an inverted population in the Nd^{3+} can result with subsequent laser action.

Most of the laser crystals are alkaline earth tungstate, molybdate or fluoride crystals containing small amounts of rare earth ions. Data for these systems are presented in Chapter 5, while data for other oxide and fluoride systems are given throughout the book.

In semiconductor lasers the radiation originates in the region of a *p–n* junction and is due to transitions of injected electrons and holes between the low-lying levels of the conduction band and the uppermost levels of the valence band. Donor levels lying just below the conduction band edge and shallow acceptor levels also can be involved in these transitions. The wavelength of the emitted radiation corresponds approximately to the band gap energy. The efficiencies of these lasers are much higher than those of optically pumped lasers.

Mechanical Properties

In the following chapters, reference is often made to the hardness, ultimate tensile strength, and the modulus of elasticity of materials. The latter two values can be obtained for engineering purposes by pulling on the ends of a bar (tensile specimen) while monitoring the load applied and measuring the corresponding change in the length of the bar. The ratio of the applied load to the original cross-sectional area is called the stress. The strain is defined as the fractional change in length.

The first part of the plot of stress versus strain (see Fig. 1.2h) is a straight line whose slope is the Young's modulus or modulus of elasticity of the material. Up until the proportional limit the curve remains a straight line. When a load or strain is not too high the specimen returns to its original length upon the removal of the load. However, above the elastic limit the material behaves in a plastic manner and will remain partly deformed even after the load is removed. If the stress is increased the specimen will eventually break; the maximum stress is called the ultimate tensile strength of the material. Most of the values given in this book are for brittle materials which break almost at the elastic limit so that the Young's modulus is approximately equal to the slope of the entire stress–strain curve. In ductile materials the true elastic limit is only slightly higher than the proportional limit.

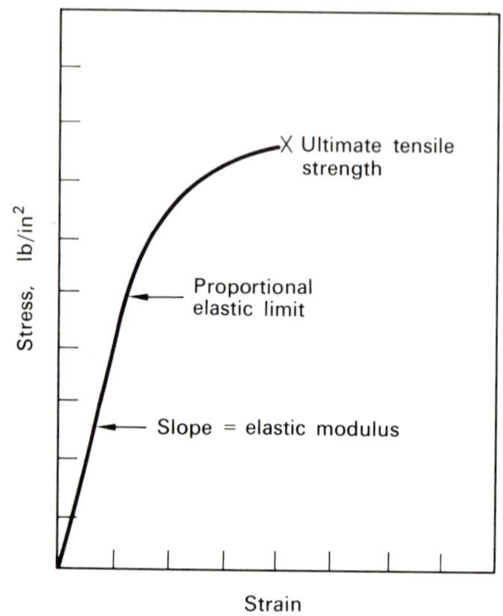

FIG. 1.2h Stress-strain diagram

The modulus of elasticity is a measure of the stiffness of a material. Since it is determined by the binding force between atoms, it is dependent on the structure of the material, and therefore can be different in various crystallographic directions. On the other hand, it is not very sensitive to alloying, cold-working or heat treatment. Thus, the values obtained from pure or nearly pure polycrystalline samples of randomly oriented grains are reproducible. This is not true of ultimate tensile strength values which can vary considerably from sample to sample.

Although hardness is easy to measure it is a very complicated mechanical property and is rather arbitrarily defined. In effect it is the resistance of a material to indentation: an applied load forces a small pyramid, ball or cone into the material and the dimension of the impression is used to indicate the hardness. Microhardness tests allow measurements on small regions of the material. The hardness values given in the following chapters are in kg/mm^2.

Emphasis in this book has been placed on the mechanical properties of high temperature, high modulus materials, such as boron, carbon, the borides, carbides, nitrides and oxides. The low density, high modulus materials which have been prepared or are being considered as fibers have been treated separately. This was done because of the interest in these fibers, for use in composite materials for aerospace applications.

Table 1.1*b References*

1. J. C. Slater, *Quantum Theory of Molecules and Solids,* Vol. 2, p. 55, *Symmetry and Energy Bands in Crystals*, McGraw-Hill, N.Y. (1965).
2. L. V. Azaroff, *Introduction to Solids*, p. 441(McGraw-Hill, N.Y., (1960).
3. V. M. Goldschmidt, *Skrifter Jet Norske. Videnskaps-Akad. Oslo,* 1. *Mat.-Naturik* **1926,** 7 (1927).
4. L. Pauling, *The Nature of the Chemical Bond,* 3rd ed., p. 514, Cornell Univ. Press, Ithaca, N.Y. (1960).
5. L. H. Ahrens, *Geochim. et Cosmochim. Acta* **2**, 155 (1952).
6. B. S. Gourary and F. J. Adrian, *Adv. Solid State Phys.* **10**, 127 (1960).
7. D. H. Templeton and C. H. Dauben in F. I. Spedding and A. H. Daane (eds.), *The Rare Earths,* p. 20, John Wiley, N.Y. (1961).
8. K. W. Herrmann, A. H. Daane and F. H. Spedding, in F. I. Spedding and A. H. Daane (eds.), *The Rare Earths,* p. 20, John Wiley, N.Y. (1961).
9. W. H. Zachariasen, *Structure Reports* **13**, 435 (1950).
10. M. P. Tosi and F. G. Fumi, *J. Phys. Chem. Solids* **25**, 45 (1964).

CHAPTER 2

COMMON STRUCTURES

THE properties of elemental solids are so well known that in this chapter they will be summarized only briefly and the accent placed on structural relationships.

Polonium is the only element which crystallizes in the simple cubic structure which can be described by a cubic unit cell containing a single atom at the corners (0, 0, 0) (see Fig. 2.0). Many of the other elements crystallize in one of three common structures: body-centered cubic, face-centered cubic and hexagonal close-packed structures. While the terms body-centered

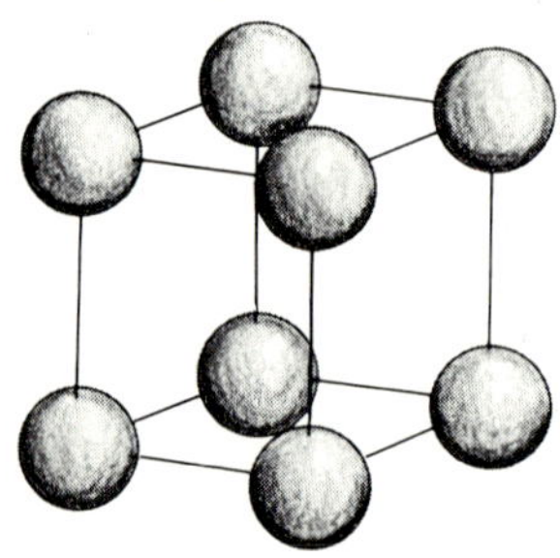

FIG. 2.0 The simple cubic structure

cubic and face-centered cubic refer to lattice types, they have been used to describe the former two structures which can be generated by placing a single atom at each lattice point. In fact, these terms are so commonly used, it is probably best to define these two special structures in this way.

2.1. Body-centered Cubic, W Type, A2, *Im*3*m*

The body-centered cubic structure can be derived from the simple cubic structure by adding other atoms of the same element at the center of each cube, so that there are two atoms per unit cell—one at 0, 0, 0, and the other at $\frac{1}{2}, \frac{1}{2}, \frac{1}{2}$ (see Fig. 2.1a). In this structure the atoms are not close-packed and

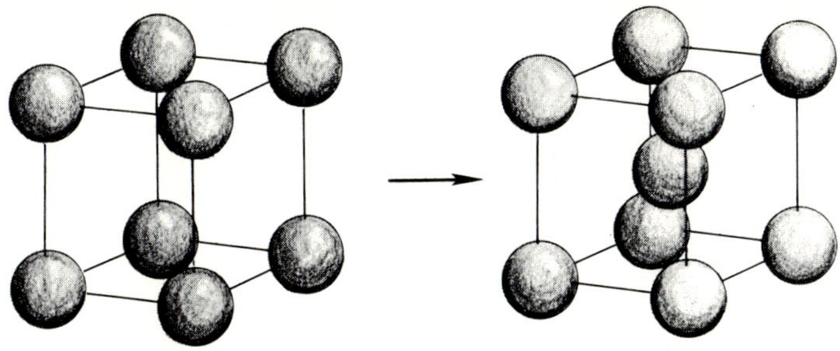

Simple cubic structure (a) Body centered cubic structure

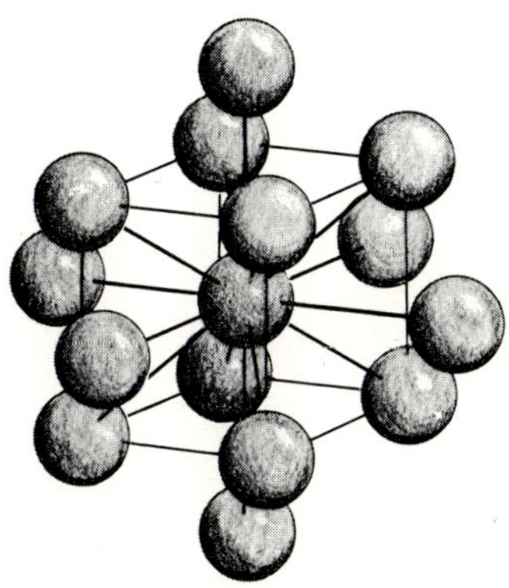

(b) Co ordination about each atom

FIG. 2.1 Body-centered cubic structure

the volume per atom is 6.16 r^3, where r = the radius of the spherical atoms. Each atom has eight equidistant nearest neighbors and six others 15% farther away. This is illustrated in Fig. 2.1b. Elements which adopt this structure and the cell size for each are listed in Table 2.1. Other elements such as Ca, Sr, Ti, Zr and Hf also form the body-centered cubic structure under certain conditions. It has been predicted theoretically that hydrogen at very high pressures also is a body-centered cubic metal.

TABLE 2.1. *Elements with the Body-centered Cubic Structure*

Element	Cell size a_0 (Å)	m.p. (°C)	Density (g/cm^3)
Ba	5.025	704	3.59
γ-Ca (~500°C)	4.38	850	1.55
δ-Ce (730° to m.p.)	4.12	795	6.67
α-Cr	2.884	1550	7.19
Cs (78°K)	6.067	28.5	1.87
Eu	4.578	826	5.259
α-Fe	2.8606	1535	7.89
β-Fe (800°C)	2.91		
δ-Fe (1425°C)	2.44		
β-Hf		1700	11.4
K (78°K)	5.247	62.3	0.86
γ-La (868° to m.p.)	4.26	920	5.98
Li	3.5093	186	0.53
Mo	3.147	2625	10.2
Na	4.291	97.5	0.97
Nb	3.300	2415	8.57
β-Nd (868° to m.p.)	4.13	1024	6.80
γ-Np (~600°C)	3.52		
β-Pr (798° to m.p.)	4.13	935	6.64
ε-Pu (500°C)	3.638		
Rb (75°K)	5.605	38.5	1.53
β-Sm (917° to m.p.)	4.07	1072	7.40
γ-Sr (614°C)	4.85	757	2.6
Ta	3.306	2996	16.6
β-Th (1450°C)	4.11	1800	11.5
β-Ti (900°C)	3.307	1820	4.43
β-Tl	3.882	300	11.85
γ-U	3.474		
V	3.024	1735	6.0
W	3.165	3410	19.26
β-Y (1490° to m.p.)	4.11	1509	
β-Yb (798° to m.p.)	4.44	824	6.54
β-Zr (850°C)	3.62	1750	6.49

2.2. Cubic Closest-Packing, Face-centered Cubic Structure, Cu Type, A1, *Fm3m*

In the unit cell of the face-centered cubic structure, as the name implies, atoms are located at the centers of the faces of the cube as well as at the corners. Thus, the unit cell contains four atoms at 0, 0, 0; $\frac{1}{2}$, $\frac{1}{2}$, 0; $\frac{1}{2}$, 0, $\frac{1}{2}$ and 0, $\frac{1}{2}$, $\frac{1}{2}$ (see Fig. 2.2a). The volume per atom is 5.66 r^3 and each atom has twelve equidistant nearest neighbors. Figure 2.2b depicts this twelvefold coordination by including atoms from neighboring unit cells which form the coordination polyhedron about each atom.

Looking at the cubic unit cell in the ⟨111⟩ directions shows the hexagonal close-packed layers clearly (Fig. 2.2c). Each atom is in contact with six others in each layer (Figs. 2.2d, 2.2e). The next layer is identical to the first, but is displaced so that the atoms fit over positions B. In the third identical layer

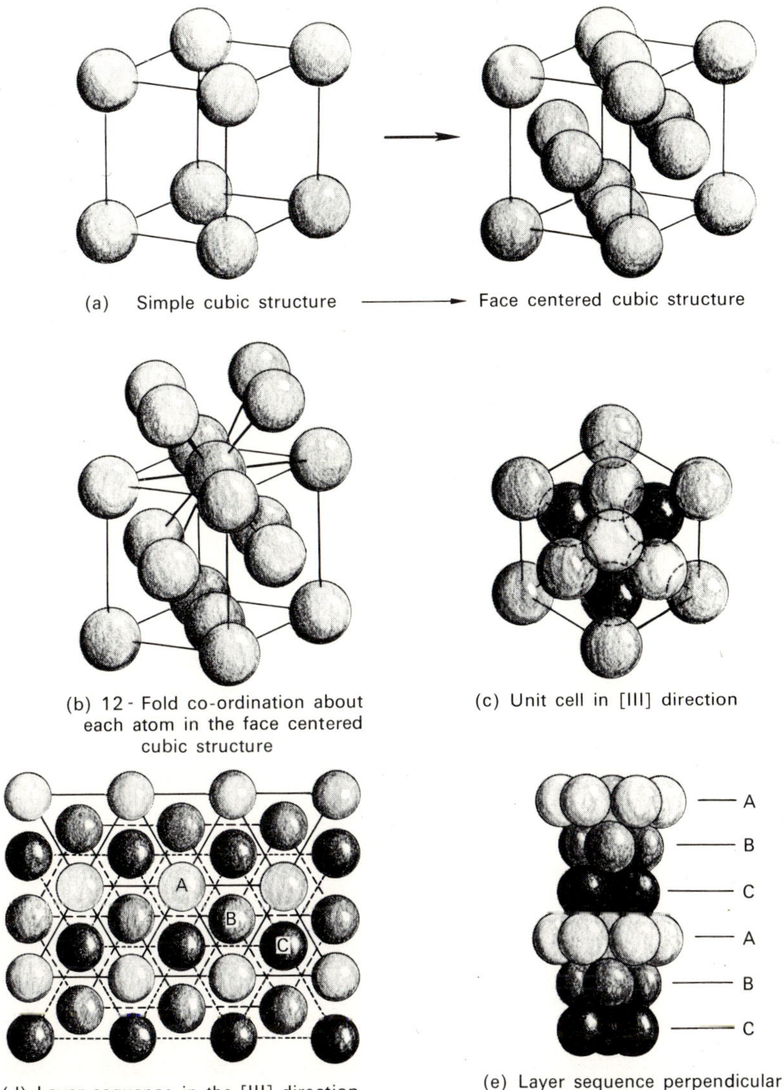

(a) Simple cubic structure ⟶ Face centered cubic structure

(b) 12 - Fold co-ordination about each atom in the face centered cubic structure

(c) Unit cell in [III] direction

(d) Layer sequence in the [III] direction

(e) Layer sequence perpendicular to the [III] direction

FIG. 2.2 The face-centered cubic structure

the atoms are set on positions C, and in the next layer the atoms are over the atoms in layer A. Thus, the repeat sequence is . . . ABC ABC . . . Elements which adopt this structure are listed in Table 2.2.

TABLE 2.2. *Elements with Face-centered Cubic Structure*

Element	Cell size a_0 (Å)	m.p. (°C)	Density (g/cm³)
Ac	5.311	1600	10.07
Ag	4.0862	960.5	10.49
Al	4.050	660	2.70
Am	4.894		
Ar (4.2°K)	5.256	−184	1.40
Au	4.078	1063	19.3
Ca	5.582	850	1.55
(< −150°)	4.85		
Ce (−10° to 730°)	5.1604	795	6.771
Co	3.548	1490	8.70
β-Cr	3.68	1550	7.20
Cu	3.615	1083	8.92
γ-Fe	3.591	1535	7.87
Ir	3.839	2454	22.5
Kr (58°K)	5.721	−157	
β-La (310° to 868°)	5.303	920	6.186
Ne (4.2°K)	4.429	−249	1.20
Ni	3.524	1455	8.90
Pb	4.951	327	11.34
Pd	3.890	1554	12.0
Pt	3.923	1774	21.45
δ-Pu (320°C)	4.637		
Rh	3.8031	1966	12.44
β-Sc (1000° to m.p.)	4.541	1539	
α-Sr	6.085	770	2.6
α-Th	5.0843	1800	11.5
Xe (58°K)	6.197	−112	5.5×10^{-3}
α-Yb (to 798°)	5.481	824	6.977

2.3. Close-packed Hexagonal, Mg type, A3, *P6mmc*

Many of the elements (Table 2.3) form the hexagonal close-packed structure. The shape of a hexagonal unit cell is shown in the Introduction, and also in Fig. 2.3. The angle between the *a*- and *b*-axes is 120° and the angle between the *ab*-plane and the *c*-axis is 90°. In the unit cell of the hexagonal close-packed structure, one atom is located at $\frac{1}{4}$ of a unit cell height on a line with x,y equal to $\frac{1}{3}, \frac{2}{3}$ and another atom at $\frac{3}{4}$ of a unit cell height on a line with x,y equal to $\frac{2}{3}, \frac{1}{3}$ (Fig. 2.3a). Thus, the atoms are located at $\frac{1}{3}, \frac{2}{3}, \frac{1}{4}$ and $\frac{2}{3}, \frac{1}{3}, \frac{3}{4}$ in the unit cell. Figure 2.3b shows another orientation of the unit cell for the hexagonal close-packed structure which is obtained by placing the origin at an atom. One of the atoms then is at the corner or 0, 0, 0 and the other at $\frac{1}{3}, \frac{2}{3}, \frac{1}{2}$.

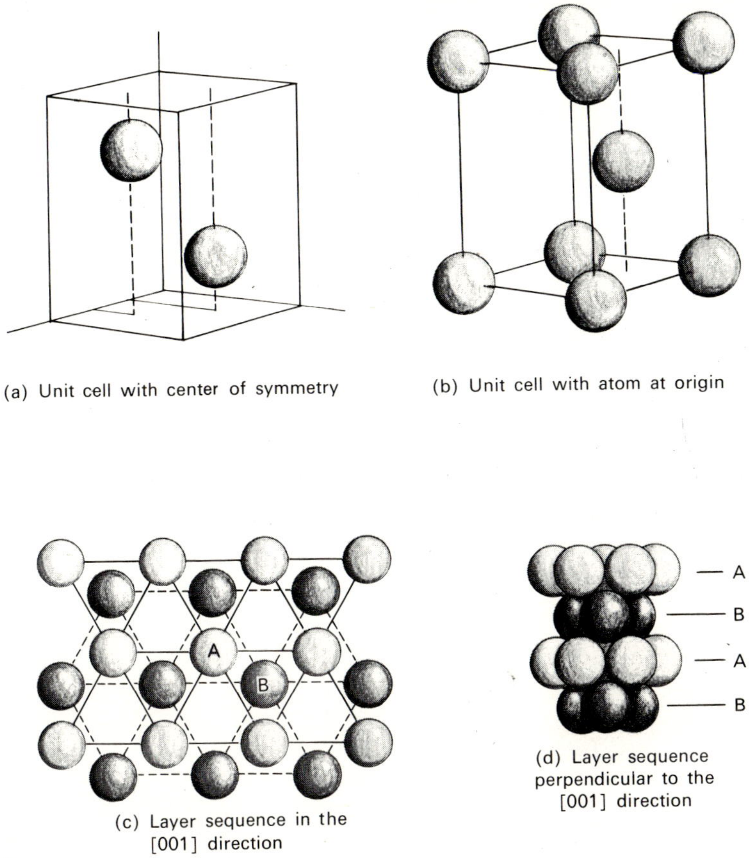

(a) Unit cell with center of symmetry

(b) Unit cell with atom at origin

(c) Layer sequence in the [001] direction

(d) Layer sequence perpendicular to the [001] direction

FIG. 2.3 The Hexagonal close-packed structure

As in the cubic closest-packed structure, the volume per atom is 5.66 r^3, and each atom is surrounded by twelve nearest neighbors. The layers of atoms in the hexagonal close-packed structure, which can be seen by looking down the *c*-axis, are identical to those in the {111} planes of the cubic close-packed structure but the packing sequence is different. The atoms in the second layer fit over the B holes in the first layer as they do the cubic closest-packed structure, but the atoms in the third layer are over those in the first layer, so the sequence is . . . ABAB . . . instead of . . . ABCABC . . . as it is in cubic closest-packed structure. (See Fig. 2.3(c) and (d).)

TABLE 2.3. *Elements with the Hexagonal Close-packed Structure*

Element	Cell size (Å) a_0	Cell size (Å) c_0	m.p. (°C)	Density (g/cm^3)
Be	2.287	3.583	1280	1.86
β-Ca (450°C)	3.98	6.52	842	1.55
Cd	2.979	5.618	3209	8.65
α-Co	2.507	4.069	1490	8.9
γ-Cr	2.722	4.427	1550	8.56
Dy	3.5925	5.6545	1407	8.536
Er	3.5590	5.592	1497	9.051
Gd	3.6315	5.777	1312	7.895
He	3.57	5.83	−272	0.126
α-Hf	3.197	5.058	1700	11.4
Ho	3.5761	5.6174	1461	8.803
Li (78°K)	3.111	5.093	186	0.53
Lu	3.5050	5.5486	1652	9.842
Mg	3.209	5.210	650	1.74
Na (5°K)	3.657	5.902	97.5	0.97
Ni	2.65	4.33	1455	8.90
Os	2.735	4.319	2700	22.5
Re	2.761	4.458	3170	20
Ru	2.704	4.282	2500	12.2
α-Sc (to 1000°)	3.3080	5.2653	1539	2.992
β-Sr (248°C)	4.32	7.06	757	2.6
Tb	3.5990	5.696	1356	8.272
Te	2.735	4.388	2700	
α-Ti	2.950	4.686	1820	4.54
α-Tl	3.456	5.525	300	9.35
Tm	3.5372	5.5619	1545	9.332
α-Y (to 1490°)	3.6451	5.7305	1509	4.478
Zn	2.665	4.947	419	7.13
α-Zr	3.232	5.147	1750	6.49

2.4. Structure of Other Elements

The elements in the periodic table are divided into A and B types, and the elements from group IIIA up to group IB are called transition metals. The structures of these elements are indicated in Fig. 2.4. Note that many of the elements have more than one structure. The structure depends on the temperature and/or pressure to which the elements are subjected. One of the most important examples of polymorphism is that for iron, which has the body-centered cubic structure below 906°C and above 1401°C to the melting point. From 906°C to 1401°C, it has the face-centered cubic structure and it is hexagonal at very high pressures. Cobalt, on the other hand, adopts the face-centered cubic structure above 500°C and becomes partially hexagonal on cooling.

Neon, Ar, Kr and Xe form the cubic close-packed structure and molecular hydrogen (H_2) the hexagonal close-packed structure in the solid state. The other elements in this area of the long form of the periodic table have structures in which the number of nearest neighbors for each atom is 8-N, where

N is the group number of the element. The halogens exist as diatomic molecules so that each halogen has one neighbor. Sulfur and selenium exist as eight-member rings and selenium and tellurium metal as puckered chains in which the coordination number is two. Phosphorus forms a P_4 molecule in the gas and Bi, Sb and As form layered structures so that the coordination number is three. Carbon, silicon, germanium and grey tin atoms have four nearest neighbors in the diamond cubic structure.

The types and atomic positions of many elements are listed below for easy reference. Solid Hg has a rhombohedral structure, boron crystallizes in complex tetragonal and rhombohedral structures, Ga has a complex structure and Al, In, Tl and Pb form nearly close-packed structures. Manganese has three polymorphs and Pu, Pa and U also have several. Other elements which logically fall into one of the structural groups of each chapter will be described later.

Diamond type—A4 discussed later in Chapter 5.

White Sn type—A5, $I4_1/amd$, tetragonal. $a = 5.8197$ Å, $c = 3.1750$ Å.
4Sn at $0, 0, 0; \frac{1}{2}, \frac{1}{2}, \frac{1}{2}; 0, \frac{1}{2}, \frac{1}{4}; \frac{1}{2}, 0, \frac{3}{4}$.

In type—A6, $F4/mmm$, tetragonal. $a = 4.588$ Å, $c = 4.9377$ Å.
4In at $0, 0, 0; 0, \frac{1}{2}, \frac{1}{2}; \frac{1}{2}, 0, \frac{1}{2}; \frac{1}{2}, \frac{1}{2}, 0$.

As type—A7, $R\bar{3}m$, rhombohedral. $a = 4.123$ Å, $\alpha = 54°10'$.
6As at $0, 0, z; \frac{1}{3}, \frac{2}{3}, \frac{2}{3} + z; \frac{2}{3}, \frac{1}{3}, \frac{1}{3} + z; 0, 0, \bar{z}; \frac{1}{3}, \frac{2}{3}, \frac{2}{3} - z; \frac{2}{3}, \frac{1}{3}, \frac{1}{3} - z$.
$z = 0.226$.

Se type—A8, $P3_121$, trigonal. $a = 4.36$ Å, $c = 4.95$ Å.
3Se at $x, 0, \frac{1}{3}; 0, x, \frac{2}{3}; \bar{x}, \bar{x}, 0$. $x = 0.217$.

Hg type—A10, $R\bar{3}m$, rhombohedral, hexagonal cell. $a = 3.464$ Å. $c =$ 6.677 Å.
3Hg at $0, 0, 0; \frac{1}{3}, \frac{2}{3}, \frac{2}{3}; \frac{2}{3}, \frac{1}{3}, \frac{1}{3}$.

Ga type—A11, $Cmca$, orthorhombic. $a = 4.523$ Å, $b = 7.661$ Å, $c =$ 4.524 Å.
8Ga at $0, y, z; 0, \bar{y}, \bar{z}; \frac{1}{2}, y, \frac{1}{2} - z; \frac{1}{2}, \bar{y}, \frac{1}{2} + z$. $z = 0.0785$.
$\frac{1}{2}, y + \frac{1}{2}, z; \frac{1}{2}, \frac{1}{2} - y, \bar{z}; 0, y + \frac{1}{2}, \frac{1}{2} - z; 0, \frac{1}{2} - y, \frac{1}{2} + z$. $y = 0.0785$,
$z = 0,1525$.

Mn type—A12, $I\bar{4}3m$, cubic. $a = 8{\cdot}894$ Å.
Atoms at $(000, \frac{1}{2}\frac{1}{2}\frac{1}{2})$ +
2Mn at 0, 0, 0.
8Mn at $x, x, x; x, \bar{x}, \bar{x}; \bar{x}, x, \bar{x}; \bar{x}, \bar{x}, x$. $x = 0.317$.
24Mn at $x, x, z; z, x, x; x, z, x; \bar{x}, x, \bar{z}; \bar{z}, x, \bar{x}; \bar{x}, z, \bar{x}; x, \bar{x}, \bar{z}; z, \bar{x}, \bar{x}; x, \bar{z}, \bar{x}; \bar{x}, \bar{x}, z; \bar{z}, \bar{x}, x; \bar{x}, \bar{z}, x$. $x = 0.356$, $z = 0.042$.
24Mn as above with $x = 0.089$, $z = 0.278$.

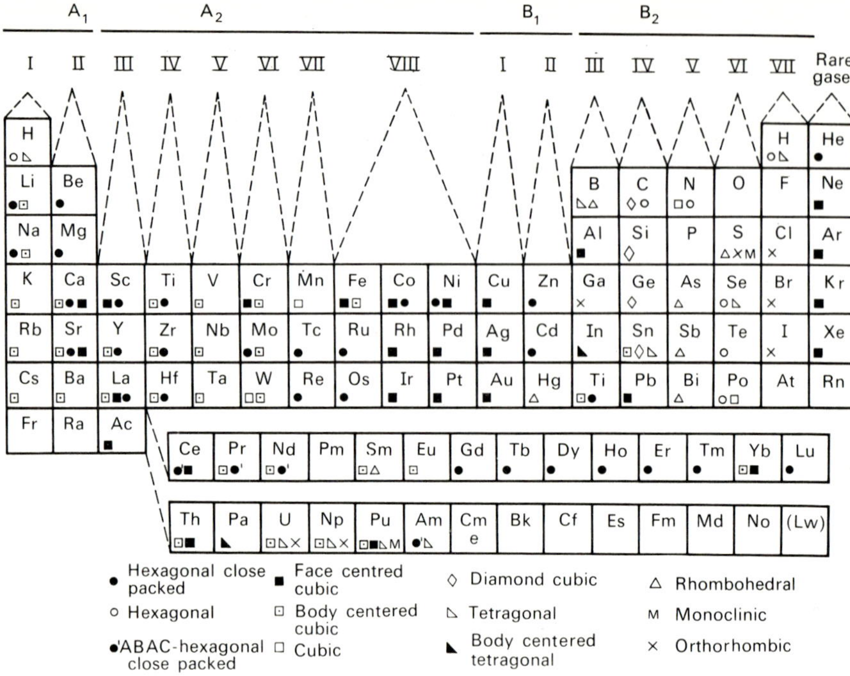

FIG. 2.4 Periodic chart of the elements indicating atomic structures

Pu There are six modifications of plutonium.

α-form—$P2_1/m$—monoclinic. $a_0 = 6.1835$ Å, $b_0 = 4.8244$ Å, $c_0 = 10.973$ Å, $\beta = 101°48'$.

16Pu at = (0.332, $\frac{1}{4}$, 0.155; 0.774, $\frac{1}{4}$, 0.175; 0.144, $\frac{1}{4}$, 0.341; 0.658, $\frac{1}{4}$, 0.457; 0.016, $\frac{1}{4}$, 0.621; 0.465, $\frac{1}{4}$, 0.644; 0.337, $\frac{1}{4}$, 0.926; 0.892, $\frac{1}{4}$, 0.897).

β-form monoclinic. $a_0 = 0.284$ Å, $b_0 = 10.463$ Å, $c_0 = 7.859$ Å, $\beta = 92°8'$.

γ-form—*Fddd*—orthorhombic. $a_0 = 3.1587$ Å, $b_0 = 5.7682$ Å, $c_0 = 10.162$ Å.

8Pu at $(0, 0, 0; \frac{1}{4}, \frac{1}{4}, \frac{1}{4}) + 0, 0, 0; \frac{1}{2}, \frac{1}{2}, 0; \frac{1}{2}, 0, \frac{1}{2}; 0, \frac{1}{2}, \frac{1}{2}$.

There also are a face-centered cubic delta phase, a body-centered cubic phase and a tetragonal delta prime phase.

Mn type—A13, $P4_332$, cubic. $a = 6.3018$ Å.

8Mn at $x, x, x; x+\frac{1}{2}, \frac{1}{2}-x, \bar{x}; \bar{x}, x+\frac{1}{2}, \frac{1}{2}-x; \frac{1}{2}-x, \bar{x}, x+\frac{1}{2}; \frac{1}{4}-x, \frac{1}{4}-x, \frac{1}{4}-x; x+\frac{3}{4}, \frac{3}{4}-x, x+\frac{1}{4}; \frac{3}{4}-x, x+\frac{1}{4}, x+\frac{3}{4}; x+\frac{1}{4}, x+\frac{3}{4}, \frac{3}{4}-x$. $x = 0.061$.

12Mn at $\frac{1}{4}-y, y, \frac{1}{8}; \frac{3}{4}-y, \frac{1}{2}-y, \frac{7}{8}; y+\frac{3}{4}, y+\frac{1}{2}, \frac{3}{8}; y+\frac{1}{4}, \bar{y}, \frac{5}{8}; \frac{1}{8}, \frac{1}{4}-y, y; \frac{7}{8}, \frac{3}{4}-y, \frac{1}{2}-y; \frac{3}{8}, y+\frac{3}{4}, y+\frac{1}{2}; \frac{5}{8}, y+\frac{1}{4}, \bar{y}; y, \frac{1}{8}, \frac{1}{4}-y; \frac{1}{2}-y, \frac{7}{8}, \frac{3}{4}-y; y+\frac{1}{2}, \frac{3}{8}, y+\frac{3}{4}; \bar{y}, \frac{5}{8}, y+\frac{1}{4}$. $y = 0.206$.

α-U type—*A*20, *Cmcm*, orthorhombic. $a = 2.858$ Å, $b = 5.877$ Å, $c = 4.955$ Å.

4U at $0, y, \frac{1}{4}; 0, \bar{y}, \frac{3}{4}; \frac{1}{2}, \frac{1}{2} + y, \frac{1}{4}; \frac{1}{2}, \frac{1}{2} - y, \frac{3}{4}$. $y = 0.105$.

α-Np type—*Pnma*, orthorhombic. $a = 6.663$ Å, $b = 4.723$ Å, $c = 4.887$ Å.

4Np at $x, \frac{1}{4}, z; \bar{x}, \frac{3}{4}, \bar{z}; \frac{1}{2} - x, \frac{3}{4}, \frac{1}{2} + x; \frac{1}{2} + x, \frac{1}{4}, \frac{1}{2} - z$. $x = 0.036$, $z = 0.208$.

4Np at same position with $x = 0.319$, $z = 0.842$.

β-Np type—*P*422, tetragonal. $a = 4.897$ Å, $c = 3.388$ Å.

2Np at $0, 0, 0; \frac{1}{2}, \frac{1}{2}, 0$.

2Np at $0, \frac{1}{2}, z; \frac{1}{2}, 0, \bar{z}$. $z = 0.375$.

Sm Type—$R\bar{3}m$, rhombohedral. $a = 8.996$ Å, $\alpha = 23°13'$.

1Sm at 0, 0, 0.

2Sm at $x, x, x; \bar{x}, \bar{x}, \bar{x}$. $x = 0.222$.

2.5. Alloys (Ordering)

When two metals are melted together and cooled, the amount of one element which will dissolve in another depends on the chemical similarity, and the relative sizes of the two types of atoms. For example, it would be expected that elements in the same group in the periodic table would be

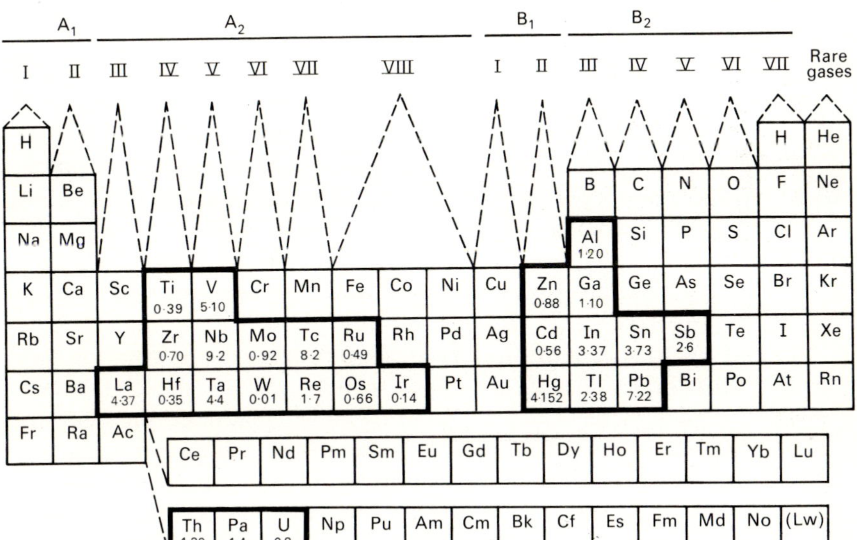

FIG. 2.5 Superconducting elements

extensively soluble in one another, provided that the size difference of the two atoms was not too great. Further, when solid solution is found between two metals of different valency, a greater degree of solution is obtained when the lower valency element is the solvent for the higher valence metal rather than dissolving in it.

In some cases, where this solid solution is cooled slowly, the atoms of the two elements take on certain preferred positions within the structure and are ordered to various degrees.

2.6. Discussion

Properties of the Elements

The melting point and density of elements with the body-centered cubic, face-centered cubic and hexagonal close-packed structures are listed in Tables 2.1, 2.2 and 2.3 respectively. Cubic close-packed metals tend to be more ductile and easier to work mechanically than the hexagonal ones. Because of the high symmetry, cubic structures contain more possible slip directions than the hexagonal structures and therefore can slip in many directions.

The electrical resistivities of metals are small in comparison with semi-conductors. Aluminum, gold, copper and silver are the best conductors, having resistivities of 2.8, 2.4, 1.7 and 1.5 times 10^{-6} ohm-cm, respectively. These resistivities, as those of all metals, increase with increasing temperature. The temperature variation of the resistivities can be stabilized by alloying, but the alloys usually have higher resistivities than the pure metal.

A number of metals have been found to be perfect electrical conductors below certain temperatures called the superconducting transition temperature. Among the transition metals, those with three, five or seven valence electrons have the highest superconducting transition temperatures (see Fig. 2.5). Some of the others have to be extremely pure to become superconducting even at temperatures below 1°K. The ferromagnetic elements and those which are highly conducting are not superconductors.

The principal ferromagnetic elements are iron, cobalt and nickel. The Curie points are 770°C, 1131°C and 358°C, the coercive forces are 0.05, 10 and 0.7 oersted and the saturation magnetization values are 1710, 1425 and 550 gauss, respectively.

Some of the metals with other interesting properties are iron, the most versatile structural material, the liquid metal mercury, nickel the most versatile alloying metal with corrosion-resistance, platinum, rhodium and iridium, with the greatest resistance to oxidation at high temperatures, and tungsten, molybdenum, tantalum and niobium, with the highest melting points. In addition, titanium, aluminum, beryllium, and boron are light structural materials which are extremely important in aerospace applications. Germanium and silicon are semiconducting elements, gold and silver are jewelry materials, zinc is used in batteries, and chromium is used as a protective coating on other metals. Cobalt, chromium, vanadium and molybdenum are used in alloying nickel to form high-temperature materials.

Properties of Alloys

When one element is added to another, materials with interesting properties can result. If nickel is added to copper, a complete range of solid solution is obtained (see Fig. 2.6a). Note that if any melt is cooled, the temperature at which it begins to solidify is the point where a vertical line intersects the liquidus curve and it is completely solidified at the temperature where the

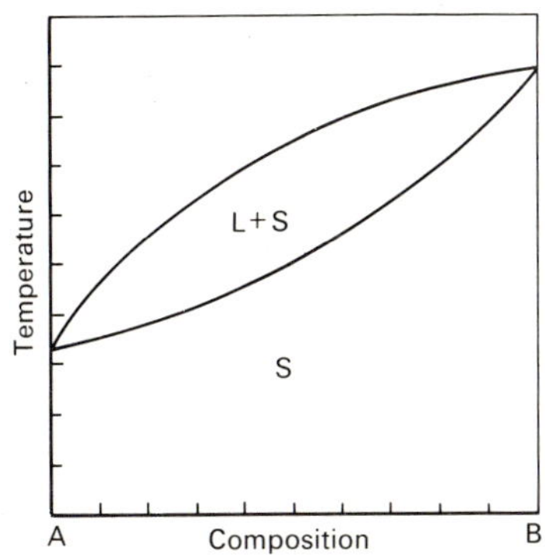

(a) Phase diagram showing solid solution

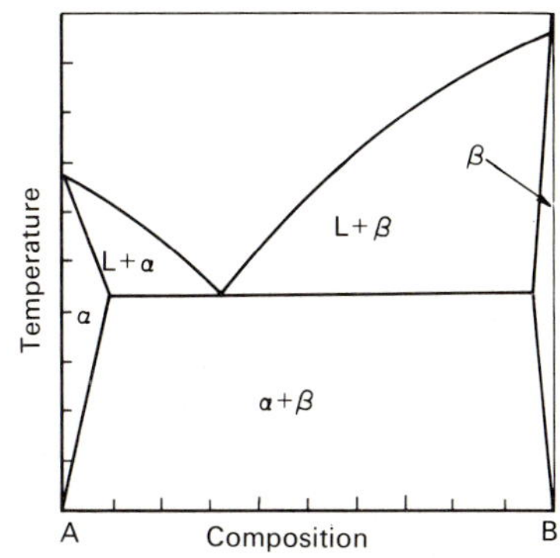

(b) Phase diagram showing eutectic formation

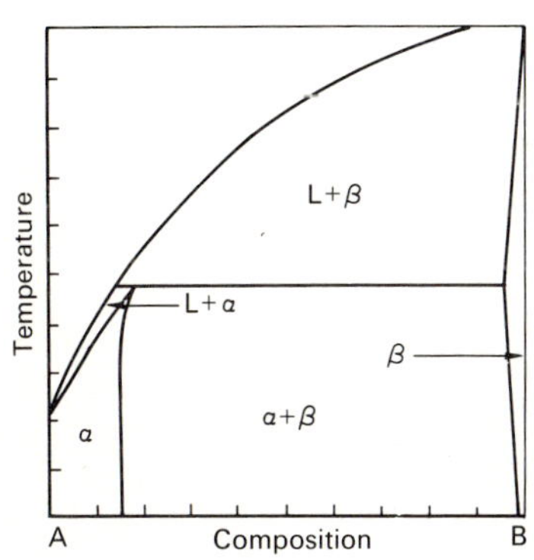

(c) Phase diagram showing peritectic reaction

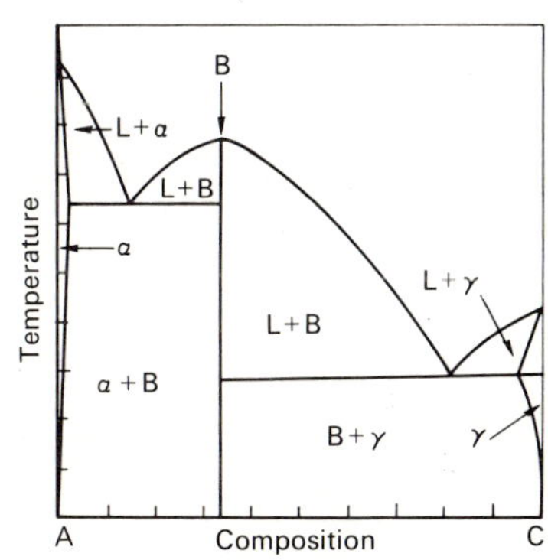

(d) Phase diagram showing congruently melting intermediate compound formation

FIG. 2.6

vertical line crosses the solidus curve. At this point a single phase with the cubic close-packed structure is present. The lattice constant, thermal expansion, specific heat and specific volume vary nearly linearly with composition. On the other hand, the tensile strength, hardness and electrical resistivity pass through a maximum while the ductility goes through a minimum.

In a eutectic system, one liquid crystallizes to form two solid phases (see Fig. 2.6b). Contrary to what might be expected, such properties as hardness, elongation and electrical resistivity do not vary linearly with composition. Often the largest irregularities in properties can be observed at the eutectic point. The low melting point of the eutectic composition makes some of these alloys useful as solders.

Recently, eutectics have been unidirectionally solidified to produce whiskers of one material aligned in the other. When the whiskers have a high modulus and the matrix is ductile, strengths several times as great as those of the as-cast structures have been obtained.

When one metal is added to another and forms a peritectic system, as shown in Fig. 2.6c, the occurrence of nonequilibrium structures make the properties of the phases unpredictable.

In other systems, intermetallic compounds with sharp compositions are formed (see Fig. 2.6d). These are true intermetallic compounds. In other cases, intermediate phases with ranges of composition are formed and these are called secondary solid solutions. Both of these type phases can melt congruently or decompose into two phases on melting. The properties of these intermediate phases can be quite different from those of the elements which make them up.

In the chapters that follow, the structures of alloys, intermetallics, halides, hydrides, oxides, etc., will be described in structurally related groups. In the tables included in these chapters, borides, carbides, halides, hydrides, nitrides, oxides and sulfides, selenides and tellurides are treated separately from the intermetallics while alloys and metal compounds have been included with the intermetallics.

CHAPTER 3

THE CsCl TYPE AND RELATED STRUCTURES

IN ARRANGING structures into related groups, the CsCl type structures have been designated as those which can be described by unit cells with an atom of one element at the corner and an atom of another element or structural group in the center. Also included are related structures with unit cells which contain several unit cells of the CsCl type. The relationship between these structures is shown in the flow sheet, Table 3.0.

TABLE 3.0. CsCl *Type Structures*

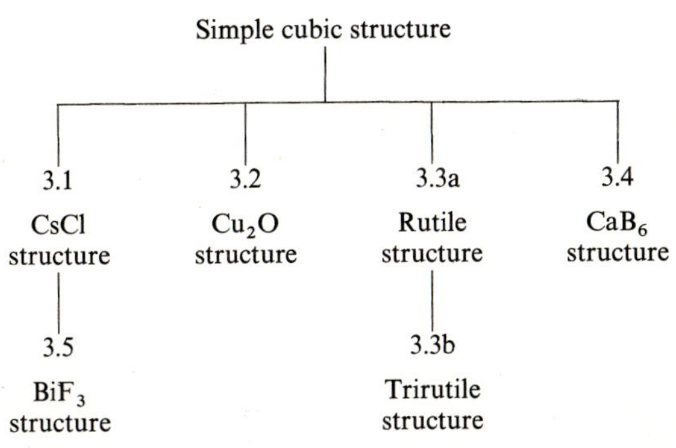

The CsCl, Cu_2O, rutile and CaB_6 structures can be derived from the simple cubic structure, the BiF_3 unit cell from four unit cells with a CsCl type arrangement and four unit cells of the body-centered cubic structure and the trirutile unit cell from three unit cells of the rutile structure with an ordered arrangement of metal atoms.

3.1. CsCl Structure, B2, *Pm3m*, Cubic

The CsCl structure is the simplest of this group because it can be described by a cesium ion at the corner (0, 0, 0) and a chlorine ion at the center ($\frac{1}{2}, \frac{1}{2}, \frac{1}{2}$)

of a cubic unit cell (see Fig. 3.1). Note that the atomic positions are the same as those in the body-centered cubic structure, but the ions at the corner and center of the unit cell are different. For example, β-brass, CuZn, has the body-centered cubic structure when the atoms are randomly arranged, but when the atoms order it has the CsCl structure.

Many of the AB type intermetallic and a few halide compounds have the CsCl structure (see Table 3.1). In general, compounds with the largest radius ratios (R_A/R_B) tend to crystallize with this structure because of the large

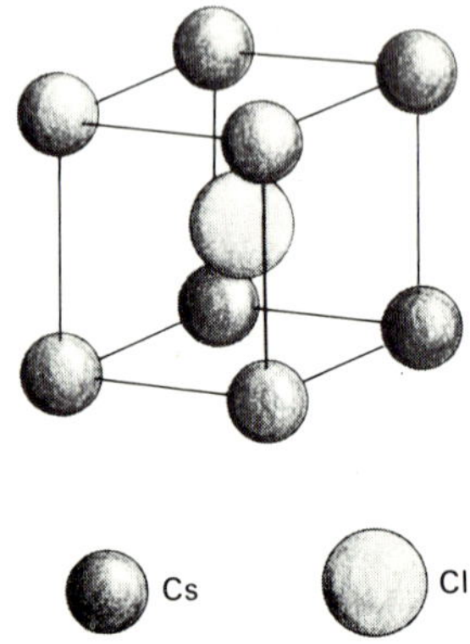

FIG. 3.1 The cesium chloride structure

coordination number (8) of the atoms. Of the alkali halides, only CsCl, CsBr and CsI have the CsCl structure under standard conditions, while the rest, most of which have R_A/R_B ratios between 0.41 and 0.73, have the sodium chloride structure.

Properties

Some of the halides listed in Table 3.1 have outstanding optical properties. Cesium bromide, cesium iodide, thallium chloride and thallium bromide transmit in part of the ultraviolet, all of the visible and the near infrared. This will be discussed further in Chapter 5.

Aside from β-brass, which has been widely studied, AlNi is one of the most interesting metallic phases. It has gained considerable attention because it is deformable above 1300°C, has a melting point of 1650°C and has excellent resistance to oxidation. Two other intermetallics, CrAl and CoAl, are not quite as oxidation resistant, but have the advantage of higher melting points.

TABLE 3.1. *Phases with the CsCl Structure*

Phase	Cell size a_0(Å)	Refs.
Intermetallics		
β-AgCd	3.332	1
AgCe	3.731	2
AgLa	3.760	2
β-AgLi	3.174	3
β-AgMg	3.28	4
AgNd	3.714	6
AgY	3.6196	5
AgYb	3.6787	5
β-AgZn	3.150	7, 8
β-AlCo	2.862	9
AlFe	2.903	10
AlNd	3.73	11
β-AlNi	2.881	12, 13
AlPd (H.T.)	3.036	14
AlSc	3.450	15
β-AuCd	3.34 (400°C)	16
AuMg	3.259	17
β-AuMn	3.255	18
AuYb	3.5634	5
β-AuZn	3.19	19
BaCd	4.215	20
BaHg	4.133	20
BeCo	2.606	19
BeCu	2.698	21
BeNi	2.616	22
BePd	2.813	19
CaTi	3.847	3
CdCe	3.86	23
CdEu	3.960	24
CdLa	3.90	23
CdPr	3.82	23
CdSr	4.011	20
CeIIg	3.815	25
CeMg	3.898	26
CeZn	3.70	23
CoFe	2.857	27
CoSc	3.16	28
CuEu	3.479	5
β-CuPd	2.988	29
CuY	3.4757	5
β-CuZn	2.945	7
ErTl	3.704	30
EuZn	3.808	24
β-GaNi	2.879	31
GaRh	3.01	32
HgLi	3.287	3
HgMg	3.44	17
HgMn	3.315	32
HgNd	3.78	25
HgPr	3.80	25

TABLE 3.1 (*cont.*)

Phase	Cell size a_0(Å)	Refs.
HgSr	3.930	20
HoIn	3.774	30
HoTl	3.720	30
InLa	3.985	5
InPd	3.53	33
InPr	3.955	5
InTm	3.739	5
InYb	3.8138	5
IrLu	3.330	34
LaMg	3.965	26
LaZn	3.75	23
β-LiPb (H.T.)	3.529	35
LiTi	3.424	3
LuRh	3.334	34
MgPr	3.88	36
MgSc	3.597	15
MgSr	3.900	25
MgTi	3.628	3
β-MnPd	~3.16	37
β-MnRh	3.051	38
OsTi	3.07	39
PrZn	3.67	23
RhY	3.410	34
α-RuSi	2.90	49
RuTi	3.06	39
SbTi	3.84	41
ScRh	3.204	42
SrTi	4.024	3
TeTh	3.827	43
TlTm	3.711	44
TmRh	3.358	42
Halides		
CsBr	4.286	45, 46, 47
CsCl	4.123	47
CsI	4.567	48, 49, 50
NH_4Br	4.0594	51
NH_4Cl	3.8756	51
NH_4I (L.T.)	4.37 (−17°C)	52
RbCl (H.T.)	3.74 (190°C)	53
TlBr	3.97	50
TlCl	3.8340	50
TlI	4.198	54
Sulfides, Selenides, Tellurides		
ThTe	3.827	55
Other phases		
CsCN	4.25	56
$CsNH_2$ (H.T.)	4.063 (50°C)	57
CsSH	4.30	58
CsSeH	4.437	58
TlCN	3.82	59

3.2. Cu_2O, Cuprite, C3, *Pn*3*m*, Cubic

The structure of Cu_2O can be described by a cubic unit cell containing an oxygen ion at the corners, and a Cu_4O unit in the center. In the Cu_4O unit, a linear Cu–Cu unit above the oxygen ion is perpendicular to a linear Cu–Cu unit below the oxygen ion, forming a tetrahedron. The orientation of the Cu_4O unit can be seen in Fig. 3.2.

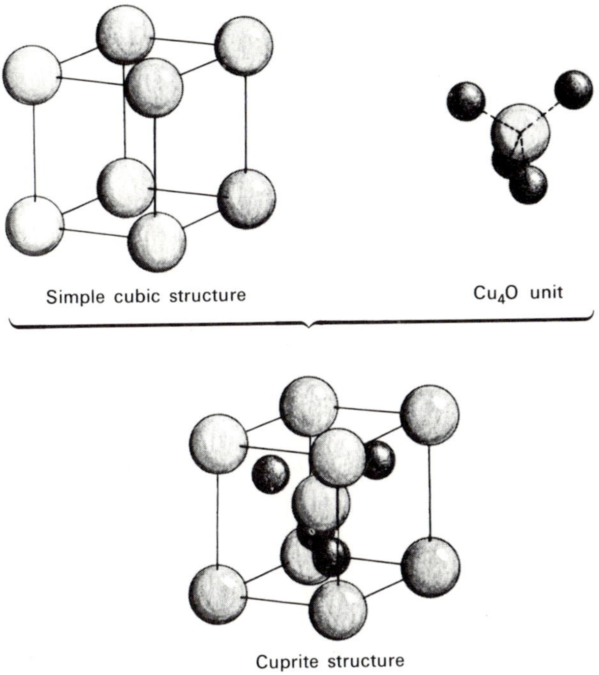

FIG. 3.2 The cuprite structure

The oxygen ions are at 0, 0, 0 and $\frac{1}{2}, \frac{1}{2}, \frac{1}{2}$ and the copper ions are at $\frac{1}{4}, \frac{1}{4}, \frac{1}{4}$; $\frac{1}{4}, \frac{3}{4}, \frac{3}{4}$; $\frac{3}{4}, \frac{1}{4}, \frac{3}{4}$ and $\frac{3}{4}, \frac{3}{4}, \frac{1}{4}$.

The oxides Ag_2O and Pb_2O and the cyanides of cadmium and zinc, listed in Table 3.2, also have the cuprite structure. However, in the structure of the cyanides, the cyanide ions are in the copper ion positions and the metal ions are in the oxygen ion positions. They are said to have the anticuprite structure.

The coordination number four for silver and copper ions in the cuprite structure is quite unusual for these ions. Since it certainly is not dictated by their sizes it must be attributed to the tendency to form covalent bonds.

Properties

As was discussed in Chapter 1, the oxides Cu_2O and Ag_2O are *p*-type semiconductors because they contain excess oxygen atoms. The energy gap in

TABLE 3.2. *Phases with the Cuprite Structure*

Phase	Cell size, a_0(Å)	Refs.
Oxides		
Ag_2O	4.72	1
Cu_2O	4.2696	2
Pb_2O	5.38	3
Others		
$Cd(CN)_2$	6.32	4
$Zn(CN)_2$	5.905	5

Cu_2O is approximately 1.5 eV, and the impurity levels (acceptors) are about 0.3 to 0.6 eV above the edge of the valence band. At room temperature the thermoelectric power of Cu_2O is 700 μV/°C.

The semiconductor properties of Cu_2O have been studied for over twenty-five years. This interest stemmed from the fact that it was used in the construction of the first rectifiers and photocells. These rectifiers contain a layer of Cu_2O with excess copper ions next to the *p*-type Cu_2O, which is a condition for rectification.

3.3a. Rutile, TiO_2 Cassiterite, SnO_2, C4, $P4_2/mnm$, Tetragonal

The rutile structure also can be visualized as having a structure related to that of CsCl. It can be described by a tetragonal unit cell containing titanium atoms at the corner and a distorted TiO_6 octahedra at the center (see Fig. 3.3a). The octahedra are distorted so that each titanium ion is surrounded by four nearest oxygen ions and two more at a slightly longer distance. The titanium ions are located in the tetragonal unit cell at $0, 0, 0; \frac{1}{2}, \frac{1}{2}, \frac{1}{2}$ and the oxygen ions are $x, x, 0; \bar{x}, \bar{x}, 0; \frac{1}{2} + x, \frac{1}{2} - x, \frac{1}{2}; \frac{1}{2} - x, \frac{1}{2} + x, \frac{1}{2}$, with $x = 0.3$ for TiO_2.

The dioxides and difluorides which assume the rutile structure are listed in Table 3.3a.

Properties

The mineral rutile, which has the chemical formula TiO_2, is probably the best known compound with this structure. It is the simplest compound of a family of titanates with high dielectric constants ranging from ($\varepsilon_{\parallel} = 167$, $\varepsilon_{\perp} = 83$) for rutile to several thousand for $BaTiO_3$. All of these compounds are related in that they have structures which contain metal ions in octahedrally coordinated positions.

The energy gap for rutile is 3.05 eV, and the intrinsic resistivity at room temperature is more than 10^{13} ohm-cm. Other properties of rutile are a melting point of 1840°C, a thermal conductivity of 0.016 cal/sec-cm-°C, and a coefficient of thermal expansion of $8.8 \times 10^{-6}/°C$. The melting points of TeO_2 and SnO_2 are 733°C and 1630°C.

Of the other oxides in Table 3.3a, CrO_2 is ferromagnetic with a Curie

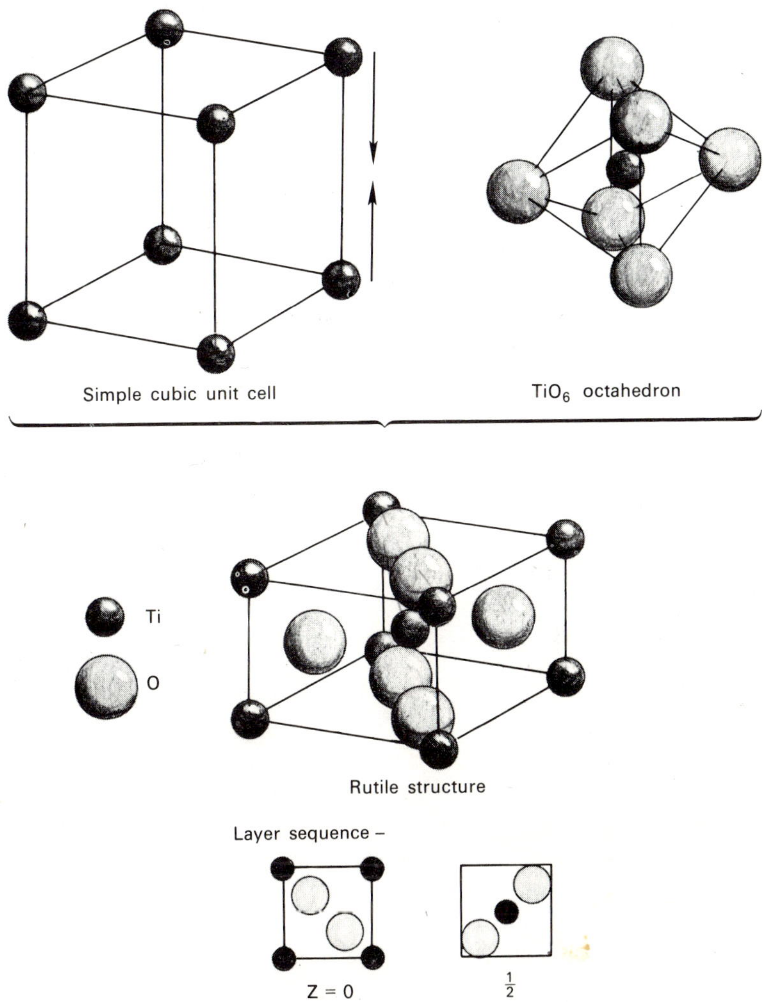

FIG. 3.3a The rutile structure

point of 116°C, and VO_2 and MnO_2 are antiferromagnetic with Néel points of 343°K and 84°K. Several fluorides, MnF_2, FeF_2, CoF_2 and NiF_2, also are antiferromagnetic materials with Néel points of 75°K, 90°K, 45°K and 85°K respectively.

The electrical resistivity of CrO_2 decreases as a function of temperature between −150°C and 350°C except for a maximum at 116°C. Anomalies in resistivity for these types of materials at Néel points have been attributed to the uncoupling of electrons at these temperatures. Because these materials exhibit resistance anomalies at specific temperatures they have been considered for thermistor applications.

TABLE 3.3a. *Phases with the Rutile Structure*

Phases	Cell size (Å)		Refs.
	a_0	c_0	
Halides			
CoF_2	4.695	3.180	1
FeF_2	4.697	3·309	1
MgF_2	4.623	3.052	2
MnF_2	4.873	3.310	3
NiF_2	4.651	3.084	1, 3
PdF_2	4.931	3.367	4
ZnF_2	4.703	3.134	1, 3
Hydrides			
$Mg(D_{0.9}H_{0.1})$	4.5025	3.0123	5
Nitrides			
Ti_2N	4.9452	3.0342	6
Oxides			
δ-CrO_2	4.41	2.91	7
GeO_2	4.395	2.859	8
IrO_2	4.49	3.14	9
α-MnO_2	4.395	2.86	10
MoO_2†	4.86	2.79	11
NbO_2	4.77	2.96	9
OsO_2	4.51	3.19	9
PbO_2	4.946	3.379	12
RuO_2	4.51	3.11	9
SiO_2 (H.P.)			13
SnO_2	4.737	3.185	8
δ-TaO_2	4.709	3.065	14
TeO_2	4.79	3.77	13
TiO_2 (rutile)	4.5929	2.9591	15
WO_2†	4.86	2.77	11
$AlAsO_4$ (H.P.)	4.359	2.815	16
$CrVO_4$ (H.P.)	4.551	2.844	17

† Distorted form.

3.3b. Trirutile, $ZnSb_2O_6$, *P4/mmm*, Tetragonal

The unit cell of $ZnSb_2O_6$ can be described as three unit cells similar to that of the rutile structure stacked on one another with the zinc and antimony atoms filling the titanium positions so that the vertical edges have a zinc atom at the base and two antimony atoms above it, while the $\frac{1}{2}, \frac{1}{2}$, *z*-line (vertical line up the center of the cell) has an antimony atom at the center of the top and bottom unit cells and a zinc atom at the center of the middle cell (see Fig. 3.3b).

2 Zn atoms are at 0, 0, 0 and $\frac{1}{2}, \frac{1}{2}, \frac{1}{2}$,
4 Sb atoms are at $\pm(0, 0, z; \frac{1}{2}, \frac{1}{2}, z + \frac{1}{2})$ with $z = \frac{1}{3}$,
4 O atoms are at $\pm(x, x, 0; x + \frac{1}{2}, \frac{1}{2} - x, \frac{1}{2})$,
and 8 O atoms are at $\pm(x, x, y; \bar{x}, \bar{x}, y; x + \frac{1}{2}, \frac{1}{2} - x, y + \frac{1}{2}; \frac{1}{2} - x, x + \frac{1}{2}, y + \frac{1}{2})$ with $x = 0.303$, $y = 0.328$.

Other oxides with this structure are listed in Table 3.3b.

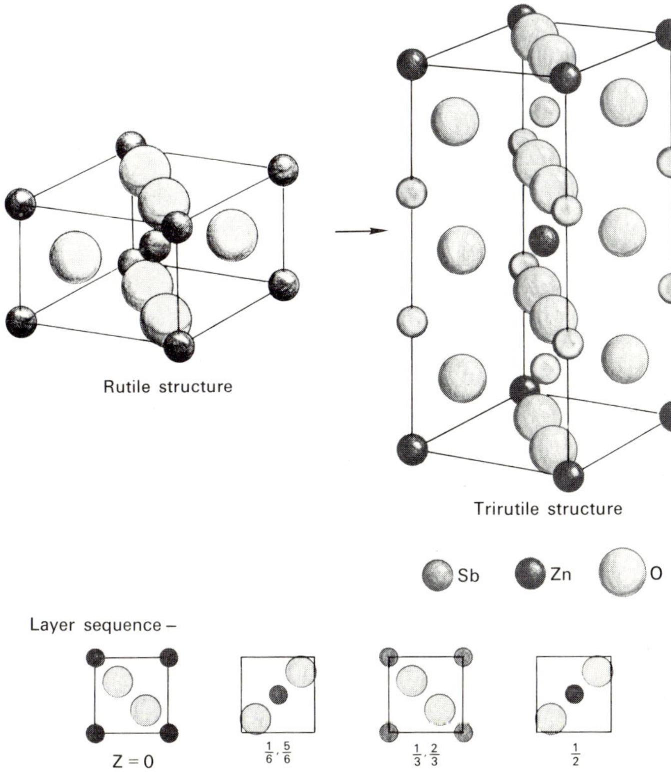

FIG. 3.3b The trirutile structure

The size of the cations in compounds which form this structure has to be quite similar for them to retain sixfold coordination. The ordering of these ions, therefore, must be attributed to the difference in charge. The importance of size and charge differences in causing ordering of cations has been clearly demonstrated in a study of complex perovskite type compounds (see Section 7.4).

TABLE 3.3b. *Compounds with the Trirutile Structure*

Compound	Cell size (Å)		Refs.
	a_0	c_0	
Oxides			
$CoSb_2O_6$	4.64	9.25	1
$CoTa_2O_6$	4.73	9.16	1
$FeSb_2O_6$	4.62	9.19	1
$MgSb_2O_6$	4.63	9.21	1
$MgTa_2O_6$	4.713	9.187	1, 2
$NiSb_2O_6$	4.63	9.18	1
$NiTa_2O_6$	4.70	9.10	1
$ZnSb_2O_6$	4.66	9.24	1

3.4. CaB_6 Type, $D2_1$, *Pm3m*, Cubic

The CaB_6 structure can be represented by a cubic unit cell with a calcium atom at the corner of the cube and an octahedron of boron atoms in the center. The octahedron is arranged so that each atom in it is approximately 0.2a from the center of one of the cube faces (see Fig. 3.4). Thus a calcium atom is at 0, 0, 0 and the boron atoms are at $x, \frac{1}{2}, \frac{1}{2}; \frac{1}{2}, x, \frac{1}{2}; \frac{1}{2}, \frac{1}{2}, x; \bar{x}, \frac{1}{2}, \frac{1}{2}; \frac{1}{2}, \bar{x}, \frac{1}{2}; \frac{1}{2}, \frac{1}{2}, \bar{x}$ where $x \sim 0.2$. If the B_6 unit is placed at the origin of the unit cell, then the calcium atom is at the center. In this arrangement it is easily seen that the calcium atom is surrounded by twenty-four boron atoms.

TABLE 3.4a. *Compounds with the* CaB_6 *Structure*

Compound	Cell size (Å)	*x*, B	Refs.
Borides			
BaB_6	4.268	0.207	1
CaB_6	4.144	0.207	1
CeB_6	4.1410		1
DyB_6	4.0976		2
ErB_6	4.102	0.20	3
GdB_6	4.112	0.207	4
LaB_6	4.156	0.207	1
LuB_6	4.11		3
NdB_6	4.128	0.207	1
PrB_6	4.121	0.20	5
SiB_6	4.150		6
SmB_6	4.1333		4
SrB_6	4.1984	0.207	1
TbB_6	4.1020		2
ThB_6	4.113	0.207	1
TmB_6	4.110		7, 8
YB_6	4.113	0.207	9
YbB_6	4.144	0.207	1

Other important higher boride structures, UB_4 and UB_{12}, contain B_6 and B_{12} units respectively. The B_6 units in the UB_4 structure lie in planes and are joined together with pairs of boron atoms, while the uranium atoms lie in planes above and below. The UB_{12} structure contains uranium atoms and B_{12} units in a sodium chloride arrangement, and the coordination number of the uranium atom is the same as it is in CaB_6.

The hexaborides which have the CaB_6 structure and the lattice parameter for their unit cells are listed in Table 3.4a.

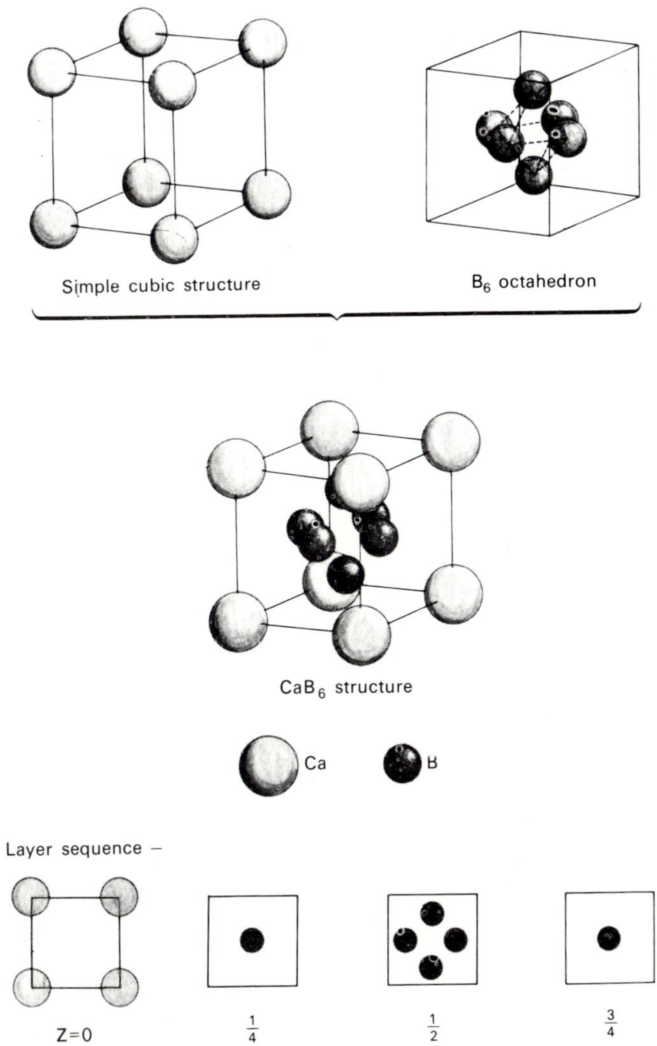

FIG. 3.4 The CaB_6 structure

Properties

These compounds have been widely investigated as thermionic emitters. They are considered high temperature materials. Their properties are listed in Table 3.4b.

TABLE 3.4b. *Properties of* CaB_6 *Type Compounds*

Compound	Melting point (°C)	Electrical resistivity (μohm-cm)	Thermal e.m.f. (μV/°C)	Micro-hardness (kg/mm²)	Young's modulus (10^6 psi)	Thermal conductivity cal/cm-deg-sec	Coeff. of thermal expansion (10^{-6}/°C)
			CaB_6 Structure Type				
BaB_6	>2100[1,2,3]	77[9]	−26.2[9]	3000[13]	55.9[18]		6.47[20]
BeB_6	2300[4]	25×10^{8}[10]					
CaB_6	2100[1,2,3]	222[9]	−32[9]	2700[13]	65·5[18]		6.5[21]
CeB_6	2190[5]	29.4[9]	2.8[9]		54.9[18]	0.081[19]	6.67[21]
EuB_6		85[9]	−17.7[9]			0.055[19]	6.9[20]
GdB_6	>2100[6]	45[9]	0.1[9]			0.049[19]	8.7[21]
LaB_6	>2100[1,2,3]	15[9]		2770[13]	69.4[18]	0.114[19]	5.75[20]
NdB_6	2540[7]	20[9]	0.4[9]	2540[7]		0.113[19]	7.3[21]
PrB_6		20[9]	−0.6[9]			0.098[19]	7.5[21]
ScB_6			−7.7[9]				
SmB_6	2400[8]	207[9]	7.6[9]	2500[14]		0.033[19]	6.47[20]
SrB_6	>2100[1,2,3]	192[11]	30.3[9]	2920[15]		0.048[19]	6.7[21]
TbB_6		37[9]	−1.1[9]				7.8[21]
ThB_6	>2100[1]	15[9]	−0.6[9]	1740[16]			7.8[21]
YB_6		41[12]	−0.5[9]	3264[17]		0.060[19]	6.2[21]
YbB_6	1538[8]	47[9]	−25.5[9]				

3.5. BiF_3, $D0_3$, *Fm3m*, Cubic

The BiF_3 structure can be described by eight small simple cubic unit cells (octants) having fluorine atoms at the corners which are joined together to form a large cubic cell, 5.853 Å on edge. Bismuth atoms are in the centers of four of the small cells (type I octants) and fluorine atoms at the center of the four other cells (type II octants) (see Fig. 3.5). The bismuth atoms are located in diagonally opposite octants to one another in the two top and two bottom octants, and are not in the octants directly above one another.

The Bi atoms are at $x, x, x; x, \bar{x}, \bar{x}; \bar{x}, x, \bar{x}; \bar{x}, \bar{x}, x$ with $x \simeq \frac{3}{4}$.
Four F atoms are at $x, x, x; x, \bar{x}, \bar{x}; \bar{x}, x, \bar{x}; \bar{x}, \bar{x}, x$ with $x = \frac{1}{4}$
and eight more F atoms are at $0, 0, 0; \frac{1}{2}, \frac{1}{2}, \frac{1}{2}; 0, \frac{1}{2}, \frac{1}{2}; \frac{1}{2}, 0, \frac{1}{2}; \frac{1}{2}, \frac{1}{2}, 0; \frac{1}{2}, 0, 0; 0, \frac{1}{2}, 0; 0, 0, \frac{1}{2}$.

Table 3.5 lists a number of phases which have this structure.

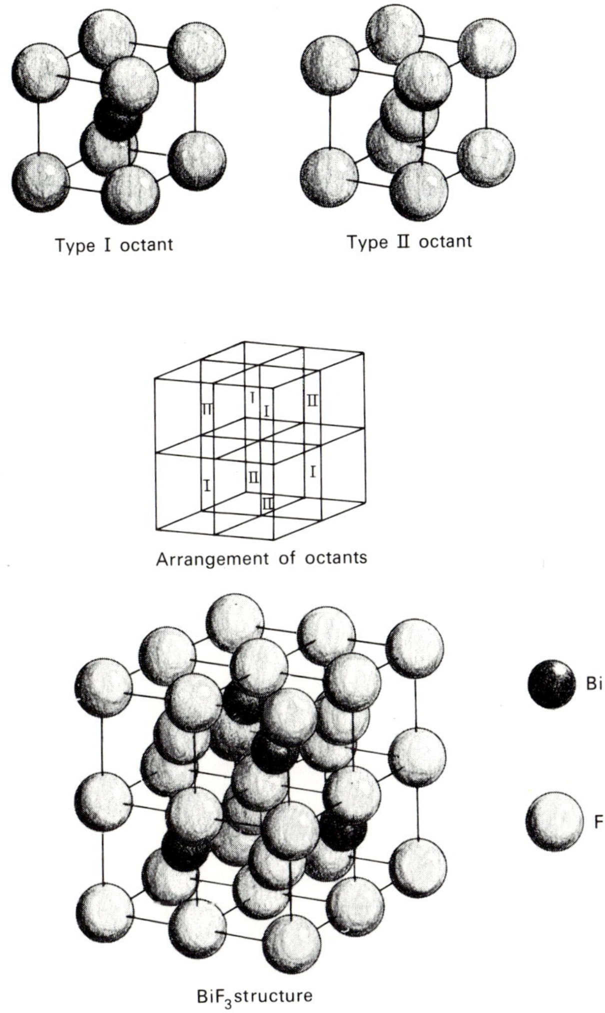

FIG. 3.5 The BiF_3 structure

Properties

If two elements were randomly arranged in this structure, only one octant of the large unit cell would be necessary to describe the structure. This is what was observed when less than 18 at.% Al was heated with iron and slowly cooled. However, upon standing the alloy began to order. Materials such as these are valuable in studying the effect of ordering on various properties. The compound Fe_3Al is a particularly interesting material for these studies because it is ferromagnetic. It has a Curie point of 500°C. The silicide, Fe_3Si, has also been well studied. It has a melting point of 1300°C, a thermal conductivity of 0.04 cal/cm-deg-sec, a coefficient of thermal expansion of 11×10^{-6}/°C and a resistivity of 130×10^{-6} ohm-cm.

TABLE 3.5. *Phases with the* BiF_3 *Structure*

Phases	Cell size a_0(Å)	Refs.
Intermetallics		
$AlFe_3$	5.792	1
BiK_3	8.805	2
$BiLi_3$	6.721	3
$CeMg_3$	7.436	4
$(Cu, Ni)_3Sb$	5.869	5
$(Cu, Ni)_3Sn$	5.958	5
Fe_3Si	5.655	6
$GdMg_3$	7.31	7
$HgLi_3$	6.5610	8
$LaMg_3$	7.480	4
β-Li_3Sb	6.572	9
Mg_3Nd		10
Mg_3Pr	7.388	11
Mg_3Tb		10
Hydrides		
$LaH_{3.0}$	5.60	11

The NaTl *Structure*

The NaTl structure also is similar to the BiF_3 and Cu_2AlMn structures. The sodium atoms are in the aluminum atom positions and the thallium atoms are in the manganese atom positions. In the center of the octants of the unit cell, the sodium and thallium atoms alternate in the same way as the bismuth and fluorine atoms do in the BiF_3 structure. Since the coordination number about each atom is eight, four of each kind, intermetallics such as LiGa and LiZn containing atoms with large differences in atomic radii tend to form this structure in preference to those in which the coordination number is six.

3.6. Discussion

Structure (Relations)

The unit cell of the CsCl structure is the simplest cell described in this chapter. It contains cesium ions at the corners of a cube and chlorine ions at the center, or vice versa. In the unit cell of the cuprite structure, a Cu_4O tetrahedron is at the center of a cube containing oxygen atoms at the corners. The CaB_6 structure contains an octahedron of boron atoms in a cubic unit cell with calcium atoms at the corners and the rutile structure can be described by a TiO_6 octahedra situated in a flattened unit cell containing a titanium atom at the corners. The trirutile structure is described by an ordered arrangement of metal atoms in unit cells of the rutile structure stacked on one another and the unit cell of the BiF_3 structure is made up of four small unit cells containing BiF in a CsCl type arrangement and four fluorine small unit cells in a body-centered cubic arrangement.

Intermetallics

Studies of the phases with each of the structures show that intermetallics with large atomic radius ratios tend to form either the CsCl or NaTl structure where the atoms are in eightfold coordination. In the former structure, each atom is surrounded by eight neighboring atoms of a different element, while in the latter each atom is coordinated with four of its own atoms as well as four atoms of the other element.

The intermetallic CuZn adopts the CsCl structure on ordering and the phase Fe_3Al forms an ordered BiF_3 structure. Ordering in these metallic phases is normally an extremely sluggish process and therefore requires slow cooling and/or annealing of the hot ingots. Since alloy samples with different degrees of ordering can be obtained by quenching at various cooling times, the effects of ordering on a number of properties have been investigated. The metallic phases Cu_3Au and Fe_3Al have been the subject of many of these studies. In the NaTl structure thallium atoms form a three-dimensional network with sodium atoms in between. Phases such as these (i.e. NaTl, LiZn and LiCd), where the B post-transition metal atoms determine the dimensions of the unit cell and electro-positive elements are placed in the holes, are called Zintl phases.

Other phases of particular interest are the β-phases, such as AgZn, AgCl, AuZn, CoAl, CuZn, FeAl and NiAl, which have the CsCl structure and electron-to-atom ratios (e/a) of three to two. Hume-Rothery's rules state that when a metal of higher valency is added to a face-centered cubic metal of lower valency a series of intermediate phases may be formed at (e/a) ratios of $\frac{3}{2}$, $\frac{7}{4}$ and $\frac{21}{13}$. As the composition is changed a phase having a face-centered cubic lattice is followed by a phase with a body-centered cubic lattice or a β-manganese-type phase with an (e/a) ratio of $\frac{3}{2}$. This phase is followed by a complex cubic γ-phase with a $\frac{7}{4}$ (e/a) ratio and then a ε-phase with a close-packed hexagonal structure and a $\frac{21}{13}$ (e/a) ratio. The β-phases mentioned above are the second phases to appear in a succession of phases predicted by Hume-Rothery. Of course, when they are ordered they have the CsCl structure and not the body-centered cubic structure. A typical phase diagram is shown in Fig. 3.6. Note the dotted lines in the β-phase region, indicating the range of temperature where the β-phase begins to order and also note that each phase has a region of homogeneity.

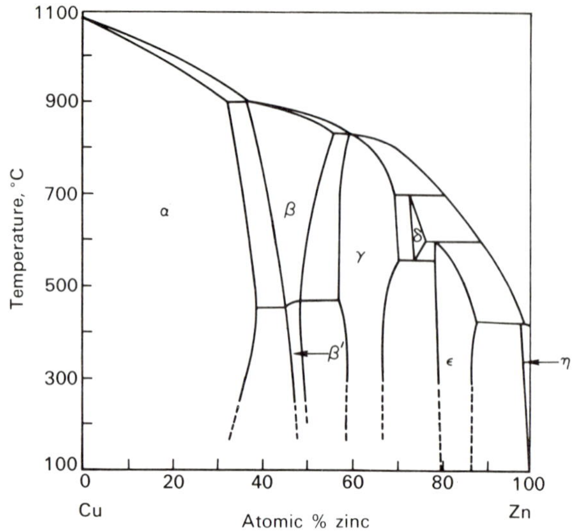

FIG. 3.6 Cu–Zn phase diagrams

Borides

The higher borides discussed in this chapter form the UB_4, CaB_6 and UB_{12} structures which contain three-dimensional arrangements of boron atoms, B_6 units joined by B atoms in UB_4, B_6 units in CaB_6 and B_{12} units in the UB_{12} structure. In subsequent chapters it will be seen that large B units are present in the structure of boron itself, isolated B atoms in the $CuAl_2$ and CaF_2 structures, and as B layers in the AlB_2 structure.

Halides

Of the alkali halides CsCl, CsBr and CsI are the only ones which form the CsCl structure. The ammonium halides NH_4Cl and NH_4Br, and the thallous bromide, chloride and iodide also have this structure. Most of the other halides crystallize with the NaCl structure.

The fluorides with the MF_2 formula, aside from BeF_2, adopt either the rutile or CaF_2 structure. Like the oxides, the smaller metal atoms form fluorides with the rutile structure and the larger ones prefer the eightfold coordination in the fluorite structure. Thus, most of the rutile type difluorides are those of the transition metals, i.e. CoF_2, FeF_2 and MnF_2.

The structure of BiF_3 is an unusual one for any of the trihalides.

Oxides

All of the oxides included in this chapter have the rutile or cuprite structure.

Most of the dioxides have the rutile structure, although a few of the larger metal ions form dioxides with the fluorite structure where the coordination number of the cation is eight instead of six. For example, TiO_2 has the rutile structure while ZrO_2 and HfO_2 have distorted and cubic CaF_2 structures respectively. Other dioxides of transition metals are shown below under the group number of the element. As would be expected, the dioxides of groups V and VI are not stable in air and will oxidize upon heating to

IV	V	VI	VII	VIII
TiO_2		CrO_2	MnO_2	
	NbO_2†	MoO_2†		RuO_2
	TaO_2	WO_2†		OsO_2 IrO_2

† Distorted or defect structure.

Nb_2O_5, Ta_2O_5, MoO_3 and WO_3 in which the valence state of the metal ion is the same as the group number. The stable oxide of chromium in air is the sequioxide Cr_2O_3. The tendency to form higher oxides from the top to bottom transition metal in any group is characteristic of these metals. In the post-transition metals the reverse is true so that while GeO_2, SnO_2 and PbO_2 form the rutile structure, PbO is more stable than PbO_2 at high temperatures.

When two elements are substituted for titanium in TiO_2 the average valence of the ions must equal 4+. In addition, the sizes of the ions must be nearly the same or the rutile structure will not be formed. Compounds such as $CrNbO_4$ and $FeTaO_4$ meet these requirements. On the other hand, the structure of compounds such as Sb_2ZnO_6 and Ta_2MgO_6 contain an ordered arrangement of cations, and a unit cell with three times the *c*-axis of the unit cell of the rutile structure is required to describe the structure. Since these cations are also similar in size, the ordering must be attributed to the charge differences.

The structure of Ag_2O, Cu_2O and Pb_2O are not typical of oxides in general. The fourfold coordination for the Cu ions is quite unusual.

Table 3.1. *References*

1. E. A. Owen, J. Rogers and J. C. Guthrie, *J. Inst. Met.* **65**, 457 (1939).
2. H. Bommer and E. Krose, *Z. Anorg. Allgem. Chem.* **252**, 62 (1943).
3. E. Zintl and G. Brauer, *Z. Physik. Chem.* **20B**, 245 (1933).
4. E. A. Owen and G. D. Preston, *Phil. Mag.* **2**, 1266 (1926).
5. J. L. Moriarty, J. E. Humphreys, R. O. Gordon and W. C. Baenziger, *Acta Cryst.* **21**, 840 (1966).
6. R. Ferro, *Gazz. Chim. Ital.* **85**, 888 (1955).
7. A. Westgren and G. Phragmen, *Phil. Mag.* **50**, 311 (1925).
8. A. Roux and J. Cournot, *Compt. Rend.* **188**, 1399 (1929).
9. A. J. Bradley and G. C. Seager, *J. Inst. Met.* **64**, 81 (1939).
10. A. J. Bradley and A. H. Jay, *J. Iron and Steel Inst.* **125**, 339 (1932).
11. C. W. Stillwell and E. E. Jukkola, *J. Am. Chem. Soc.* **56**, 56 (1934).
12. R. W. G. Wyckoff and A. H. Armstrong, *Z. Krist.* **72**, 319 (1929).

13. J. BRADLEY and A. TAYLOR, *Proc. Roy. Soc. (London)* **159** A, 56 (1937).
14. H. PFISTERER and K. SCHUBERT, *Naturwiss.* **37**, 112 (1950).
15. E. PARTHE, D. HOHNKE, W. JEITSCHKO and O. SCHOB, *Naturwiss.* **52**, 155 (1965).
16. A. OLANDER, *Z. Krist.* **83**, 145 (1932).
17. G. BRAUER and W. HAUCKE, *Z. Physik. Chem.*, **33B**, 304 (1936).
18. E. RAUB, U. ZWICKER, and H. BAUR, *Z. Metallk.* **44**, 312 (1953).
19. L. MISCH, *Metallwirtschaft* **15**, 163 (1936).
20. R. FERRO, *Acta Cryst.* **7**, 781 (1954).
21. L. MISCH, *Z. Physik. Chem.* **29B**, 42 (1935).
22. L. LOSANA and C. GLORIEA, *Alluminio* **11**, 17 (1942).
23. A. IANDELLI and R. BOTTI, *Rend. Gazz. Chim. Ital.* **67**, 638 (1937).
24. A. IANDELLI and A. PALENZONA, *Atti. Accad. Naz. Lincei, Rend. Classe Sci. Fis.* **37**, 165 (1964).
25. A. IANDELLI and R. FERRO, *R. C. Accad. Naz. Lincei* **10**, 48 (1951).
26. H. NOWOTNY, *Z. Metallk.* **34**, 247 (1942).
27. W. C. ELLIS and E. S. GREINER, *Trans. ASM* **29**, 475 (1941).
28. P. I. KRIPYAKEVICH, V. S. PROTASOV and YU B. KUZ'MA, *Dopovidi Akad. Nauk Uhr. RSR* **2**, 212 (1964).
29. C. H. JOHANSSON and J. O. LINDE, *Ann. Physik.* **82**, 449 (1927).
30. J. L. MORIARTY, R. O. GORDON and J. E. HUMPHREYS, *Acta Cryst.* **19**, 285 (1965).
31. E. HELLNER, *Z. Metallk.* **41**, 480 (1950).
32. J. DEWET, *Angew. Chem.* **67**, 208 (1955).
33. E. HELLNER and F. LAVES, *Z. Naturforsch.* **A2**, 177 (1947).
34. T. H. GEBALLE, B. T. MATTHIAS, V. B. COMPTON, T. CORENZWIT, G. W. HULL, JR. and L. D. LONGINATTI, *Phys. Rev.* **137**, A119 (1965).
35. H. NOWOTNY, *Z. Metallk.* **33**, 388 (1941).
36. A. ROSSI and A. IANDELLI, *Rend. Accad. Lincei* **18**, 156 (1933).
37. E. RAUB and W. MAHLER, *Z. Metallk.* **45**, 430 (1954).
38. E. RAUB and W. MAHLER, *Z. Metallk.* **46**, 282 (1955).
39. C. B. JORDAN, *J. Metals* **7**, 832 (1955).
40. J. H. BUDDERY and A. J. E. WELCH, *Nature* **167**, 362 (1951).
41. E. PERSSON and A. WESTGREN, *Z. Physik. Chem. (Leipzig)* **136**, 208 (1928).
42. A. E. DWIGHT, R. A. CONNER JR. and J. W. DOWNEY, *Acta Cryst* **18**, 837 (1965).
43. R. FERRO, *R. C. Accad. Naz. Lincei* **18**, 641 (1955).
44. A. IANDELLI and A. PALENZONA, *J. Less Common Metals* **9**, 1 (1965).
45. R. W. G. WYCKOFF, *J. Wash. Acad. Sci.* **11**, 429 (1921).
46. W. P. DAVEY, *Phys. Rev.* **21**, 143 (1923).
47. V. HORI, *Ann. Univ. Turko.*, Ser. AI, **26**, 8 (1957).
48. W. P. DAVEY and F. G. WICK, *Phys. Rev.* **17**, 403 (1921).
49. T. B. RYMER and P. G. HAMBLING, *Acta Cryst.* **4**, 565 (1951).
50. A. SMAKULA and J. KALNAJS, *Phys. Rev.* **99**, 1737 (1955).
51. V. C. ANSELMO and N. O. SMITH, *J. Phys. Chem.* **63**, 1344 (1959).
52. A. SMITS and D. TOLLENAAR, *Z. Physik. Chem.* **52B**, 222 (1942).
53. L. F. VERESHCHAGIN and S. S. KABALKINA, *Dokl. Akad. Nauk SSSR* **113**, 797 (1957).
54. L. HELMHOLZ, *Z. Krist.* **95A**, 129 (1936).
55. R. FERRO, *Acta Cryst.* **8**, 360 (1955).
56. G. NATTA and L. PASSERINI, *Gazz. Chim. Ital.* **61**, 191 (1931).
57. R. JUZA and A. MEHNE, *Z. Anorg. Allgem. Chem.* **299**, 33 (1959).
58. W. TEICHERT and W. KLEMM, *Z. Anorg. Allgem. Chem.* **243**, 86 (1939).
59. M. STRADA, *Rend. Acad. Lincei* **19**, 809 (1934).

Table 3.2. *References*

1. G. R. LEVI and A. QUILICO, *Gazz. Chim. Ital.* **54**, 598 (1924).
2. T. OKADA, *Proc. Phys. Soc. Japan* **4**, 140 (1949).
3. A. FERRARI, *Gazz. Chim. Ital.* **56**, 630 (1926).
4. E. A. SHUGAM and G. S. ZHDANOV, *Acta Physicochim. URSS* **20**, 247 (1945).
5. G. S. ZHDANOV, *Compt. Rend. Acad. Sci. URSS* **31**, 352 (1941).

Table 3.3*a*. *References*

1. W. H. BAUR, *Acta Cryst.* **11**, 488 (1958).
2. A. DUNCANSON and R. W. H. STEVENSON, *Proc. Phys. Soc. (London)* **72**, 1001 (1958).
3. W. H. BAUR, *Naturwiss.* **44**, 349 (1957).
4. N. BARTLETT and M. A. HEPWORTH, *Chem. Ind. (London)* **1956**, 1425 (1956).
5. W. H. ZACHARIASEN, C. E. HOLLEY, JR. and J. F. STAMPER, JR., *Acta Cryst.* **16**, 352 (1963).
6. B. HOLMBERG, *Acta Chem. Scand.* **16**, 1253 (1962).
7. O. GLEMSER, U. HAUSCHILD and F. TRUPEL, *Z. Anorg. Allgem. Chem.* **277**, 113 (1954).
8. W. H. BAUR, *Acta Cryst.* **9**, 515 (1956).
9. V. M. GOLDSCHMIDT, *Skrifter Norske Videnskaps-- Akad. Oslo I. Mat.-- Naturv. Kl.* No. 1 **1926**, 7 (1927).
10. A. M. BYSTROM, *Acta Chem. Scand.* **3**, 163 (1949).
11. A. MAGNELI and G. ANDERSSON, *Acta Chem. Scand.* **9**, 1378 (1955).
12. T. KATZ, *Ann. Chim.* **5**, 5 (1950).
13. N. A. BENDELIANI and L. F. VERESHCHAGIN, *Dokl. Akad. Nauk SSSR* **158**, 819 (1964).
14. N. SCHÖNBERG, *Acta Chem. Scand.* **8**, 620 (1954).
15. W. H. BAUR, *Acta Cryst.* **14**, 214 (1961).
16. A. P. YOUNG, C. B. SCLAR and C. M. SCHWARTZ, *Z. Krist.* **118**, 223 (1963).
17. A. P. YOUNG and C. M. SCHWARTZ, *Acta Cryst.* **15**, 1305 (1962).
18. R. S. ROTH and J. L. WARING, *Am. Mineralogist* **48**, 1348 (1963).
19. G. BAYER, *Z. Krist.* **118**, 158 (1963).

Table 3.3*b*. *References*

1. A. BYSTROM, B. HOK and B. MASON, *Arkiv. Kemi, Mineral Geol.* **15B**, No. 4 (1941).
2. Y. BASKIN and D. C. SCHELL, *J. Am. Ceram. Soc.* **46**, 174 (1963).

Table 3.4*a*. *References*

1. P. BLUM and F. BERTAUT, *Acta Cryst.* **7**, 81 (1954).
2. H. A. EICK and P. W. GILLES, *J. Am. Chem. Soc.* **81**, 5030 (1959).
3. V. S. NESHPOR and G. V. SAMSONOV, *Zh. Fiz. Khim.* **32**, 1328 (1958).
4. N. N. ZHURAVLEV, A. A. STEPANOVA, Y. B. PADERNO and G. V. SAMSONOV, *Kristallografiya* **6**, 791 (1961).
5. M. V. STACKELBERG and F. NEWMANN, *Z. Physik. Chem.* **19B**, 314 (1932).
6. N. N. ZHURAVLEV, *Kristallografiya* **1**, 666 (1956).
7. Y. B. PADERNO and G. V. SAMSONOV, *Zh. Strukt. Khim.* **2**, 213 (1961).
8. G. V. SAMSONOV, V. S. FOMENKO and YU B. PADERNO, *Upr. Fiz. Zh.* **8**, 700 (1963).
9. G. ALLARD, *Bull. Soc. Chim. France* **51**, 1213 (1932).

Table 3.4*b*. *References*

1. W. G. BRADSHAW and C. O. MATHEWS, *Properties of Refractory Materials*, LMSD-2466 (June 24, 1958).
2. N. K. HIESTER, F. A. FERGUSON and N. FISHMAN, *Chem. Eng.* 237 (March 1957).
3. C. E. POWELL, I. E. CAMPBELL and B. W. GONSER, *Vapor Plating*, Wiley, New York (1955).
4. A. SEARCY and A. THARP, *J. Phys. Chem.* **64**, 1939 (1960).
5. J. LAFFERTY, *J. Appl. Phys.* **22**, 299 (1951).
6. J. BINDER and R. STEINITZ, *Planseeber Pulvermet* **7**, 18 (1959).
7. G. V. SAMSONOV, YU. B. PADERNO and V. S. FOMENKO, *Izd. An. Ukr. SSR, Kiev* **8**, 66 (1960).
8. L. BREWER, *J. Am. Ceram. Soc.* **33**, 273 (1950).
9. YU. B. PADERNO and G. V. SAMSONOV, *Dokl. Akad. Nauk SSSR* **137**, 646 (1961).
10. G. S. MARKEVICH, Chemical Faculty, Lenningrad State Univ., Lenningrad (1961).
11. G. V. SAMSONOV and T. I. SEREBRYAKOVA, A. S. BOLGAR, *Zhur. Neorg. Khim.* **6**, 2243 (1961).
12. R. JOHNSON and A. DAANE, *J. Chem. Phys.* **38**, 425 (1963).
13. G. V. SAMSONOV and A. E. GRODSHTEIN, *Zhur. Fiz. Khim.* **30**, 379 (1956).
14. G. V. SAMSONOV, W. N. ZHURAVLEV, YU. B. PADERNO and V. R. MELIK ADAMYAN, *Kristallografiya* **4**, 538 (1959).

15. G. V. SAMSONOV, T. I. SEREBRYCKOVA and A. S. BOLGAR, *Zhur. Neorg. Khim.* **6**, 2243 (1961).
16. G. V. SAMSONOV and O. I. ZORINA, *Zhur. Neorg. Khim.* **1**, 2260 (1956).
17. G. A. KUDINTSEVA, M. D. POLYAKOVA, G. V. SAMSONOV and B. M. TSAREV, *Fiz. Metalli Metalloved* **6**, 272 (1958).
18. G. V. SAMSONOV, V. S. NESHPOR, Problems of Powder Metallurgy and the Strength of Materials, No. 5, *Izd. An. Ukr. SSR, Kiev* **3** (1958).
19. S. N. L'VOV, V. F. NEMCHENKO and YU. B. PADERNO, *Dokl. Akad. Nauk SSSR* **142**, 1371 (1963).
20. O. H. KRIKORIAN, *Thermal Expansion of High Temperature Materials,* UCRL-6132 (September 1960).
21. N. N. ZHURAVLEV, A. A. STEPANOVA, YU. B. PADERNO and G. V. SAMSONOV, *Kristallografiya* **6**, 791 (1961).

Table 3.5. *References*

1. J. BRADLEY and A. H. JAY, *J. Iron and Steel Ind.,* **125**, 339 (1932).
2. D. E. SANDS, D. H. WOOD and W. J. RAMSEY, *Acta Cryst.* **16**, 316 (1963).
3. E. ZINTL and G. BRAUER, *Z. Elektrochem.* **41**, 297 (1935).
4. R. VOGEL and T. HEUMANN, *Z. Metallk.* **38**, 1 (1947).
5. RAHLFS, *Metallwirt.* **16**, 640 (1937).
6. M. C. FARQUHAR, H. LIPSON and A. R. WEILL, *J. Iron and Steel Inst.* **152**, 457 (1945).
7. P. I. KRIP'YAKEVICH, V. F. TEREKHOVA, O. S. ZARECHNYUK and J. V. BUROV, *Kristallografiya* **8**, 268 (1963).
8. E. ZINTL and A. SCHNEIDER, *Z. Electrochem.* **41**, 771 (1935).
9. G. BRAUER and E. ZINTL, *Z. Phys. Chem.* **B37**, 323 (1937).
10. P. I. KRIPYAKEVICH and V. I. EVDOKIMENKO, *Vapr. Teorii i Primeneiya Redoze Metal. Akad. I Nauk, SSSR,* 146 (1964).
11. A. ROSSI and A. IANDELLI, *R. C. Accad. Naz. Lincei* **19**, 415 (1934).

CHAPTER 4

THE NaCl TYPE AND RELATED STRUCTURES

THE sodium chloride type structures have been grouped into those which have unit cells with atoms of one element at the corners and at the center of every face, and atoms or groups of atoms of the other element at the center of the cube and the center of the edges. These unit cells are sometimes distorted from cubic symmetry. In a related structure, Cu_2AlMn, the octants of these cubes are filled with additional atoms. The flow diagram of the structures is shown in Table 4.0. This diagram shows that the FeS_2, CaC_2 and the Cu_2AlMn structure can be derived from the NaCl structure.

TABLE 4.0. NaCl *Type Structures*

4.1
NaCl
structure

4.2	4.3	4.4
FeS_2	CaC_2	Cu_2AlMn
structure	structure	structure

4.1. NaCl, Rock Salt, B1, *Fm3m*, Cubic

In the sodium chloride structure the sodium ions are at the points of a face-centered cubic lattice. They are at the corners and face centers of a unit cell and chlorine ions are at the center of the cube and center of the edges (see Fig. 4.1a). The layer sequence at different levels of z is also shown.

Many halides, oxides, sulfides, selenides, nitrides and intermetallics which combine in a 1:1 ratio form this structure. These are listed in Table 4.1a.

When the radius of the A atom is much larger than that of the B atom in AB compounds, the CsCl structure is adopted instead. In the case of intermetallics, either the CsCl or NaTl structures are formed.

The sodium ions are surrounded with six chlorine ions and each chlorine ion is surrounded with six sodium ions. Formally, the sodium ions are at

$$0, 0, 0; 0, \tfrac{1}{2}, \tfrac{1}{2}; \tfrac{1}{2}, 0, \tfrac{1}{2}; \tfrac{1}{2}, \tfrac{1}{2}, 0$$

and the chlorine ions at

$$\tfrac{1}{2}, \tfrac{1}{2}, \tfrac{1}{2}; \tfrac{1}{2}, 0, 0; 0, \tfrac{1}{2}, 0; 0, 0, \tfrac{1}{2}.$$

Properties

Some of the nitrides and carbides such as NbC, TaC and NZr with this structure have been found to have high superconducting transition temperatures (see Fig. 4.1b). While it is suspected that this property may be related to structure, no direct evidence has been obtained to date. However, the electron per atom ratio does appear to be important. Note that the best superconductors with this structure have an electron per atom ratio of 4.5.

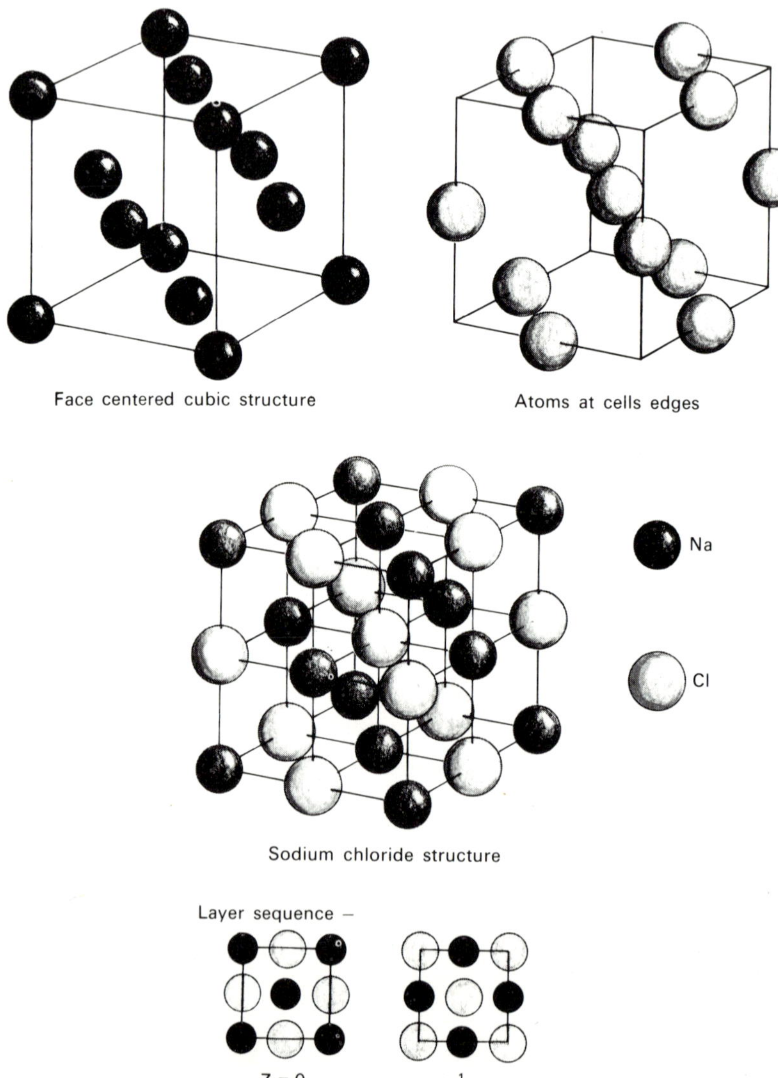

FIG. 4.1a. The sodium chloride structure

Many of the oxides, carbides and nitrides with the sodium chloride structure are often nonstoichiometric. This affects their electrical properties, and in the case of superconductors can reduce their transition temperatures drastically. Titanium monoxide exists over the range $Ti_{0.85}O$ to TiO, while ferrous oxide has a range of from $Fe_{0.90}O$ to $Fe_{0.96}O$. As a consequence of

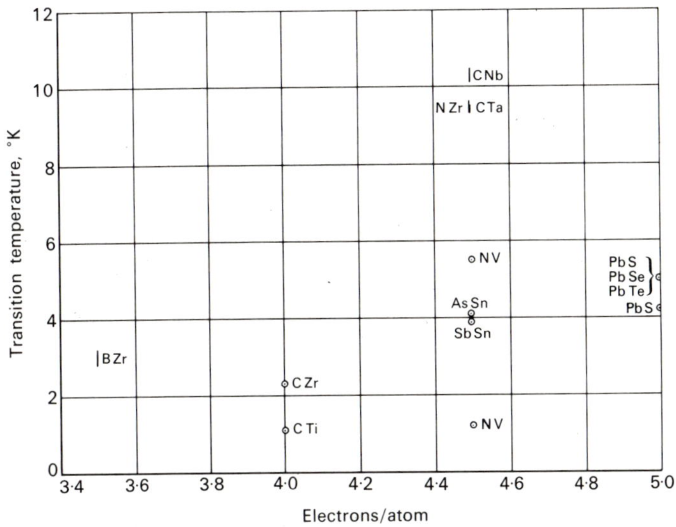

FIG. 4.1b. T_c vs. $\bar{e}/A$ for superconducting compounds with the sodium chloride structure

these vacancies, the transition metals exist in two valence states, causing the material to exhibit semiconductor properties. The electrons are conducted from one metal ion to another through an oxygen ion. The same type of conductivity is observed in oxides such as NiO after substituting Li^+ for Ni^{2+}, which results in the same number of Ni^{3+} ions being formed as Li^+ ions added. The carrier mobility in this type of semiconductor is much lower than that in Cu_2O and ZnO.

A number of compounds listed in Table 4.1a also have interesting magnetic properties. Compounds TbN, DyN, HoN, ErN, EuO, EuS and EuSe are ferromagnetic with Curie points of 42°, 26°, 18°, 5°, 77°, 16° and 6°K, and MnS, MnSe, FeO, CoO and MnO are antiferromagnetic with Néel temperatures of 11°, 165°, 247°, 198°, 273° and 122°K. The oxide, MnO, is a classic case in which superexchange causes the d-electron spins of manganese ions connected linearly through oxygen ions to be oriented antiparallel.

Carbides with the sodium chloride structure are characterized by hardness and refractoriness. They are chemically inert and have melting points which range from 2475°C to 4000°C. Titanium carbide has a melting point of

TABLE 4.1a. *Phases with the Sodium Chloride Structure*

Phases	Cell size, a_0 (Å)	Refs.
Intermetallics		
AsCe	6.072	1
AsDy	5.780	2
AsEr	5.732	2
AsGd	5.854	2
AsHo	5.771	2
AsLa	6.137	1, 2
AsNd	5.970	1
AsPr	6.009	1
AsPu	5.855	3
AsSc	5.487	2
AsSm	5.921	4
AsSn	5.727	5
AsTb	5.827	2
AsTh	5.972	6
AsTm	5.711	2
AsU	5.766	7
AsY	5.786	2
AsYb	5.698	2
BiCe	6.500	1
BiHo	6.228	8
BiLa	6.578	1
BiNd	6.424	1
BiPr	6.461	1
BiPu	6.350	9
BiSm	6.362	1
BiTb	6.280	8
BiU	6.364	10
CeP	5.909	11
CeSb	6.411	1
DySb	6.153	2
ErSb	6.106	2
GdSb	6.217	2
HoP	5.626	8
HoSb	6.130	7
LaP	6.025	1
LaSb	6.488	1
NdP	5.838	1
NdSb	6.322	1
PPr	5.860	11
PPu	5.664	3
PSc	5.312	12
PSm	5.760	4
PTb	5.688	2
PTh	5.818	13
PY	5.661	12
PZr	5.27	14
PrSb	6.366	1
SbSc	5.859	2
SbSm	6.271	4
SbSn	6.130	15

TABLE 4.1a *(cont.)*

Phases	Cell size a_0 (Å)	Refs.
SbTb	6.181	1
SbTh	6.318	16
SbTm	6.083	2
SbU	6.191	10
SbYb	5.922	2
Borides		
PuB	4.92	17
ZrB	4.65	18
Carbides		
HfC	4.4578	19
NbC	4.4691	20, 21
NpC	5.004	22
PuC	4.920	1
TaC	4.4540	23
ThC	5.338	24
TiC	4.3186	25, 26
UC	4.9591	24
VC	4.182	27
ZrC	4.6828	25
Halides		
AgBr	5.7745	28
AgCl	5.547	29
AgF	4.92	30
CsF	6.008	31
KBr	6.5982	32, 33
KCl	6.29294	34
KF	5.347	35
KI	7.06555	32
LiBr	5.5013	31
LiCl	5.12954	36
LiF	4.0173	37
LiI	6.000	31
NaBr	5.97324	36
NaCl	5.64056	37
NaF	4.620	38
NaI	6.4728	31
NH_4I	7.2613	39
RbBr	6.854	31
RbCl	6.5810	40
RbF	5.64	31
RbI	7.342	40
Hydrides		
KH	5.700	41
LiH	4.085	41
NaH	4.880	41
PdH	4.02	42
RbH	6.037	41

TABLE 4.1a *(cont.)*

Phases	Cell size a_0 (Å)	Refs.
Nitrides		
CeN	5.011	1
CrN	4.140	43
DyN	4.905	44
ErN	4.839	44
EuN	5.014	44
GdN	4.999	44
HoN	4.874	44
LaN	5.301	44
LuN	4.766	44
$NbN_{0.98}$(H.T.)	4.702	45
NdN	5.151	44
PrN	5.155	1
PuN	4.905	46
ScN	4.44	47
SmN	5.0481	48
TbN	4.933	44
ThN	5.20	49
TiN	4.235	50
$Ti_{0.9}N$	4.231	51
TmN	4.809	44
UN	4.884	10
VN	4.128	52
YN	4.877	53
YbN	4.7852	48
ZrN	4.61	54
Oxides		
AmO	5.05	55
BaO	5.523	56
CaO	4.8105	56
CdO	4.6953	57
CoO†	4.2667	58
EuO	5.141	48
FeO(H.T.)†	4.28 to 4.309	60
MgO	4.2112	61
MnO†	4.445	62
NbO†	4.2101	63
NiO†	4.1684	64
NpO	5.01	46
PaO	4.961	65
PuO	4.959	46
SmO	4.9883	66
SrO	5.1602	67
TaO	4.22–4.439	68
TiO†	4.1766	69
UO	4.92	70
VO†	4.062 (800°C)	71
YbO	4.86	72
ZrO	4.62	73

TABLE 4.1a *(cont.)*

Phases	Cell size a_0 (Å)	Refs.
Sulfides, Selenides and Tellurides		
BaS	6.3875	74
BaSe	6.600	75
BaTe	6.986	75
BiSe	5.99	68
BiTe	6.47	68
CaS	5.6903	76
CaSe	5.91	76
CaTe	6.345	76
CdS (H.P.)	5.464	77
CdSe (H.P.)	5.49	77
CdTe (H.P.)	5.81	77
CeS	5.778	78
CeSe	5.982	78
CeTe	6.346	78
DyTe	6.092	2
ErTe	6.063	2
EuS	5.968	59
EuSe	6.197	59
EuTe	6.603	59
GdSe	5.771	79
HoS	5.465	8
HoSe	5.680	8
HoTe	6.049	2
InTe (H.P.)	6.1060	80
LaS	5.854	81
LaSe	6.066	81
LaTe	6.429	81
MgS	5.2033	74
MgSe	5.451	82
MnS	5.2236	83
MnSe	5.456	84
NdS	5.681	78
NdSe	5.907	81
NdTe	6.282	81
PbS	5.9362	85
PbSe	6.1243	86
PbTe	6.454	87
PrS	5.727	78
PrSe	5.944	81
PrTe	6.317	81
PuS	5.536	46
PuTe	6.183	3
SmS	4.970	4
SmSe	6.200	79
SmTe	6.594	4
SnSe	6.020	88
SnTe	6.313	88
SrS	6.0198	74
SrSe	6.23	89
SrTe	6.47	90

TABLE 4.1a (*cont.*)

Phases	Cell size a_0 (Å)	Refs.
TbS	5.516	8
TbSe	5.741	8
TbTe	6.102	2
ThS	5.682	46
ThSe	5.875	91
TmTe	6.042	2
US	5.484	46
USe	5.750	92
UTe	6.163	92
YTe	6.095	2
YbSe	5.867	93
YbTe	6.353	93
ZrS	5.250	94
Others		
BaNH	5.84	95
CaNH	5.006	95
KCN	6.527	96
NaCN	5.893	97
RbCN	6.82	98
$RbNH_2$	6.395 (50°C)	99
SrNH	5.45	95

† Defect or distorted structures.

3147°C, a high hardness and modulus of elasticity and is resistant to oxidation up to 1200°C. It has been used in cermets for high-temperature devices. Zirconium carbide has a higher melting point (3530°C) and has been employed in crucibles for melting metals and in nuclear power engineering. Uranium carbide has also been used in nuclear engineering as heat evolving elements in an atomic reactor. A list of phases with the NaCl structure and their properties are tabulated in Table 4.1b.

The sulfide, selenide and telluride of lead in Table 4.1a form layers which are photosensitive in the infrared region of the spectrum. Lead sulfide can be used as a detector out to 3 microns at room temperature and 4.5 microns at liquid hydrogen temperature, while lead telluride is effective out to 4.5 and 7 microns at these temperatures. The energy gaps of lead sulfide, selenide and telluride are 1.04 eV, 0.83 eV and 0.62 eV, respectively.

Because the electron mobility in PbS and PbTe is so high (500 cm^2/V-sec and 2000 cm^2/V-sec, respectively) and the thermal conductivity is so low, these compounds have been considered for thermoelectric application. However, they are not as good as Bi_2Te_3 which is commonly used for thermoelectric elements. (See discussion of thermoelectrics in the Introduction.)

Another structure which is closely related to the NaCl structure is that of NbO. In the NbO structure one NbO molecule in the unit cell is missing, Nb from 0, 0, 0 and O from $\frac{1}{2}, \frac{1}{2}, \frac{1}{2}$.

TABLE 4.1b. *Properties of Phases with the* NaCl *Structure*

Compound	Melting point(°C)	Electrical resistivity (μ ohm-cm)	Hardness (kg/mm^2)	Modulus (10^6 psi)	Thermal conductivity (cal/cm-sec-°C)	Coeff. of therm. expan. (10^{-6}/°C)
HfB	3060[1]					
ZrB	2922[1]	30[21]				
HfC	3890[2]	37–65[22]	2500[22]	46[22]	0.053[22]	6.6[22]
NbC	3480[3]	35–74[22]	1490[25]	49[22]	0.034[22]	6.2[22]
TaC	3880[4]	30–41[22]	1600[25]	41[22]	0.053[22]	5.5[22]
ThC	d 2625[5]					
TiC	3147[5]	60[22]	3200[22]	45[22]	0.041[22]	10.2[22]
UC	2315[6]					
VC	d 2810[7]					
ZrC	3530[8]	42–67[22]	2600[22]	69[22]	0.049[22]	9.1[22]
TiN	3205[7]	22–30[21]	1770[22]	11.4[22]	0.070[22]	9.3[22]
ThN	2630[9]					
UN	2650[9]					
VN	2360[7]					
ZrN	2980[7]					
BaO	1923[10]					
CaO	2572[11]		560[26]		0.037[26]	12.9[27]
FeO	1370[12]					
MgO	2800[13]	10^8[23]		36 [26]	0.097[26]	13.6[28]
MnO	1775[14]					
NiO	1950[14]				0.029[23]	
SrO	2415[15]					13.5[28]
TiO	1760[14]					12.3[29]
CeS	2450[7]					
EuS	1667[16]					14.2[16]
EuSe	1307[16]					18.2[16]
LaS		30[24]				
LaSe		50[24]				
LaTe		50[24]				
NdS	2140[17]					
NdSe		60[24]				
NdTe		60[24]				
PrS	2230[18]	50[24]				
PrSe		50[24]				
PrTe		50[24]				
ThS	2400[19]					
US	2000[20]					
Phases						
TaC–11M%HfC	4050[30]					
TiC–40M%HfC			4250[29]			
NbC–43M%HfC		77.5[29]				

Ordered Rock Salt Structure

$LiFeO_2$ and a number of other compounds have an ordered form of the sodium chloride structure and also a random form in which the Li and Fe ions are in the sodium ion positions (see Table 4.1c). $LiNiO_2$ and $LiInO_2$ also have superstructures of the NaCl structure.

TABLE 4.1c. *Compounds with the* $LiFeO_2$ *Structure*

Compounds	Cell size, a_0 (Å)	Refs.
Oxides		
$LiFeO_2$	4.141	1, 2
$LiTiO_2$	4.140	3
γ-$LiTiO_2$	4.568	4
Complex Oxides		
$Li(Co_{0.5}Mn_{0.5})O_2$	4.150	1
$Li(Co_{0.5}Ti_{0.5})O_2$	4.166	1
$Li(Fe_{0.5}Mn_{0.5})O_2$	4.184	1
$Li(Fe_{0.5}Ti_{0.5})O_2$	4.137	1
$Li(Mn_{0.5}Ti_{0.5})O_2$	4.216	1
$Li(Ni_{0.5}Mn_{0.5})O_2$	4.137	1
$Li(Ni_{0.5}Ti_{0.5})O_2$	4.144	1
Sulfides, Selenides, Tellurides		
$AgBiS_2$	5.648	5
$AgBiTe_2$	6.155	6
$AgSbS_2$	5.647	7
$AgSbSe_2$	5.786	5
$AgSbTe_2$	6.078	5
$CuBiSe_2$	5.69	5
$KBiS_2$	6.04	5, 8
$KBiSe_2$	5.922	5, 8
$LiBiS_2$	5.603	8, 9
$NaBiS_2$	5.775	10
$NaBiSe_2$	5.852	8
$TlBiS_2$	6.18	6

4.2. FeS_2, Pyrite, C2, *Pa*3, Cubiv

The FeS_2 type structure is a sodium chloride type structure with iron atoms at the sodium ion positions and the center of linear S_2 molecules at the chlorine ion positions (see Fig. 4.2). The layers at different values of z are also shown. The four Fe atoms are at

$$0, 0, 0; 0, \tfrac{1}{2}, \tfrac{1}{2}; \tfrac{1}{2}, 0, \tfrac{1}{2}; \tfrac{1}{2}, \tfrac{1}{2}, 0.$$

and the eight S atoms are at

$$x, x, x; \tfrac{1}{2}+x, \tfrac{1}{2}-x, \bar{x}; \bar{x}, \tfrac{1}{2}+x, \tfrac{1}{2}-x; \tfrac{1}{2}-x, \bar{x}, \tfrac{1}{2}+x; \bar{x}, \bar{x}, \bar{x};$$
$$\tfrac{1}{2}-x, \tfrac{1}{2}+x, x; x, \tfrac{1}{2}-x, \tfrac{1}{2}+x; \tfrac{1}{2}+x, x, \tfrac{1}{2}-x \text{ for } FeS_2 \;\; x = 0.386.$$

For the compounds ZrS_2, SnS_2 and PtS_2 with the CdI_2 structure and in the MoS_2 structure the S atoms surround each metal atom. In the FeS_2 structure, the S_2 exist as discrete groups.

In the PdS_2 structure which is similar to the FeS_2 structure the unit cell is elongated in one direction so that Pd has four nearest neighbors instead of six as it does in FeS_2. As a consequence, PdS_2 has a layer type structure.

A list of compounds with the FeS_2 structure and the cell parameters are given in Table 4.2.

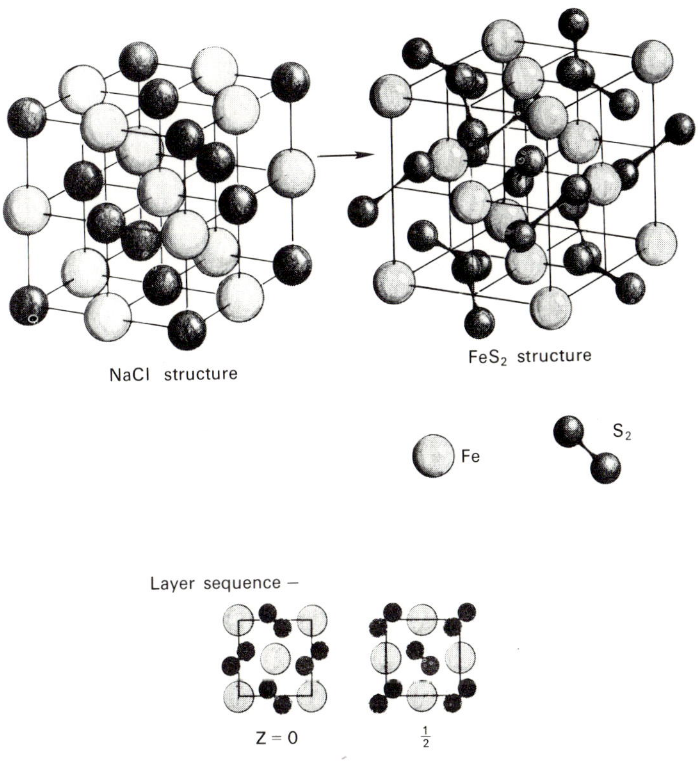

FIG. 4.2. The FeS_2 structure

Properties

The coefficients of thermal expansion for two compounds, CoS_2 and FeS_2, listed in Table 4.2 have been measured and found to be 12.74×10^{-6} and $9.25 \times 10^{-6}/°C$ respectively. In addition, CoS_2 has been shown to be ferromagnetic below $-180°C$.

TABLE 4.2. *Phases with the* FeS_2 *Structure*

Phase	Cell parameter a_0 (Å)	Atomic parameter, x	Refs.
Intermetallics			
As_2Pd	5.98	0.38	1, 2
As_2Pt	5.9665	0.38	3
$AuSb_2$	6.660	0.375	4, 5
Bi_2Pd	6.68		6
Bi_2Pt	6.7022	0.38	7
P_2Pt	5.6956	0.38	3
$PdSb_2$	6.459	0.38	8
$PtSb_2$	6.4400	0.38	3
$RuSn_2$			2
Oxides			
CdO_2	5.313	0.4192	9
α-KO_2	6.12 (300°C)		10
β-NaO_2	5.490		11
Sulfides, Selenides, Tellurides			
AsCoS (H.T.)	5.61	(As & Co randomly distributed; 0.385)	12
AsNiS	5.66	(As & S randomly distributed; 0.385)	13
CoS_2	5.52318	0.389	14, 15
$CoSe_2$	5.857	0.380	16
FeS_2 (pyrite)	5.4080	0.386	14, 15
$IrTe_{2+x}$	6.411		17
MnS_2 (haverite)	6.095	0.401	18
$MnSe_2$	6.430	0.393	19
$MnTe_2$	6.951	0.386	5
NiS_2	5.676	0.395	14, 20
$NiSe_2$	5.9604	0.384	21
OsS_2	5.6188	0.375	5, 22
$OsSe_2$	5.945	0.38	2
$OsTe_2$	6.382	0.38	2
RhS_2	5.585	0.38	2
$RhSe_2$	6.002	0.38	23
$RhTe_2$ (L.T.)	6.441	0.362	24
RuS_2	5.58	0.375	5
$RuSe_2$	5.933	0.38	2
$RuTe_2$	6.403	0.38	2

4.3. CaC_2, C11, *I4/mmm*, Tetragonal

Another sodium chloride type structure is the CaC_2 structure with calcium atoms at the sodium ion positions and the centers of the C_2 molecules at the chlorine ion positions. The long axes of the C—C molecule are aligned in one direction as shown in Fig. 4.3 so that the unit cell is tetragonal instead of cubic. The base of the smallest tetragonal cell which can describe this structure is delineated by the dotted lines in Fig. 4.3. The a axis of this cell is $\sqrt{2}/2$ times that of the large one. In the unit cell the calcium atoms are at $0, 0, 0; \frac{1}{2}, \frac{1}{2}, \frac{1}{2}$ and the carbon atoms are at $0, 0, z; \frac{1}{2}, \frac{1}{2}, \frac{1}{2} + z; 0, 0, \bar{z}; \frac{1}{2}, \frac{1}{2}, \frac{1}{2} - z$.

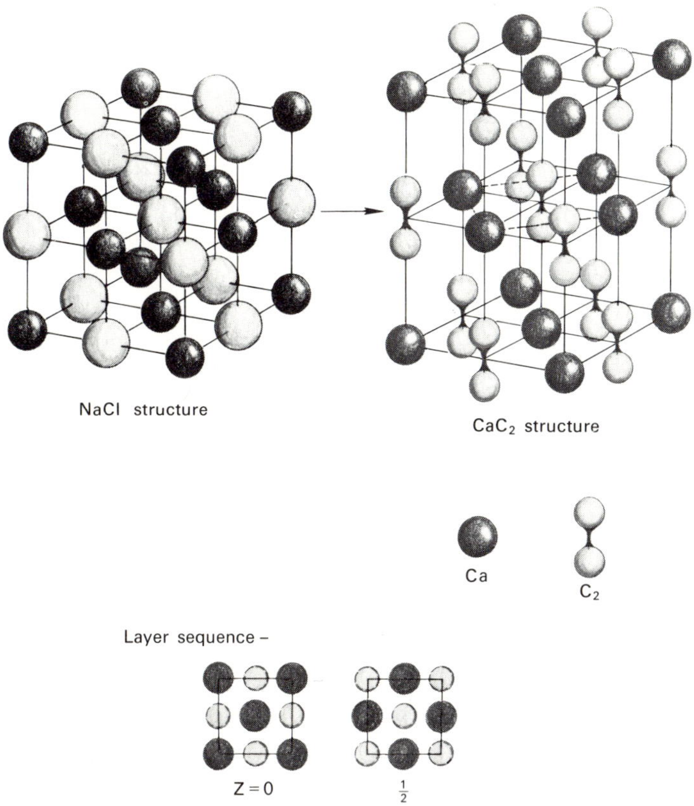

FIG. 4.3. The CaC_2 structure

Table 4.3a includes other carbides and several metal peroxides with the CaC_2 structure.

Properties

Alkaline earth, rare earth, yttrium and uranium carbides adopt this structure. They are high-temperature materials, as can be seen from the melting point data in Table 4.3b. The alkaline earth carbides react with water to yield acetylene. The other carbides hydrolyze to produce several gaseous products.

Silicides in general have good oxidation resistant properties. The compound $MoSi_2$ is excellent in this respect. However, it does have a low impact strength and therefore has limited use as an oxidation preventive coating.

The alloy Cr_2Al forms an ordered structure on cooling from the body centered cubic structure. The ordered structure is the CaC_2 structure. However, the Cr atoms are not as close together as the C atoms are, as shown in Fig. 4.3.

TABLE 4.3a. *Phases with the* CaC_2 *Structure*

Phase	Cell size (Å)		Atomic parameter, z	Refs.
	a_0	c_0		
Intermetallics				
Ag_2Er	3.668	9.135		1
Ag_2Ho	3.682	9.172		1
Ag_2Yb	3.624	8.88		2
$AlCr_2$	3.0045	8.6477	Cr: 0.3192	3
Au_2Er	3.665	8.932		1
Au_2Ho	3.676	8.934		1
Au_2Yb	3.6274	8.889		2
β-Ge_2Mo	3.313	8.195	Ge: 0.333	4
Hg_2Mg	3.838	8.799	Hg: 0.333	5
$MoSi_2$	3.203	7.89	Si: 0.333	6
$ReSi_2$	3.129	7.764	Si: 0.333	7
Si_2W	3.211	7.868	Si: 0.333	6
Carbides				
BaC_2	4.41	7.07	C: 0.39	8
CaC_2	3.633	6.036	0.395	8, 9
CeC_2	3.814	6.485		10
DyC_2	3.669	6.176		11
ErC_2	3.620	6.094		11
GdC_2	3.718	6.275		11
HoC_2	3.643	6.139		11
LaC_2	3.93	6.56		11, 12
LuC_2	3.563	5.964		11
MgC_2	4.86	5.76		13
NdC_2	3.91	6.29		11
PrC_2	3.86	6.39		11
SmC_2	3.770	6.331		11
SrC_2	4.12	6.69	C: 0.39	8
TbC_2	3.690	6.217		11
TmC_2	3.600	6.047		11
UC_2	3.525	6.005	C: 0.4	12, 14, 10
YC_2	3.664	6.169		15
YbC_2	3.637	6.109		11
Oxides				
BaO_2	5.384	6.841	O: 0.3911	16
CaO_2	3.54	5.92		17
CsO_2	6.29	7.21	O: 0.405	18
KO_2 (L.T.)	5.704	6.699	O: 0.4047	19
RbO_2	6.01	7.04	O: 0.405	20
SrO_2	5.03	6.56		21

TABLE 4.3b. *Melting Points of Phases with the* CaC_2 *Structure*

Phase	Melting point (°C)
BaC_2	d 2300[1]
CaC_2	2300[2]
CeC_2	d 2540[3]
GdC_2	d 2200[3]
LaC_2	d 2440[4]
NdC_2	d 2000[3]
PrC_2	d 2535[3]
SmC_2	d 2200[3]
SrC_2	>1900[2]
UC_2	2260[5]
YC_2	2300[6]

4.4. Cu_2AlMn Type, Heusler Alloy Type, $L2_1$, *Fm3m*, Cubic

A more complex structure is adopted by Cu_2AlMn, which can be described by a cubic unit cell in which manganese and aluminum atoms are arranged as the sodium and chlorine ions are in the rock salt structure, with copper atoms situated in the center of the octants which make up the cell. Thus, the Mn atoms are at the corners and centers of each of the faces of a cube, the Al atoms are at the center and edges and the Cu atoms are in the corners of the cube (see Fig. 4.4). More formally, the Mn atoms are at 0, 0, 0; 0, $\frac{1}{2}$, $\frac{1}{2}$; $\frac{1}{2}$, 0, $\frac{1}{2}$ and $\frac{1}{2}$, $\frac{1}{2}$, 0 the Al atoms are at $\frac{1}{2}$, $\frac{1}{2}$, $\frac{1}{2}$; $\frac{1}{2}$, 0, 0; 0, $\frac{1}{2}$, 0 and 0, 0, $\frac{1}{2}$ and the Cu atoms are at $\frac{1}{4}$, $\frac{1}{4}$, $\frac{1}{4}$; $\frac{1}{4}$, $\frac{3}{4}$, $\frac{3}{4}$; $\frac{3}{4}$, $\frac{1}{4}$, $\frac{3}{4}$; $\frac{3}{4}$, $\frac{3}{4}$, $\frac{1}{4}$; $\frac{3}{4}$, $\frac{3}{4}$, $\frac{3}{4}$; $\frac{3}{4}$, $\frac{1}{4}$, $\frac{1}{4}$; $\frac{1}{4}$, $\frac{3}{4}$, $\frac{1}{4}$ and $\frac{1}{4}$, $\frac{1}{4}$, $\frac{3}{4}$. Phases which have this structure are listed in Table 4.4.

TABLE 4.4. *Phases with the* Cu_2AlMn *Structure*

Phase	Cell size, a_0 (Å)	Refs.
Intermetallics		
$AlAu_2Mn$	6.358	1
$AlCu_2Mn$ (H.T.)	5.947	2
$AlNi_2Ti$	5.87	3
$CoCu_2Sn$	5.98	4
Co_2MnSn	5.989	5
(Cr, Ni)Cu_2Sn (H.T.)	5.98	4
Cu_2FeSn (H.T.)	5.93	4
Cu_2InMn	6.186	6
Cu_2MnSn (H.T.)	6.161	7
Cu_2NiSn	5.92	4
$MgNi_2Sb$	6.06	8, 9
$MgNi_2Sn$	6.109	9
$MnNi_2Sb$	6.013	10
$MnNi_2Sn$	6.048	5

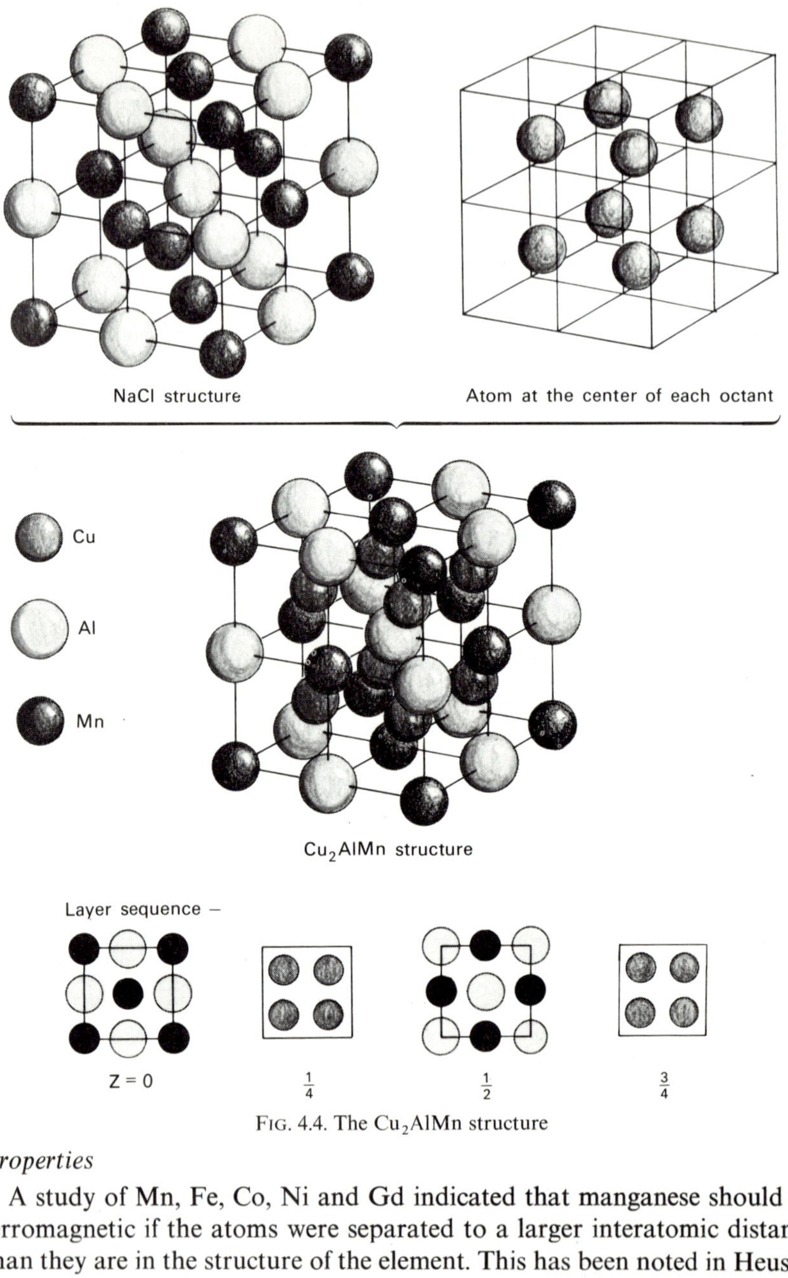

FIG. 4.4. The Cu_2AlMn structure

Properties

A study of Mn, Fe, Co, Ni and Gd indicated that manganese should be ferromagnetic if the atoms were separated to a larger interatomic distance than they are in the structure of the element. This has been noted in Heusler alloy and some Mn compounds with the NiAs structure. Ferromagnetism has been observed in Cu_2MnAl, Cu_2MnIn, Cu_2MnGa and Cu_2MnSn. The Curie points of the first two compounds are 630° and 560°K.

The NaY_3F_{10} *Structure*

The crystal structure of NaY_3F_{10} can be considered to be the same as that of Cu_2AlMn. In the unit cell the Na and Y ions are ordered in the manganese atom positions, with Na ions at the corners and the Y ions in the face centers. The fluorine ions are randomly distributed in the aluminum and the copper atom positions filling ten of the twelve available sites.

4.5. Discussion

Structures (Relations)

The sodium chloride structure is made up of two interpenetrating face-centered cubic lattices, one with sodium ions at each lattice point and the other with chlorine ions. In the random $LiFeO_2$ structure the Li and Fe ions are situated in the sodium ion sites and oxygen ions on the chlorine ion sites and the NbO structure is a NbO deficient NaCl structure. In the FeS_2 and CaC_2 structure, molecules of S_2 and C_2 replace the halide ions. In the latter structure the C_2 molecules are aligned so that the unit cell is tetragonal. Then in Heusler's alloy, and the NaY_3F_{10} structures, the same positions are occupied as those in the NaCl structure, and additional atoms are at the center of each octant of the unit cell.

Intermetallics

Among the intermetallic AB phases as well as sulfides, selenides and tellurides, those with radius ratios $(R_A/R_B) = 0.41–0.73$ tend to form the NaCl structure. The ones with the larger ratios $(R_A/R_B > 0.73)$ normally form either the CsCl or NaTl structure while the more covalent intermetallics with appropriate ratios for having the sodium chloride structure adopt the NiAs structure instead. The structure of Cu_2AlMn is not a common one; however, it is important since some of the Mn compounds with this structure are ferromagnetic.

Borides, Carbides and Nitrides

Most of the carbides and nitrides, with the formula AC and AN respectively where A is a metal, and ZrB have the sodium chloride structure, while WC, MoC have the WC structure. They are hard, have high melting points and high conductivities. For any metal the resistivity of the carbide nitride and boride is higher than that of the metal. Niobium carbide and zirconium nitride have high superconducting transition temperatures (~10°K), although deficiencies in carbon or nitrogen reduce the transition temperature drastically.

The alkali metal and alkaline earth carbides usually form transparent crystals and are not electrically conductive. They are very reactive with water and are good reducing agents. Calcium carbide, whose structure is described in this chapter, has C_2 units in its structure. It is prepared by reacting C with $CaCN_2$, and yields acetylene when reacted with water. Alkali metals form A_2C_2 type carbides, beryllium forms Be_2C, aluminium, Al_4C_3 and iron, manganese and nickel, Fe_3C, Mn_3C and Ni_3C. All of these are attacked by water or acids. The carbides Be_2C and Al_4C_3 yield methane on hydrolysis. The carbide Mn_3C reacts with water, and Fe_3C and Ni_3C with acids, to give mixtures of hydrocarbons. There also are a large number of chromium, manganese, iron and cobalt carbides with complicated structures which will not be described here.

Halides

All of the alkali metal halides, except CsCl, CsBr and CsI, and AgF, AgBr and AgCl, and NH_4I form the NaCl structure. In general, the AX halides with the smaller R_A/R_X ratios (0·41–0·73) tend to adopt the NaCl structure rather than the CsCl structure. Melting point and coefficient of thermal expansion data for the alkali metal halides are given below. Temperature given in (°C) and coefficients of thermal expansion (10^{-6}/°C).

	F	Cl	Br	I
Li	840	606	535	450
	34	44	50	59
Na	992	803	740	653
	36	40	42	47
K	846	768	748	693
	37	38	40	43
Rb	755	717	681	638
		36	38	42
Cs	684	638	627	621
	32	46	47	48

Hydrides

The alkali and alkaline earth hydrides are salt-like hydrides. They can be formed by direct combination of hydrogen and the metal. The alkali metal hydrides LiH, NaH, KH, RbH and CsH have the sodium chloride structure; MgH_2 has the rutile structure.

The hydrides of Be, Zn, Cd and Al are formed by reacting the halides with $LiAlH_4$ in ether solution. The transition metal hydrides, like the alkali and alkaline earth metal hydrides, form by direct action of the metal and hydrogen. These are called interstitial hydrides and are often nonstoichiometric.

Oxides

Of the AO oxides, those which are most ionic form the sodium chloride structure. A number of them are listed in table form with group I and other AO_2 oxides, which have related structures.

I	II	IV	V	VI	VII	VIII	IIB
NaO_2	MgO	TiO†	VO†	FeO†	MnO†	CoO† NiO†	
	CaO						
KO_2	CaO_2						
	SrO						
RbO_2	SrO_2		NbO†				CdO
	BaO						
CsO_2	BaO_2	ZrO					

† Defect or distorted structures.

Other AO oxides form the zinc-blende or wurtzite structures (BeO), PtS structure (PdO, PtO), PbO structure (PbO, SnO) and the tenorite structure (CuO, AgO). The superoxides of K, Rb and Cs and the peroxides of Ca, Sr and Ba form the CaC_2 structure. The phase NaO_2 has the FeS_2 structure.

Table 4.1*a References*

1. A. Iandelli and R. Botti, *Rend. Atti R. Acad. Lincei* **67**, 638 (1937).
2. L. H. Brixner, *J. Inorg. Nucl. Chem.* **15**, 199 (1960).
3. A. E. Gorum, *Acta Cryst.* **10**, 144 (1957).
4. A. Iandelli, *Z. Anorg. Allgem. Chem.* **288**, 81 (1956).
5. W. H. Willott and E. J. Evans, *Phil. Mag.* **18**, 114 (1934).
6. R. Ferro, *Acta Cryst.* **8**, 360 (1955).
7. A. Iandelli, *R. C. Accad. Naz. Lincei* **13**, 138 (1952).
8. G. Bruzzone, *Atti. Accad. Lincei, Classe Sci. Fis. Mat. Nat.* **30**, 208 (1961).
9. A. S. Coffinberry and F. H. Ellinger, *U.S.A. Rep. A/Conf.* **8**, 826 (1955).
10. R. Ferro, *R. C. Acad. Naz. Lincei*, **13**, 401 (1952).
11. A. Iandelli and E. Botti, *R. C. R. Accad. Naz. Lincei* **24**, 459 (1936).
12. E. Parthé and E. Parthé, *Acta, Cryst.* **16**, 91 (1963).
13. K. Meisel, *Z. Anorg. Allgem. Chem.*, **240**, 300 (1939).
14. K. Bachmayer, H. Nowotny and A. Kohl, *Monatsh. Chem.* **86**, 39 (1955).
15. G. Hägg and A. G. Hybinette, *Phil. Mag.* **20**, 913 (1935).
16. R. Ferro, *Acta Cryst.* **9**, 817 (1956).
17. B. J. McDonald and W. I. Stuart, *Acta Cryst.* **13**, 447 (1960).
18. B. Post and F. W. Glaser, *J. Chem. Phys.* **20**, 1050 (1952).
19. H. Nowotny, R. Kieffer, F. Benesovsky and C. Bruki, *Monatsh. Chem.* **90**, 86 (1959).
20. C. P. Kempter, E. K. Storms and R. J. Fries, *J. Chem. Phys.* **33**, 1873 (1960).
21. J. Johnston, U. of Cal. Berk., Lawrence Rad. Lab, UARL-11390 (1964).
22. D. H. Templeton and C. H. Dauben, U.S.A.E.C. Publ. AECD 3443 (1952).
23. A. L. Bowman, *J. Phys. Chem.* **65**, 1596 (1961).
24. E. Laube and H. Nowotny, *Monatsh. Chem.* **89**, 312 (1958).
25. R. O. Elliott and C. P. Kempter, *J. Phys. Chem.* **62**, 630 (1958).
26. V. Zwicker, *Z. Metallk. Deutch.* **54**, 477 (1963).
27. H. Nowotny and R. Kieffer, *Metallforschung*, **2**, 257 (1947).
28. H. D. Keith and J. W. Mitchell, *Phil. Mag.* **42**, 1331 (1951).
29. H. Wilman, *Proc. Phys. Soc. (London)* **52**, 323 (1940).
30. H. Ott, *Z. Krist.* **63**, 222 (1926).
31. W. P. Davey, *Phys. Rev.* **21**, 143 (1923).
32. E. T. Teatum and N. O. Smith, *J. Phys. Chem.* **61**, 697 (1957).
33. A. DeCugnac and J. Pouradier, *Compt. Rend.* **258**, 4761 (1964).
34. R. E. Glover, *Z. Physik.* **138**, 222 (1954).
35. E. Broch, I. Oftedal and A. Pabst. *Z. Physik. Chem.* **3B**, 209 (1929).
36. A. Jevins, M. Straumanis and K. Karlons, *Z. Physik. Chem.* **40B**, 146 (1938).
37. C. W. Tucker, Jr. and P. Senio, U.S. At. Energy Comm. Rept. AECU-3428, 1244 (1952).
38. T. Barth and G. Lunde, *Central Mineral. Geol.* **1927A**, 57 (1927).
39. A. Smits and D. Tollenaar, *Z. Physik. Chem.* **52B**, 222 (1942).

40. L. F. VERESHEHAGIN and S. S. KABALKINA, *Dokl. Akad. Nauk SSSR* **113,** 797 (1957).
41. E. ZINTL and A. HARDER, *Z. Physik. Chem.* **14B**, 265 (1931).
42. J. E. WORSHAM, JR., M. K. WILKINSON and C. G. SHULL, *J. Phys. Chem. Solids* **3**, 303 (1957).
43. Z. G. PINSKER and S. Y. KARERIN, *Kristallografiya* **2**, 386 (1957).
44. W. KLEMM and G. WINKELMANN, *Z. Anorg. Allgem. Chem.* **288**, 87 (1956).
45. G. BRAUER and R. ESSELBORN, *Z. Anorg. Allgem. Chem.* **309**, 151 (1961).
46. W. H. ZACHARIASEN, *Acta Cryst.* **2**, 388 (1949).
47. K. BECKER and F. EBERT, *Z. Physik.* **31**, 268 (1952).
48. H. A. EICK, N. C. BAENZIGER and L. EYRING, *J. Am. Chem. Soc.* **78**, 5987 (1956).
49. R. E. RUNDLE, *Acta Cryst.* **1**, 180 (1948).
50. A. MUNSTER and K. SAGEL, *Z. Electrochem.* **57**, 571 (1953).
51. M. E. STRAUMANIS, C. A. FAUNCE and W. J. JAMES, *Acta Met.* **15**, 65 (1967).
52. E. SANDOR and W. A. WOOSTER, *Acta Cryst.* **13**, 339 (1960).
53. C. P. KEMPTER, N. H. KRIKORIAN and J. C. MCGUIRE, *J. Phys. Chem.* **61**, 1237 (1957).
54. T. W. BAKER, *Acta Cryst.* **11**, 300 (1958).
55. F. H. ELLINGER and W. H. ZACHARIASEN, *J. Inorg. Nucl. Chem.* **15**, 185 (1960).
56. H. HUBER and S. WAGENER, *Z. Tech. Physik* **23**, 1 (1942).
57. R. FAIVRE and A. MICHEL, *Compt. Rend.* **207**, 156 (1938).
58. S. GREENWALD, *Acta Cryst.* **6**, 396 (1953).
59. U. ENZ, J. F. FAST, S. VAN HOUTEN and J. SMIT, *Philips Res. Rpts.* **17**, 451 (1962).
60. W. L. ROTH, *Acta Cryst.* **13**, 140 (1960).
61. E. RIANO and J. L. AMOROŚ PORTOLÉS, *Bol. Real. Soc. Espan. Hist. Nat., Secc. Geol.* **56**, 391 (1958).
62. J. P. BORQUET, W. DORNELAS and P. LACOMBE, *Compt. Rend.* **260**, 4771 (1965).
63. A. L. BOWMAN, T. C. WALLACE, J. L. YARNELL and R. G. WENZEL, *Acta Cryst.* **21,** 843 (1966).
64. Y. SHIMOMURA, *Bull. Naniwa Univ.* **A3**, 175 (1955).
65. W. H. ZACHARIASEN, *Acta Cryst.* **5**, 19 (1952).
66. H. A. EICK, N. C. BAENZIGER and L. EYRING, *J. Am. Chem. Soc.* **78**, 5147 (1956).
67. W. G. BURGERS, *Z. Physik* **80**, 352 (1933).
68. N. SCHÖNBERG, *Acta Chem. Scand.* **8**, 240 (1954).
69. M. E. STRAUMANIS and H. W. LI, *Z. Anorg. Allgem. Chem.* **305**, 143 (1960).
70. R. E. RUNDLE, N. C. BAENZIGER, A. S. WILSON and R. A. MCDONALD, *J. Am. Chem. Soc.* **70**, 99 (1948).
71. W. KLEMM and L. GRIMM, *Z. Anorg. Allgem. Chem.* **250**, 42 (1942).
72. J. C. ARCHARD and G. TSOUCARIS, *Compt. Rend.* **246**, 285 (1958).
73. E. P. SAL'DAV, *Zap. Vses. Mineralog. Obshchestra* **86**, 324 (1957).
74. O. J. GUNTERT and A. FAESSLER, *Z. Krist.* **107**, 357 (1956).
75. E. MILLER, K. KOMAREK and I. CADOFF, *Trans. AIME* **218**, 978 (1960).
76. I. OFTEDAL, *Z. Physik. Chem. (Leipzig)* **128**, 154 (1927).
77. N. MARIANO and E. F. WAREKOIS, *Science* **142**, 1672 (1963).
78. A. IANDELLI, *Gazz. Chim. Ital.* **85**, 881 (1955).
79. C. B. SCLAR, L. C. CARRISON and C. M. SCHWARTZ, *Science* **143**, 1352 (1964).
80. M. GUITTARD and A. BENACERRAF, *Compt. Rend.* **248**, 2589 (1959).
81. A. V. GOLUBKOV, T. B. ZHUKOVA and V. M. SERGEEVA, *Inorg. Materials* **2**, 66 (1966).
82. E. BROCH, *Z. Physik. Chem. (Leipzig)* **127**, 446 (1927).
83. L. M. CORLISS, N. ELLIOTT and J. M. HASTINGS, *Phys. Rev.* **104**, 924 (1956).
84. G. I. MAKOVELSKII and N. N. SIROTA, *Dokl. Akad. Nauk Belorussk, SSR* **8**, 289 (1964).
85. B. WASSERSTEIN, *Am. Mineralogist* **36**, 102 (1951).
86. J. W. EARLEY, *Am. Mineralogist* **35**, 337 (1950).
87. A. NISHIYAMA and T. OKADA, *Mem. Fac. Sci. Kyushu Univer.* Ser. B, **3**, 3 (1960).
88. H. KREBS, K. GRÜN and D. KALLEN, *Z. Anorg. Allgem. Chem.* **312**, 307 (1961).
89. M. K. SLATTERY, *Phys. Rev.* **25**, 333 (1925).
90. V. M. GOLDSCHMIDT, *Skrifter Norske Videnskaps—Akad. Oslo I. Mat.—Naturv. Ki.* **1926**, No. 8 (1926).
91. R. W. M. D'EYE, P. G. SELLMAN and J. R. MURRAY, *J. Chem. Soc.* **1952**, 2555 (1952).
92. R. FERRO, *Z. Anorg. Allgem. Chem.* **275**, 320 (1954).
93. H. SENIFF and W. KLEMM, *Z. Anorg. Allgem. Chem.* **242**, 92 (1939).

94. F. K. McTaggart and A. D. Wadsley, *Australian J. Chem.* **11**, 445 (1958).
95. H. Hartmann, H. J. Fröhlich and F. Ebert, *Z. Anorg. Chem.* **218**, 181 (1934).
96. N. Elliott and J. Hastings, *Acta Cryst.* **14**, 1018 (1961).
97. L. A. Siegel, *J. Chem. Phys.* **17**, 1146 (1949).
98. G. Natta and L. Passerini, *Gazz. Chim. Ital.* **61**, 191 (1931).
99. R. Juza and A. Mehne, *Z. Anorg. Allgem. Chem.* **299**, 33 (1959).

Table 4.1*b*, *References*

1. B. Post, D. Moskowitz and F. Glaser, *J. Chem. Soc.* **78**, 1800 (1956).
2. P. Cotter and I. Kohn, *J. Am. Ceram. Soc.* **37**, 415 (1954).
3. M. Nadler and C. Kempter, *J. Phys. Chem.* **64**, 1471 (1960).
4. F. Ellinger, *Trans. ASM* **31**, 89 (1943).
5. P. Chiotti, *Iowa State Coll. J. Sci.* **26**, 185 (1952).
6. D. Keller, J. Fackelmann, E. Speidel and S. Paprocki, *Pulvermet. in der Atomkerntechnik*, 4, Planseesemunir, Springer-Verlag 279 (1962).
7. *Welding Engr.* No. 4 (1958).
8. L. Brewer, L. A. Bromley, P. W. Gilles and N. L. Lofgren, *Chemistry and Metallurgy of Miscellaneous Materials,* Ed. L. L. Quill, McGraw-Hill Book Co., New York (1950).
9. R. Keiffer and P. Swarzkopf, *Hard Alloys, Metallurgizdat,* Moscow (1957).
10. N. A. Lange, *Handbook of Chemistry,* Handbook Pub. Co., Sandusky, O. (1946).
11. L. J. Cronin, *Am. Ceram. Soc. Bull.* **30**, 234 (1951).
12. O. Kubaschewski and E. L. Evans, *Metallurgical Thermochemistry,* Pergamon, London (1958).
13. O. J. Wittermore, Jr., *J. Can. Ceram. Soc.* **28**, 43 (1959).
14. W. G. Bradshaw and C. O. Mathews, *Properties of Refractory Materials,* LMSD-2466 (June 24, 1958).
15. I. E. Campbell, *High Temperature Technology,* Wiley, New York (1956).
16. E. M. Dudnik, G. V. Lashkarev, Yu. B. Paderno and V. A. Obolonchik, *Izv. Akad. Nauk SSSR, Neorg. Mat.* **7**, 980 (1966).
17. M. Picon and N. Patrie, *Compt. Rend.* **242**, 1521 (1956).
18. M. Picon, J. Flahaut, M. Guittard and M. Patrie, 16*th Int. Congress Pure et Appl. Chemie* (Paris, 1957).
19. G. V. Samsonov and N. M. Popova, *Zhur. Obshcei Khim.* **27**, 3 (1957).
20. K. Siborg and D. Katz, *Actinides,* Moscow (1955).
21. W. Arbiter, *New High Temperature Intermetallic Compounds,* WADC TR-53-190 (1953).
22. *Ceramics for Advanced Technologies,* Ed. J. E. Hove and W. C. Riley, Wiley, New York (1965).
23. G. R. Finlay, *Chem. in Canada* **4**, 41 (1952).
24. A. V. Golubkov, T. B. Zhukova and V. M. Sergeeva, *Inorg. Materials* **2**, 66 (1966).
25. Ya S. Umanskii and V. I. Fadeeva, *Inorg. Materials* **2**, 70 (1966).
26. W. A. Plummer, D. E. Campbell and A. A. Comstock, *J. Am. Ceram. Soc.* **45**, 310 (1962).
27. W. A. Fisher and A. Hoffmann, *Arch. Eisenhuettenu* **35**, 27 (1964).
28. O. H. Krikorian, *Thermal Expansion of High Temperature Materials,* UCRL-6132 (September 1960).
29. G. V. Samsonov and V. N. Paderno, *Planseeber. Pulvermet.* **12**, 19 (1964).

Table 4.1*c*, *References*

1. L. H. Brixner, *J. Inorg. Nucl. Chem.* **16**, 162 (1960).
2. J. L. Anderson and M. Schieber, *J. Phys. Chem. Solids* **25**, 961 (1964).
3. A. Lecerf, *Compt. Rend.* **254**, 2003 (1962).
4. F. F. Hockings and J. G. White, *Acta Cryst.* **14**, 328 (1961).
5. S. Geller and S. H. Wernick, *Acta Cryst.* **12**, 46 (1959).
6. S. A. Semiletov and L. L. Mun, *Kristallografiya* **4**, 414 (1959).
7. W. Hofmann, *Ber. Preuss. Akad. Wiss.* **1938**, 111 (1938).
8. G. Gattow and J. Zemann, *Z. Anorg. Allgem. Chem.* **279**, 324 (1955).
9. O. Glemser and M. Filcek, *Z. Anorg. Allgem. Chem.* **279**, 321 (1955).
10. G. Gattow and J. Zeman, *Z. Anorg. Allgem. Chem.* **279**, 327 (1955).

Table 4.2, *References*

1. G. S. SAIRRI, L. D. CALVERT, R. D. HEYDING and J. B. TAYLOR, *Can. J. Chem.* **42**, 620 (1964).
2. L. THOMASSEN, *Z. Phys. Chem.* **4B**, 277 (1929).
3. A. KJEKSHUS, *Acta Chem. Scand.* **14**, 1450 (1960).
4. O. NIAL, A. ALMIN and A. WESTGREN, *Z. Phys. Chem.* **B14**, 81 (1931).
5. I. OFTEDAL, *Z. Phys. Chem. (Leipzig)* **135**, 291 (1928).
6. R. FERRO, *Acta Cryst.* **10**, 476 (1957).
7. H. J. WALLBAUM, *Z. Metallik* **35**, 200 (1943).
8. B. T. MATTHIAS, T. H. GEBALLE and V. B. COMPTON, *Ref. Mod. Phys.* **35**, 1 (1963).
9. C. W. W. HOFFMAN, R. C. ROPP and R. W. MOONEY, *J. Am. Chem. Soc.* **81**, 3830 (1959).
10. S. C. ABRAHAMS and J. KALNAJS, *Acta Cryst.* **8**, 503 (1955).
11. G. F. CARTER and D. H. TEMPLETON, *J. Am. Chem. Soc.* **75**, 5247 (1953).
12. L. S. RAMSDELL, *Am. Mineralogist* **10**, 281 (1925).
13. G. B. BOKU and L. I. TSINOBER, *Tr. Inst. Kristallogr. Akad. Nauk SSSR* **9**, 239 (1954).
14. N. ELLIOTT, *J. Chem. Phys.* **33**, 903 (1960).
15. M. E. STRAUMANIS, G. C. AMSTUTZ and S. CHAN, *Am. Minerologist* **49**, 206 (1964).
16. P. RAMDOHR, *Neues Jahrb. Mineral, Monatsh. Chem.* **6**, 133 (1955).
17. E. F. HOCKINGS and J. G. WHITE, *J. Phys. Chem.* **64**, 1042 (1960).
18. F. SGARLATA, *Rend. Ist. Super. Sanita* **22**, 851 (1959).
19. N. ELLIOTT, *J. Am. Chem. Soc.* **59**, 1958 (1937).
20. D. LUNGOVIST, *Arkiv. Kemi. Min. Geol.* **24A**, 22 (1947).
21. J. W. EARLY, *Am. Mineralogist* **35**, 337 (1950).
22. K. MEISEL, *Z. Anorg. Chem.* **219**, 141 (1934).
23. S. GELLER and B. B. CETLIN, *Acta Cryst.* **8**, 272 (1955).
24. S. GELLER, *J. Am. Chem. Soc.* **77**, 2641 (1955).

Table 4.3*a*, *References*

1. J. L. MORIARTY, R. O. GORDON and J. E. HUMPHREYS, *Acta Cryst.* **19**, 285 (1965).
2. J. L. MORIARTY, J. F. HUMPHREYS, R. O. GORDON and N. C. BAENZIGER, *Acta Cryst.* **21**, 840 (1966).
3. A. J. BRADLEY and S. S. LU, *J. Inst. Met.* **60**, 319 (1937).
4. A. W. SEARCY and R. J. PEAVLER, *J. Am. Chem. Soc.* **75**, 5657 (1953).
5. G. BRAUER and R. RUDOLPH, *Z. Anorg. Chem.* **248**, 405 (1941).
6. W. H. ZACHARIASEN, *Z. Physik. Chem. (Leipzig)* **128**, 39 (1927).
7. H. J. WALLBAUM, *Z. Metallk.* **33**, 378 (1941).
8. M. V. STACKELBERG, *Z. Physik. Chem.* **9B**, 437 (1930).
9. N. BREDRIG, *J. Phys. Chem.* **46**, 801 (1942).
10. P. STECHER, A. NECKEL, F. BENESOVSKY and H. NOWOTNY, *Planseeber Pulvermet.* **12**, 181 (1964).
11. F. H. SPEDDING, K. GSCHNEIDER, JR. and A. H. DAANE, *J. Am. Chem. Soc.* **80**, 4499 (1958).
12. M. A. BREDIG, *J. Am. Ceram. Soc.* **43**, 493 (1960).
13. W. H. C. RUEGGEBERG, *J. Am. Chem. Soc.* **65**, 602 (1943).
14. I. HIGASHI, *Rika Gaku Kenkyusho Hokoku* **37**, 271 (1961).
15. M. V. STACKELBERG, *Z. Elektrochem.* **37**, 542 (1931).
16. S. C. ABRAHAMS and J. KLANAJS, *Acta Cryst.* **7**, 838 (1954).
17. V. KOTOV and S. RAIKHSHTEIN, *J. Phys. Chem., USSR* **15**, 1057 (1941).
18. K. R. TSAI, P. M. HARRIS and E. N. LASSETTRE, *J. Phys. Chem.* **60**, 338 (1956).
19. S. C. ABRAHAMS and J. KALNAJS, *Acta Cryst.* **8**, 503 (1955).
20. A. HELMS and W. KLEMM, *Z. Anorg. Allgem. Chem.* **242**, 33 (1939).
21. J. D. BERNAL, E. DIATLOWA, J. KASARNOVSKY, S. REICHSTEIN and A. G. WARD, *Z. Krist.* **92A**, 344 (1935).

Table 4.3*b*, *References*

1. *Gmelins Handbuck der Anorg. Chem. System* No. 30, Verlag Chemie, Berlin 300 (1932).
2. L. BREWER *et al., Chemistry and Metallurgy of Miscellaneous Materials*, Ed. L. L. QUILL, McGraw-Hill Book Co., New York (1950).

3. R. Vickery, R. Sedlacek and A. Ruben, *J. Chem. Soc.* **2**, 498 (1959).
4. F. Spedding, K. Gschneider and A. Daane, *Trans. AIME* **215**, 192 (1959).
5. L. Cronin, *Am. Ceram. Soc. Bull.* **30**, 234 (1951).
6. G. V. Samsonov, T. Ya Kosolapova and G. N. Makarenko, *Zhur. Neorg. Khim.* **7**, 975 (1962).

Table 4.4, *References*

1. D. P. Morris, C. D. Price and J. L. Hughes, *Acta Cryst.* **16**, 839 (1963).
2. O. Heusler, *Ann. Phys. Lpz.* **19**, 155 (1934).
3. A. Taylor and R. W. Floyd, *J. Inst. Met.* **81**, 25 (1952).
4. A. P. Klyucharev, *Z. Eksper. Theoret. Fiz.* **9**, 1501 (1939).
5. L. Castelliz, *Monatsh. Chem.* **84**, 765 (1953).
6. B. R. Coles, W. Hume-Rothery and H. P. Myers, *Proc. Roy. Soc.* A **196**, 125 (1949).
7. S. Valentiner, *Z. Metallk.* **44**, 59 (1953).
8. H. Nowotny and B. Glatzl, *Monatsh. Chem.* **83**, 237 (1952).
9. P. Rahlfs, *Metallwirt* **16**, 640 (1937).
10. L. Castelliz, *Monatsh. Chem.* **82**, 1059 (1951).

CHAPTER 5

ZnS TYPE AND RELATED STRUCTURES

THE zinc sulfide type structures can be described as a cubic close-packed structure with all or part of the tetrahedrally coordinated holes in the eight corners inside the cell filled with atoms. Related structures have unit cells which are multiples of the basic cell. A flow sheet is shown in Table 5.0.

TABLE 5.0. ZnS *Type Structure*

- 5.1 Diamond structure
  - 5.2 Zinc blende structure
    - 5.3 Cristobalite (high form) structure
    - 5.4 Fluorite structure
      - 5.5 MgAgAs structure
      - 5.6 K_2PtCl_6 structure
      - 5.7 C-rare Earth structure
      - 5.8 Pyrochlore structure
      - 5.9 Scheelite structure

The zinc-blende structure is an ordered diamond cubic structure, the high-temperature form of cristobalite is a zinc-blende-type structure, and the CaF_2 structure is formed by adding atomic positions to the zinc-blende structure. The MgAgAs structure is an ordered CaF_2 structure, the K_2PtCl_6 has the CaF_2 structure and the C-rare earth and pyrochlore unit cells are made up of eight oxygen-deficient unit cells of the CaF_2 structure. Two unit cells of the CaF_2 structure make up the scheelite unit cell.

5.1. Diamond, A4, *Fd*3*m* Cubic

The simplest structure of the zinc sulfide type is the diamond structure which can be derived from a unit cell with atoms on points of a face-centered cubic lattice by placing four additional carbon atoms at the centers of every other of the eight tetrahedrally coordinated sites at the octant centers, as shown in Fig. 5.1. These additional carbon atoms are diagonally opposite one another in two top and two bottom octants and are not located in octants directly above one another in the unit cell. Germanium silicon and α-Sn also have this structure. The X-ray data are given in Table 5.1.

The eight atoms are located at $0, 0, 0; 0, \frac{1}{2}, \frac{1}{2}; \frac{1}{2}, 0, \frac{1}{2}; \frac{1}{2}, \frac{1}{2}, 0; \frac{1}{4}, \frac{1}{4}, \frac{1}{4}; \frac{1}{4}, \frac{3}{4}, \frac{3}{4}; \frac{3}{4}, \frac{1}{4}, \frac{3}{4}; \frac{3}{4}, \frac{3}{4}, \frac{1}{4}$.

Properties

Silicon and germanium are important semiconductor materials. Their properties have been attributed to the fourfold coordinated atoms which are

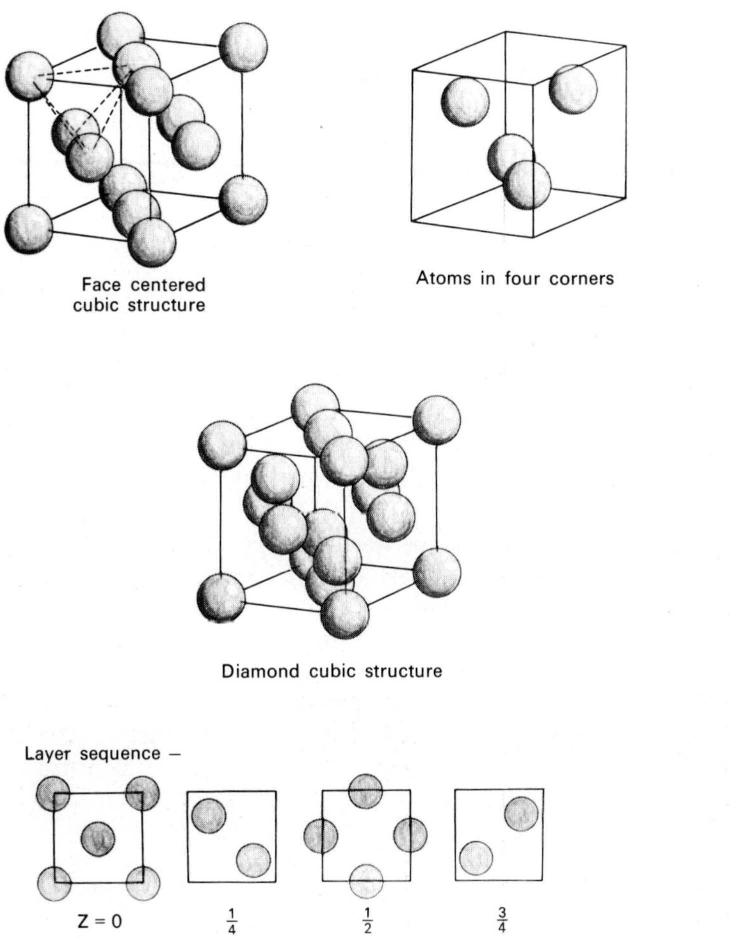

FIG. 5.1. The diamond cubic structure

covalently bonded. A list of energy gaps is given in Table 5.1 for elements with the diamond cubic structure. Note that the energy gap decreases as the cell size increases and the bonding becomes weaker.

The mobility of electrons in silicon is 1550 cm^2/V-sec and of holes is 250 cm^2/V-sec. This is smaller than the mobilities in grey tin ($\mu_- = 3000$ cm^2/V-sec) and germanium ($\mu_- = 4400$ cm^2/V-sec, $\mu_+ = 2700$ cm^2/V-sec) and larger than the mobility of electrons in diamond ($\mu_- = 900$ cm^2/V-sec). The role of germanium and silicon in electronic devices is well known.

TABLE 5.1. *Elements with the Diamond Structure*

Element	Cell size, a_0 (Å)	m.p. (°C)	E_g (eV)	σ (ohm–cm^{-1})	Coeff. thermal expansion (10^{-6}/°C)	Refs.
$C_{(d)}$	3.56679		6	10–14	1.2	1, 2
Ge	5.65754	937	0.665	10^{-2}	6.1	3, 4
Si	5.43070	1420	1.12	5×10^{-6}	4.6	5
α-Sn(13°C)	6.4912		0.08	10^3	20.9	6

5.2. Zinc Blende, Sphalerite, ZnS Type, B3, *F43m*, Cubic

The next structure in complexity is the zinc blende structure which can be visualized as a diamond cubic structure (shown in Fig. 5.1) having a unit cell with atoms of one element at the corners and at the centers of the faces of the cube and atoms of the other element in the tetrahedrally coordinated sites in the same corners as those filled in the diamond cubic structure (see Fig. 5.2a).

Thus, the four zinc atoms are located at 0, 0, 0; 0, $\frac{1}{2}$, $\frac{1}{2}$; $\frac{1}{2}$, 0, $\frac{1}{2}$; $\frac{1}{2}$, $\frac{1}{2}$, 0 and four sulfur atoms are at $\frac{1}{4}$, $\frac{1}{4}$, $\frac{1}{4}$; $\frac{1}{4}$, $\frac{3}{4}$, $\frac{3}{4}$; $\frac{3}{4}$, $\frac{1}{4}$, $\frac{3}{4}$; and $\frac{3}{4}$, $\frac{3}{4}$, $\frac{1}{4}$ positions.

Compounds of the AB type which have covalent bonds tend to adopt this structure. For phases such as SiC which are purely covalent and whose atoms fit well in fourfold coordination, the formation of the zinc blende structure is not surprising. However, the fact that CdTe has the zinc blende structure must be attributed to the polarizability of the large telluride ion. A number of compounds form both the zinc blende and the hexagonal ZnS form (see Chapter 6), but CdTe forms only the zinc blende structure. The less covalent AB compounds probably favor the formation of the wurtzite structure (see Table 5.2a).

Properties

Some of the most important semiconductor compounds form this structure. One of these, silicon carbide, has an energy gap of 3.5 eV. The III-V compounds such as AlP, AlAs, AlSb, GaP, GaAs, GaSb, InP, InAs and InSb and II-VI compounds such as ZnS, ZnSe, ZnTe and CdSe are other examples. A list of energy gaps for these compounds are listed in Table 5.2b. Note that compounds formed from elements from the bottom of the periodic table have lower E_g values and higher mobilities.

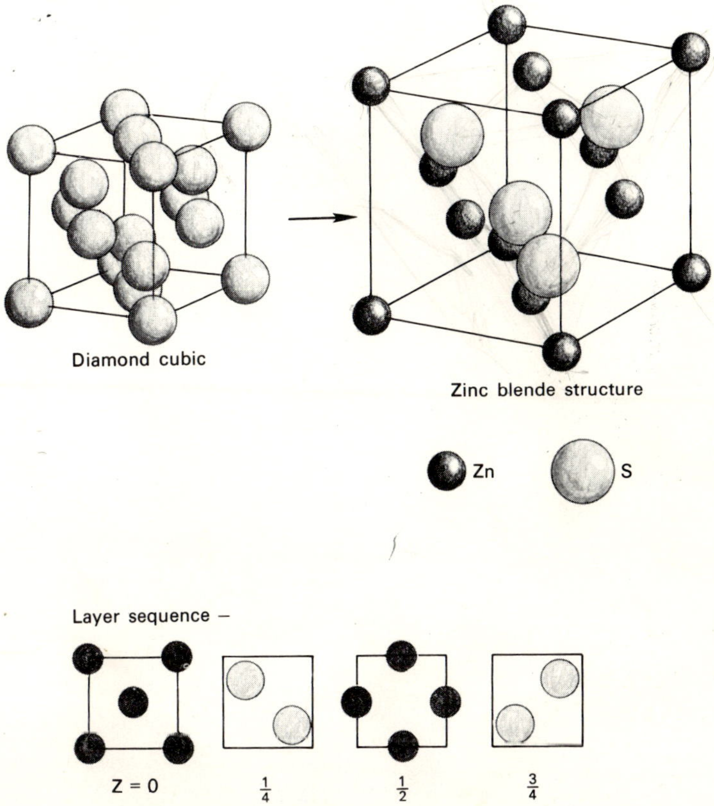

FIG. 5.2a. The zinc blende structure

Cadmium sulfide, CdSe, InSb and CdTe have a narrow forbidden energy gap and are photoconductive in the infrared. These materials along with PbS, PbSe and PbTe have been used in infrared detectors, photoresistors, photocells and transistors.

The phases HgTe, InAs and InSb have large electron and hole mobilities and are therefore used as Hall Effect devices.

High-temperature rectifiers, diodes and microwave detectors have been made from GaAs, InSb, InP, InAs, AlSb and GaP and fluorescent lamps make use of the luminescence power of ZnS and CdS doped with Cu, Mn, Sn, etc.

TABLE 5.2a. *Compounds with the Zinc Blende Structure*

Compound	Cell size, a_0 (Å)	Refs.
Intermetallics		
AlAs	5.656	1
AlP	5.451	2, 3
AlSb	6.1355	4
AsGa	5.6532	4, 5
AsIn	6.0584	4
GaP	5.450	2, 3, 4
GaSb	6.09612	6
InP	5.8688	3, 4
InSb	6.4788	6
Borides		
AsB	4.777	7
Pb	4.538	7, 8
Carbides		
β-SiC	4.3596	9
Halides		
AgI	6.473	10
CuBr	5.6905	11, 12
CuCl	5.4057	13
CuF	4.255	14
α-CuI	6.15	12
Nitrides		
BN (H.P.)	3.615	15
Oxides		
ZnO	4.62	16
Sulfides, Selenides, Tellurides		
BeS	4.8624	17
BeSe	5.139	18
BeTe	5.626	18
CdS	5.830	19
CdTe	6.48	20
γ-Ga_2S_3 (L.T.)	5.181	21, 22
Ga_2Se_3	5.429	23
Ga_2Te_3	5.901	23, 24
HgS	5.86	25
HgSe	6.074	26
HgTe	6.460, 6.429	20, 26
In_2Te_3	6.518	24
β-MnS (red)	5.61	27
β-MnSe	5.83	28
β-ZnS	5.4109	29
ZnSe	5.667	30

A recent application of these materials has been as injection lasers. The action takes place at the junction of a diode. The diode GaAs emits radiation at 0·840 μ, GaSb at 1.6 μ, InAs at 3.112 μ, InP at 0.9 μ and InSb at 5.2 μ.

The transmission ranges for some compounds with the ZnS structure as well as for some of the group IV elements are shown in Fig. 5.2b.

TABLE 5.2b. *Melting Points, E_g and Mobility Data for Compounds with the Zinc Blende Structure*

	m.p. (°C)	E_g (eV)	μ-Mobility		Refs.
			μ_- (cm²/V-sec)	μ_+ (cm²/V-sec)	
III-V compounds					
AlAs	1600	2.1			1
AlP		3		200	1, 2
AlSb	1060	1.55			1
GaAs	1280	1.35	1,200		3
GaP	1350	2.35			3, 2
GaSb	728	0.70	5,000	650	3
InAs	942	0.33	23,000	100	1
InP	1055	1.30	3,400	650	3, 2
InSb	525	0.17	80,000	700	4
II-VI compounds					
CdS	1750 (100 atm)	2.38	1,460		5
CdSe	1350	1.74			5
CdTe	1105	1.5	600		5
HgS		2.5	>250		5
HgTe	600	0.01	20,000		6, 7
ZnS	1850	3.6			5
ZnSe	1500	2.68			5
ZnTe	1254	2.25	100		5

Beta SiC has interesting mechanical properties as well as semiconducting properties. It is extremely hard, has a density of only 3.21 g/cm³, and has a modulus of approximately 60×10^6 psi. Silicon carbide has good oxidation resistant properties up to 1600°C. It is very stable in air below 1000°C and above 1150°C because of the formation of a protective silica coating. Above 1600°C it transforms to a hexagonal alpha form.

Boron phosphide and BAs have good abrasive properties as does BN with the ZnS structure which is formed at high pressures and temperatures. This form, called borazone, is almost as hard as diamond.

The compounds Ga_2S_3, Ga_2Se_3, Ga_2Te_3 and In_2Te_3 have the zinc blende structure, where only two-thirds of the A sites are filled. The energy gaps for these phases are 2.5, 1.9, 1.35 and 1.12 eV respectively.

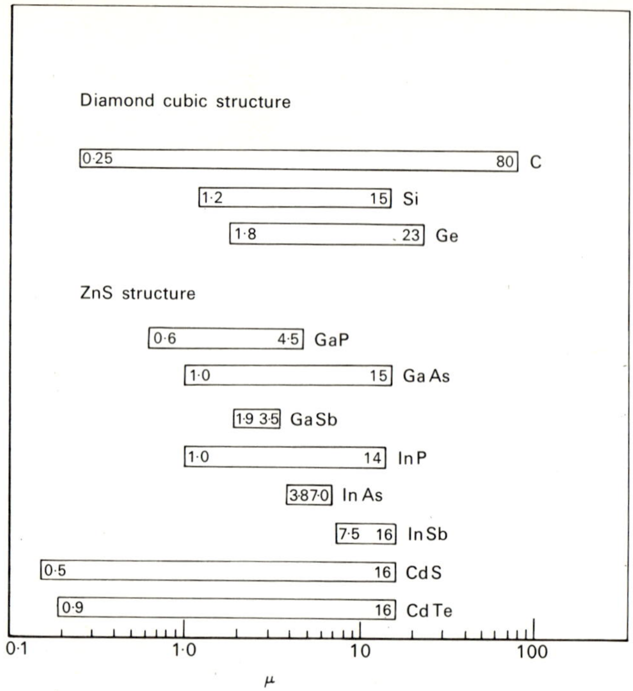

FIG. 5.2b. Range of transmittance for semiconductor samples (2mm thick sample, 10% cut off)

5.3. SiO_2, Cristobalite (High Form), *Fm3m*, Cubic

By treating the SiO_4 tetrahedra as a single unit the similarity between the structure of the high form of cristobalite and the zinc sulfide structure can be seen (Fig. 5.3). In the unit cell of the cristobalite structure, Si atoms are in the zinc atom positions and the silicon atoms in the SiO_4 tetrahedra are in the sulfur atom positions. Thus, Si atoms are at corners and centers of the faces of a cubic unit cell and SiO_4 tetrahedra are situated in diagonally opposed top octants and diagonally opposed bottom octants. These tetrahedra are not located in the octants directly above one another. The position of the oxygen atoms in each octant is the same as the arrangement of tetrahedra in the unit cell. The unit cell size is a = 7.16 Å at 290°C; however, there are indications that the cell is distorted.

In the cubic cell there are eight silicon atoms at

$$0, 0, 0; \tfrac{1}{4}, \tfrac{1}{4}, \tfrac{1}{4};$$
$$\tfrac{1}{2}, \tfrac{1}{2}, 0; \tfrac{3}{4}, \tfrac{3}{4}, \tfrac{1}{4};$$
$$\tfrac{1}{2}, 0, \tfrac{1}{2}; \tfrac{3}{4}, \tfrac{1}{4}, \tfrac{3}{4};$$
$$0, \tfrac{1}{2}, \tfrac{1}{2}; \tfrac{1}{4}, \tfrac{3}{4}, \tfrac{3}{4}$$

and sixteen oxygen atoms at

$$\tfrac{1}{8}, \tfrac{1}{8}, \tfrac{1}{8};\ \tfrac{1}{8}, \tfrac{3}{8}, \tfrac{3}{8};\ \tfrac{3}{8}, \tfrac{1}{8}, \tfrac{3}{8};\ \tfrac{3}{8}, \tfrac{3}{8}, \tfrac{1}{8};$$
$$\tfrac{5}{8}, \tfrac{5}{8}, \tfrac{1}{8};\ \tfrac{5}{8}, \tfrac{7}{8}, \tfrac{3}{8};\ \tfrac{7}{8}, \tfrac{5}{8}, \tfrac{3}{8};\ \tfrac{7}{8}, \tfrac{7}{8}, \tfrac{1}{8};$$
$$\tfrac{5}{8}, \tfrac{1}{8}, \tfrac{5}{8};\ \tfrac{5}{8}, \tfrac{3}{8}, \tfrac{7}{8};\ \tfrac{7}{8}, \tfrac{1}{8}, \tfrac{7}{8};\ \tfrac{7}{8}, \tfrac{3}{8}, \tfrac{5}{8};$$
$$\tfrac{1}{8}, \tfrac{5}{8}, \tfrac{5}{8};\ \tfrac{1}{8}, \tfrac{7}{8}, \tfrac{7}{8};\ \tfrac{3}{8}, \tfrac{5}{8}, \tfrac{7}{8};\ \tfrac{3}{8}, \tfrac{7}{8}, \tfrac{5}{8}.$$

Other forms of SiO_2 are listed in Table 5.3. In these forms, the silicon atom is surrounded with a tetrahedral arrangement of oxygen atoms.

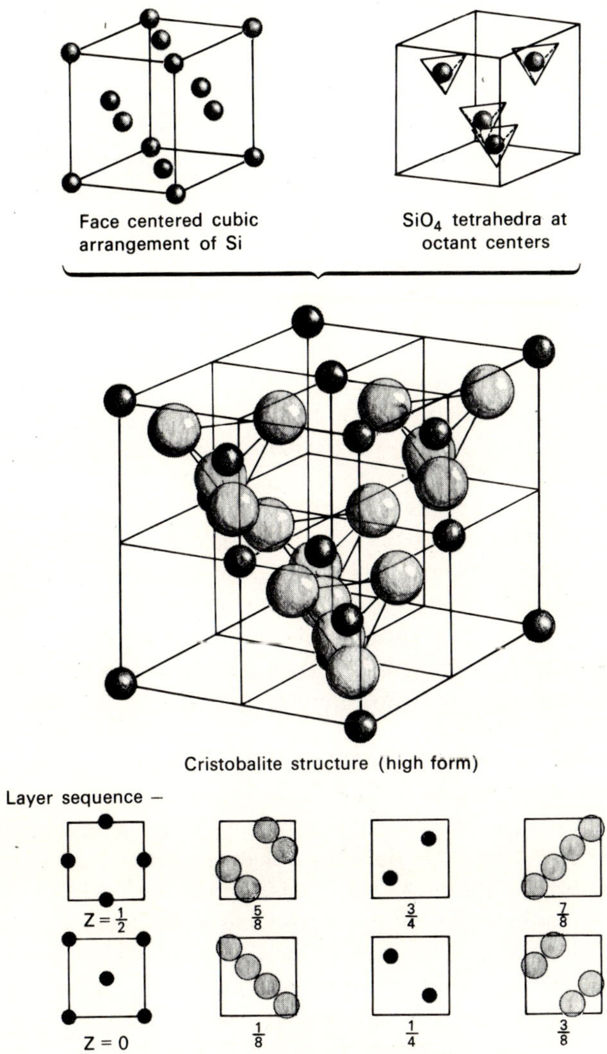

FIG. 5.3. The cristobalite structure (high form)

Recently, by using high pressures of 75–100 kilobars and temperatures greater than 800°C, a rutile form of SiO_2 has been prepared in which the silicon atom is surrounded with six oxygen atoms.

The three main forms of silica are quartz, tridymite and cristobalite. The transformation temperatures are 870° and 1470°C, respectively. In addition,

TABLE 5.3. *Forms of* SiO_2

Form	Structure	Space group	Cell size (Å)	Refs.
α-Quartz (L.T.)	trigonal	$P3_121$	$a_0 = 4.913$, $c_0 = 5.405$	1
β-Quartz (H.T.)	hexagonal	$P6_222$	$a_0 = 5.01$, $c_0 = 5.47$ (600°C)	1
Tridymite (H.T.)	hexagonal	$P6_3/mmc$	$a_0 = 5.03$, $c_0 = 8.22$ (200°C)	2
α-Cristobalite (L.T.)	tetragonal	$P4_12_12$	$a_0 = 4.973$, $c_0 = 6.926$ (200°C)	3
β-Cristobalite (H.T.)	cubic	$Fd3m$	$a_0 = 7.16$ (290°C)	3
Coestite	monoclinic	$B2/b$	$a_0 = b_0 = 7.17$, $c_0 = 12.38$, $\gamma = 8.604$	4
Keatite (SiO_2)	tetragonal	$P4_12_12$	$a_0 = 7.456$, $c_0 = 8.604$	4
Rutile (H.P.)				5

there is a high and low temperature form of these with transformation temperatures at 573°C for quartz, ~ 140°C for tridymite and ~ 250°C for cristobalite. In each structure linkage of SiO_4 tetrahedra is involved.

In the simplest silicate, the SiO_4^{2-} ion is present, i.e. Mg_2SiO_4, Be_2SiO_4. When two tetrahedral SiO_4 ions share an oxygen, $Si_2O_7^{6-}$ ions are formed as in $Ca_2MgSi_2O_7$. Three Si ions surrounded by oxygen ions form a ring, the ion being $(Si_3O_9)^{6-}$ as in $BaTiSi_3O_9$. A ring of six tetrahedra is found in beryl, which contains the ion $(Si_6O_{18})^{12-}$. The formation of chains produces the ion $[(SiO_3)_n]^{2n-}$, while a double chain contains $[(Si_4O_{11})_n]^{6n-}$ as found in the pyroxenes and the amphiboles. Sheets are formed in talc which contain the $[(Si_2O_5)_n]^{2n-}$ ion. The next step in complexity is the three-dimensional structures such as those of the zeolites and feldspars.

5.4. Fluorite and Antifluorite Structures, CaF_2 Type, C1, *Fm3m*

The fluorite structure can be considered to be a filled zinc blende structure because it can be constructed by substituting positive calcium ions in the zinc atom sites and negative fluorine ions in the sulfur atom sites and then adding four more fluorine ions in each unit cell in tetrahedrally coordinated sites not occupied by sulfur atoms so that all these sites are filled (see Fig. 5.4). The calcium ions occupy the corner and face-centered positions and the fluorine ions are situated in calcium tetrahedra. Each calcium ion is surrounded by eight fluorine ions and each fluorine ion is coordinated with four calcium ions. The 4-calcium ions are at $0, 0, 0$; $0, \frac{1}{2}, \frac{1}{2}$; $\frac{1}{2}, 0, \frac{1}{2}$; $\frac{1}{2}, \frac{1}{2}, 0$ and the 8-fluorine ions at $\frac{1}{4}, \frac{1}{4}, \frac{1}{4}$; $\frac{1}{4}, \frac{3}{4}, \frac{3}{4}$; $\frac{3}{4}, \frac{1}{4}, \frac{3}{4}$; $\frac{3}{4}, \frac{3}{4}, \frac{1}{4}$; $\frac{3}{4}, \frac{3}{4}, \frac{3}{4}$; $\frac{3}{4}, \frac{1}{4}, \frac{1}{4}$; $\frac{1}{4}, \frac{3}{4}, \frac{1}{4}$; $\frac{1}{4}, \frac{1}{4}, \frac{3}{4}$.

The oxides, sulfides, selenides and tellurides of lithium, sodium and potassium adopt the antifluorite structure which is identical to the fluorite structure except that the positive and negative ions are interchanged.

The tetravalent metal oxides of cerium, praseodymium and terbium and the actinide elements from thorium to americium form dioxides with this

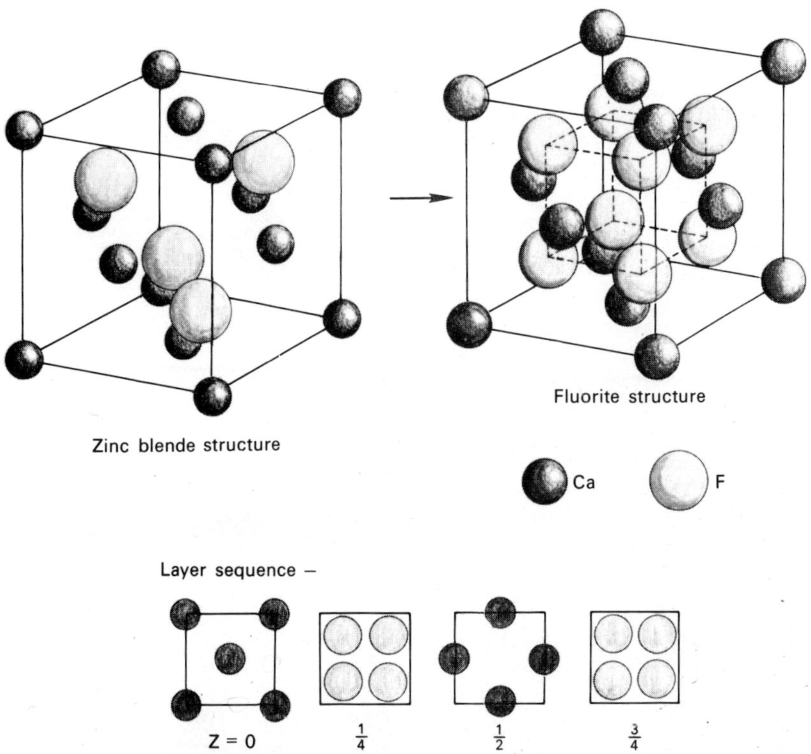

FIG. 5.4. The fluorite structure

structure. Of the transition metals, zirconium and hafnium are the only ones which have dioxides with the fluorite structure, but zirconium dioxide is monoclinic at room temperature. However, ZrO_2 in the cubic form is stabilized at room temperature by the addition of other cations such as calcium.

The fluorite structure is formed by AX_2 type compounds when the radius ratio R_A/R_X is greater than 0.73. If the ratio falls between 0.73 and 0.41, they often adopt the rutile structure where the coordination number of the A ion is 6 instead of 8. A list of compounds with the CaF_2 structure and cell data are given in Table 5.4

TABLE 5.4a. *Phases with the Fluorite Structure*

Phase	Cell size, a_0(Å)	Refs.
Intermetallics		
Al_2Au	6.01	1
Al_2Pt	5.922	2
As_3GeLi_5	6.09	3
As_3Li_5Si	6.055	3
As_3Li_5Ti	6.14	3
AsLiZn	5.924	4
$AuGa_2$	6.075	2
$AuIn_2$	6.515	2
$CoSi_2$	5.376	5
Ga_2Pt	5.923	2
$GeLi_5P_3$	5.89	3
$GeMg_2$	6.390	6
In_2Pt	6.366	2
Ir_2P	5.546	7
$IrSn_2$	6.338	8
LiMgP	6.02	4
Li_5P_3Si	5.854	3
Li_5P_3Ti	5.96	3
Mg_2Pb	6.81–6.86	6, 9, 10
$MgPu_2$	7.34	11
Mg_2Si	6.351	12
Mg_2Sn	6.7630	13
$NiSi_2$	5.406	14
PRh_2	5.516	15
$PtSn_2$	6.425	16
Borides		
Be_2B	4.661	17
Carbides		
Be_2C	4.3420	18
Halides, Oxyhalides		
AcOF	5.943	19
BaF_2	6.2001	20
CaF_2	5.46295	21
CdF_2	5.3880	22
CeOF	5.66–5.73	23
EuF_2	5.796	24
HgF_2	5.54	25
HoOF	5.523	26
LaOF	5.756	18, 27
NdOF	5.595	27
β-PbF_2	5.92732	28
PrOF	5.644	27
PuOF	5.71	18
RaF_2	6.368	29
SmOF	5.519	27
$SrCl_2$	6.9767	30
SrF_2	5.7996	31
β-YOF	5.363	18

TABLE 5.4a *(cont.)*

Phase	Cell size, a_0(Å)	Refs.
Hydrides		
CeH_2	5.590	32
DyH_2	5.201	33
ErH_2	5.123	33
GdH_2	5.297	34
HoH_2	5.165	33
LuH_2	5.033	33
NbH_2	4.563	35
NdH_2	5.470	33
PrH_2	5.517	33
ScH_2	4.78315	36
SmH_2	5.376	33
TbH_2	5.246	33
TmH_2	5.090	33
YH_2	5.199	33
Nitrides		
$GeLi_5N_3$	4.75	37
Li_5SiN_3	4.68	37
Li_5TiN_3	4.76	37
UN_2	5.32	38
Oxides		
AmO_2	5.376	39
CeO_2	5.409	40
CmO_2	5.372	39
K_2O	6.449	41
Li_2O	4.628	41
Na_2O	5.56	41
NpO_2	5.4341	39
PaO_2	5.505	42
PoO_2	5.59, 5.637	43, 44
PuO_2	5.3960	39
Rb_2O	6.742	45
TbO_2	5.220	46
ThO_2	5.586, 5.597	47, 48
UO_2	5.469	49
ZrO_2 (H.T.)		
Sulfides, Selenides, Tellurides		
K_2S	7.391	41, 50
K_2Se	7.676	41
K_2Te	8.152	41
Li_2S	5.708	41
Li_2Se	6.017	41
Li_2Te	6.517	41
Na_2S	6.526	41
Na_2Se	6.809	41
Na_2Te	7.314	41
Rb_2S	7.65	51

Properties

The oxides ThO_2 and ZrO_2 have been used extensively for high-temperature applications. The properties of thoria, zirconia and other high-temperature materials with the fluorite structure are tabulated in Table 5.4b. Thoria has been used in crucibles for melting platinum, palladium and rhodium and both materials have been employed in applications which require resistance to oxidation at very high temperatures.

TABLE 5.4b. *Properties of Phases with the* CaF_2 *Structure*

Compound	Melting point (°C)	Electrical resistivity (ohm-cm)	Micro-hardness (kg/mm^2)	Modulus (10^6 psi)	Thermal conductivity cal/cm-deg-sec	Coefficient of thermal expansion (10^{-6}/°C)
$NiSi_2$	1280[1]	0.02[7]	1019[9]		0.02	
Be_2B	1520[2]	0.02[7]				
CeO_2	2600[3]					8.9
HfO_2	2780[4]				0.002	5.9[14]
ThO_2	3205[5]		945[10]	35[11]	0.024[5]	9.4[14]
UO_2	2870[6]	10^{-4}[8]	600[10]	25[12]	0.018[13]	11.2[14]
ZrO_2	2677[6]		1200[10]	36[12]	0.004[12]	7.6[14]

The oxide UO_2 is another example of a *p*-type semiconductor. Because of its stability and high-temperature coefficient of electrical resistivity, it has been used in thermistors.

Of the intermetallics, Mg_2Si, Mg_2Ge and Mg_2Sn have been most thoroughly investigated. The energy gaps for these phases were found to be 0.77, 0.74 and 0.36 eV, respectively.

In recent years, there also has been considerable interest in optical

TABLE 5.4c. *Solid-state Lasers Using Fluorides with the* CaF_2 *Structure as Hosts*

Host	m.p.(°C)	Doping	Temperature (°K)	Wavelengths (μ)	Refs.
BaF_2	1320	U^{3+}	20	2.556	1
CaF_2	1418	Dy^{2+}	77	2.3–2.6	2, 3
		Nd^{3+}	77	1.046	4
		Sm^{2+}	20	0.708	5
		Tm^{2+}	4	1.116	6
		U^{3+}	300	2.613	7
SrF_2	870	Sm^{2+}	77	0.71	8
		U^{3+}	77	2.407	9

materials with the fluorite structure for laser application because the structure has cubic symmetry and contains a center of symmetry. Such symmetrical crystals have the least effect on the energy levels of the substituted ions. As a consequence, BaF_2, SrF_2 and CaF_2 have been used as hosts for many rare earth ions and a number of them have been lased. These laser systems are listed in Table 5.4c. Normally, these lasers are single crystals, but a study by S. F. Hatch *et al.* (*Appl. Phys. Letters* **5**, 153 (1964)) showed that a hot-pressed polycrystalline CaF_2 doped with Dy^{2+} sample which was clear could be made to lase.

5.5 MgAgAs Type, C1b, *F*43*m*, Cubic

The MgAgAs type structure can be considered to be an ordered CaF_2 structure and can be visualized by placing the arsenic atoms in the zinc atom positions in the unit cell of the ZnS structure, silver atoms in the sulfur atom position, and magnesium atoms in the four corners not occupied by the silver atoms. Thus, the arsenic atoms are at the corners and centers of the faces of a cube with silver atoms in the center of the left back octant and right front octant in the bottom octants in the cell, and in the octants not directly above the silver atoms in the top part of the cell. The magnesium atoms occupy the other four octants (see Fig. 5.5). The *x*, *y* and *z* values for these

TABLE 5.5. *Phases with the* AgAsMg *Structure*

Phase	Cell size, a_0 (Å)	Atomic positions			Refs.
		4*a*	4*c*	4*d*	
Intermetallics					
AgAsMg	6.240	As	Ag	Mg	1
AgAsZn	5.912	Ag	As, Zn	As, Zn	2
AsLiMg	6.22	As	Li	Mg	3
AsNaZn	5.912	As	Na, Zn	Na, Zn	2
BiCuMg	6.256	Cu	Bi	Mg	1
BiLiMg	6.76	Bi	Li	Mg	3
BiMgNi	6.162	Ni	Mg	Bi	4
CdCuSb	6.262	Cu	Cd	Sb	5
CoMnSb	5.900	Co	Mn, Sb	Mn, Sb	4
CuMgSb	6.152	Cu	Mg	Sb	1
CuMgSn	6.262	Cu	Mg	Sn	6
CuMnSb	6.066	Cu	Mn	Sb	4
LiMgSb	6.62	Sb	Li	Mg	3
LiPZn	5.780	P	Zn	Li	7
MgNiSb	6.048	Ni	Mg, Sb	Mg, Sb	4
MnNiSb	5.915				8
Nitrides					
LiMgN	4.970	N	Li, Mg	Li, Mg	9
LiZnN	4.877	N	Li, Zn	Li, Zn	9

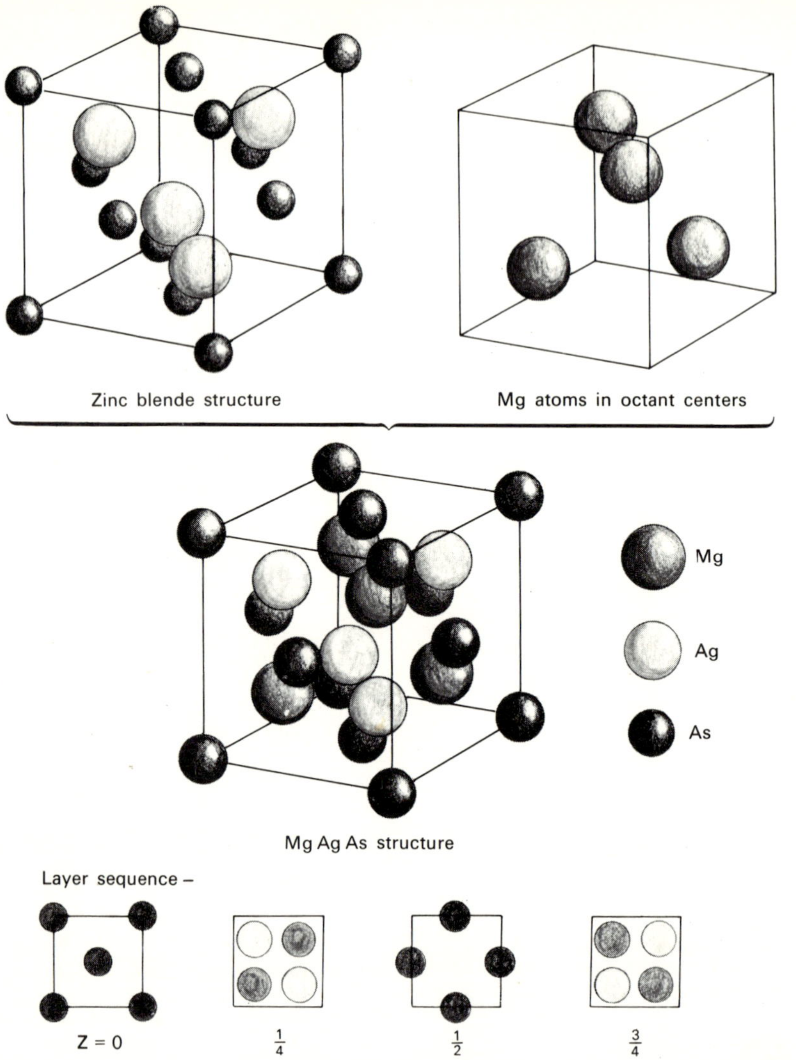

FIG. 5.5. The MgAgAs structure

atoms are 0, 0, 0; 0, $\frac{1}{2}$, $\frac{1}{2}$; $\frac{1}{2}$, 0, $\frac{1}{2}$ and $\frac{1}{2}$, $\frac{1}{2}$, 0 for the As atoms; $\frac{1}{4}$, $\frac{1}{4}$, $\frac{1}{4}$; $\frac{1}{4}$, $\frac{3}{4}$, $\frac{3}{4}$; $\frac{3}{4}$, $\frac{1}{4}$, $\frac{3}{4}$; and $\frac{3}{4}$, $\frac{3}{4}$, $\frac{1}{4}$ for the Ag atoms; and $\frac{3}{4}$, $\frac{3}{4}$, $\frac{3}{4}$; $\frac{3}{4}$, $\frac{1}{4}$, $\frac{1}{4}$; $\frac{1}{4}$, $\frac{3}{4}$, $\frac{1}{4}$ and $\frac{1}{4}$, $\frac{1}{4}$, $\frac{3}{4}$ for the Mg atoms.

Other phases with this structure are listed in Table 5.5. Note that if one element is placed in the Mg and As atom positions, the arrangement is the fluorite structure. Thus the MgAgAs type structure can be considered as the CaF_2 structure with the Mg and Ag atoms ordered.

5.6. K_2PtCl_6 Structure, *Fm*3*m*, Cubic

The K_2PtCl_6 structure is essentially the same as the CaF_2 structure with potassium ions in the fluorine ion positions and $PtCl_6$ octahedra in the calcium ion positions, as in Fig. 5.6a. Thus the $PtCl_6$ octahedra are situated at the corners and centers of the faces of the unit cell, with potassium ions in the center of the eight octants which make up the unit cell. More formally, the atoms are in the positions $(0, 0, 0; \frac{1}{2}, \frac{1}{2}, 0; \frac{1}{2}, 0, \frac{1}{2}; 0, \frac{1}{2}, \frac{1}{2})$ + for Pt: (0, 0, 0), for K: $(\frac{1}{4}, \frac{1}{4}, \frac{1}{4}; \frac{3}{4}, \frac{3}{4}, \frac{3}{4})$ and for Cl: $(x, 0, 0; 0, x, 0; 0, 0, x; \bar{x}, 0, 0; 0, \bar{x}, 0; 0, 0, \bar{x})$. (See Fig. 5.6b.)

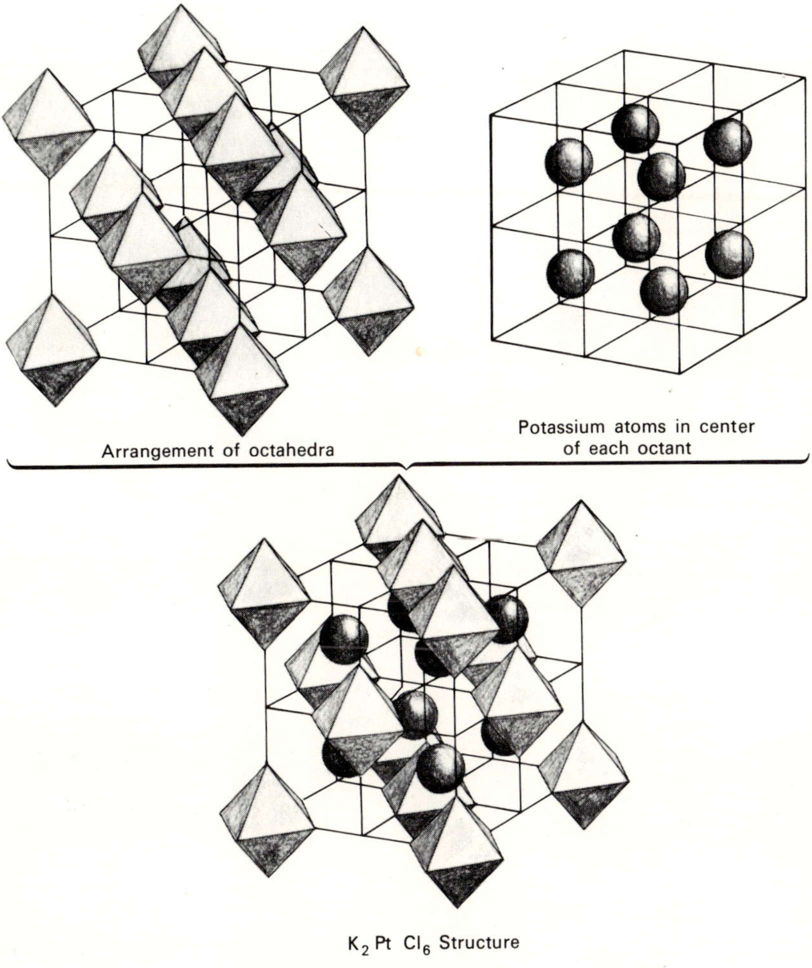

FIG. 5.6a. The K_2PtCl_6 structure (showing the octahedra)

An extremely large number of compounds with the general formula A_2BX_6 where A is NH_4^+, K^+, Rb^+, Tl^+ and Cs^+ and B is Co^{2+}, Pt^{2+}, Pd^{2+} and Sn^{2+} adopt this structure. A few of the compounds are listed in Table 5.6.

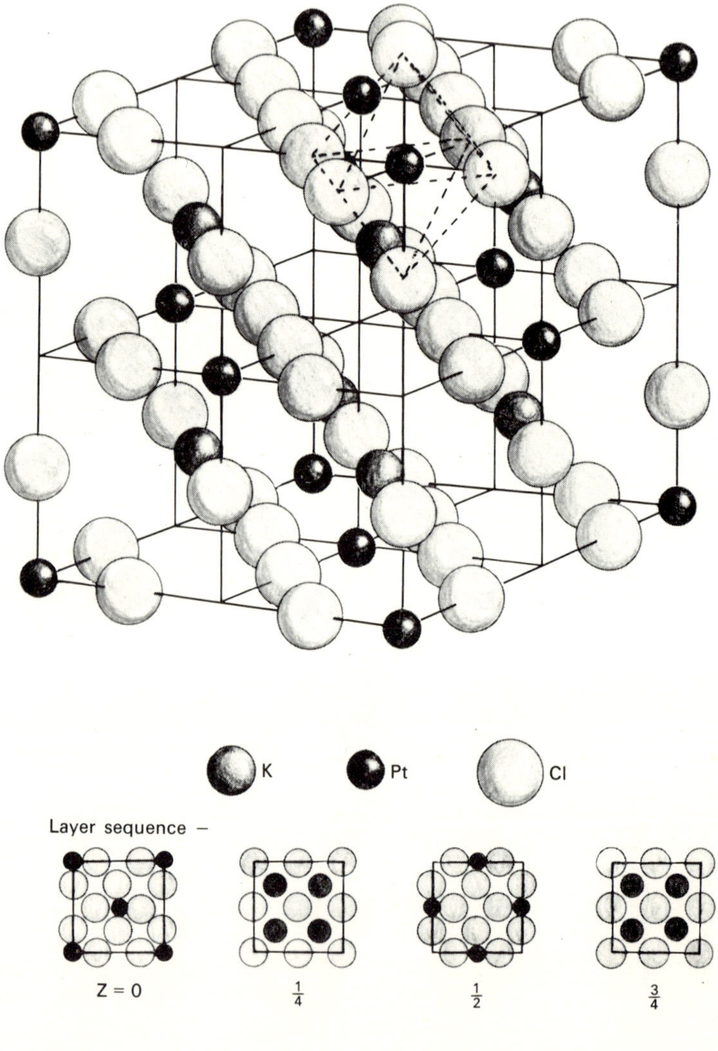

FIG. 5.6b. The K_2PtCl_6 structure (showing individual atoms)

TABLE 5.6. *Compounds with the* K_2PtCl_6 *Structure*

Compound halide	Cell size, a_0(Å)	Atomic parameter, x	Refs.
Cs_2CoF_6	8.91		1
Cs_2CrF_6	9.02	0.192	2
Cs_2GeCl_6	10.21	0.23	3
Cs_2MnCl_6	10.17		4
Cs_2PdBr_6	10.62		5
Cs_2SnCl_6	10.38	0.245	6
K_2MnCl_6	9.0445		4
K_2PtCl_6	9.76	0.25	6
K_2SnCl_6	10.002		6
$(NH_4)_2SbCl_6$†	10.66		7
$(NH_4)_2PtCl_6$	9.858	0.240	6
Rb_2MnCl_6	9.82		4

† Distorted.

5.7. C-Rare Earth Structure, Bixbyite, $D5_3$, *Ia*3, Cubic

When eight anion deficient unit cells of the CaF_2 structure are placed together in the appropriate manner a unit cell of the C-rare earth structure results. For oxides with this structure, two oxygen atoms which lie on the body diagonal are removed from each unit cell thus obtained. The unit cell which comprises an octant of the large unit cell used to describe the C-rare earth structure is obtained by taking the origin half-way up the edge of the CaF_2 unit cell (see Fig. 5.7a). This new small cell has an A atom at each edge and at the center and still has an oxygen atom in each corner except two. These two oxygen vacancies lie on body diagonals and are different in three different octants. The type I octant has the upper right front oxygen atom and the oxygen on the same body diagonal missing. In type II octants it is the lower left front oxygen and the upper right back oxygen atom which are missing, and in type III octants the lower right back and upper left front oxygen atoms are absent. A unit cell with three filled octants is shown in Fig. 5.7b along with the layer sequence for this structure.

There are eight A atoms at $(0, 0, 0; \frac{1}{2}, \frac{1}{2}, \frac{1}{2}) + \frac{1}{4}, \frac{1}{4}, \frac{1}{4}; \frac{1}{4}, \frac{3}{4}, \frac{3}{4}; \frac{3}{4}, \frac{1}{4}, \frac{3}{4}; \frac{3}{4}, \frac{3}{4}, \frac{1}{4}$; 24 more at $\pm(x, 0, \frac{1}{4}; \frac{1}{4}, x, 0; 0, \frac{1}{4}, x; \bar{x}, \frac{1}{2}, \frac{1}{4}; \frac{1}{4}, \bar{x}, \frac{1}{2}; \frac{1}{2}, \frac{1}{4}, \bar{x}) + (0, 0, 0; \frac{1}{2}, \frac{1}{2}, \frac{1}{2})$ with $x = -0.035$; and 48 O at $(0, 0, 0; \frac{1}{2}, \frac{1}{2}, \frac{1}{2}) \pm (x, y, z; x, \bar{y}, \frac{1}{2} - z; \frac{1}{2} - x, y, \bar{z}; \bar{x}, \frac{1}{2} - y, z; z, x, y; z, \bar{x}, \frac{1}{2} - y; \frac{1}{2} - z, x, \bar{y}; \bar{z}, \frac{1}{2} - x, y; y, z, x; y\bar{z}, \frac{1}{2} - y, z, \bar{x}; \bar{y}, \frac{1}{2} - z, x$; with $x \cong 0.38$, $y \cong 0.162$ and $z = 0.40$.

The smaller rare earth ions form oxides with the C-rare earth structure. See Table 5.7. The larger ones such as lanthanum ions form oxides with the

Properties

The melting points of Mn_3P_2 and the nitrides Be_3N_2 and Ca_3N_2 are

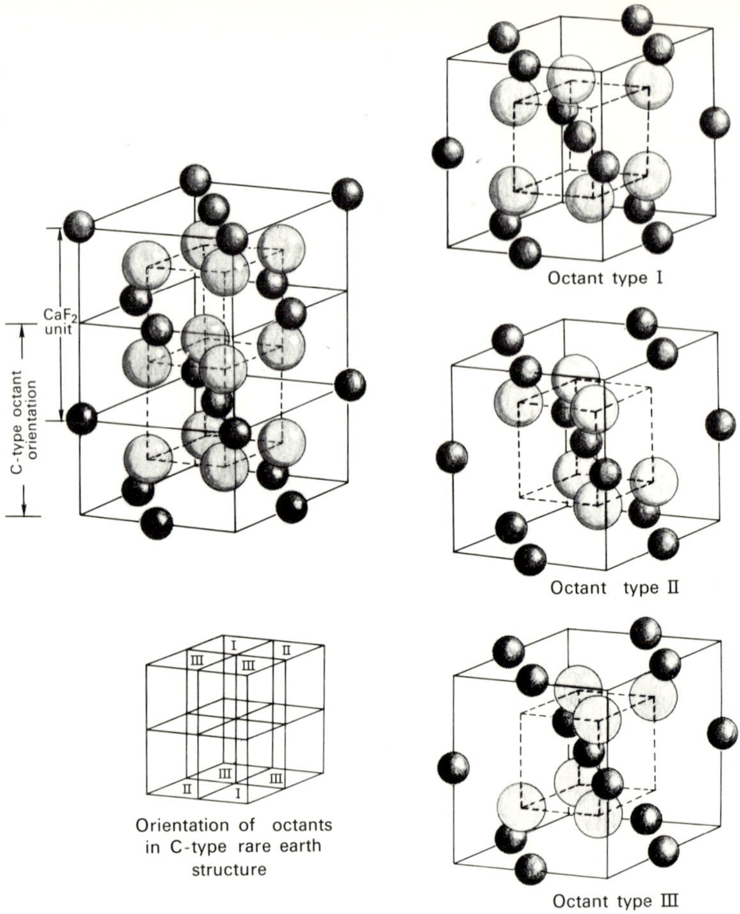

FIG. 5.7a. The C-rare earth structure. Approximate ion positions corresponding to those of fluorite structure

1200, 2200 and 1195°C, respectively. The rare earth oxides Dy_2O_3, Pr_2O_3, Sm_2O_3, Tb_2O_3, Y_2O_3 and Yb_2O_3 have melting points of 2340°C, 2200°C, 2350°C, 2390°C, 2410°C and 2350°C. The coefficients of thermal expansion for the sesquioxides of Y, Gd, Dy, Ho, Er, Tm, Yb and Lu are approximately equal to 8.1×10^{-6}/°C from room temperature to 660°C and 9.4×10^{-6}/°C from 660° to 1530°C. The coefficient for Eu_2O_3 and Sc_2O_3 is slightly higher, 8.5×10^{-6}/°C from room temperature to 428°C and 9.7×10^{-6}/°C from 428° to 1200°C. The moduli of two of the oxides, Sm_2O_3 and Y_2O_3, have been measured and found to be 27×10^6 and 17×10^6 psi, respectively.

TABLE 5.7. *Phases with the C-Rare Earth Structure*

Phase	Cell size, a_0(Å)	Atomic parameters	Refs.
Intermetallics			
As_2Mg_3	12.35	As: $x = 0.97$; Mg: $x = 0.385$, $y = 0.145$, $z = 0.380$	1
Be_3P_2	10.17	Be: $x = 0.385$, $y = 0.145$, $z = 0.380$; P: $x \cong 0$	1
Mg_3P_2	12.03	Mg: $x = 0.385$, $y = 0.145$, $z = 0.380$; P: $x = 0.875$	2
Nitrides			
Be_3N_2	8.150		1
α-Ca_3N_2	11.42		3
Cd_3N_2	10.79		4
Mg_3N_2	9.97		1
U_2N_3	10.670	U: $x = 0.982$; N: $x \cong 0.385$, $y \cong 0.145$, $z \cong -0.380$	5
Zn_3N_2	9.743		4
Oxides			
Dy_2O_3	10.665	Dy: $x \cong 0.97$; $x \cong 0.385$, $y \cong 0.145$, $z \cong 0.380$	6, 7
Er_2O_3	10.517	As in Dy_2O_3	6
Eu_2O_3	10.860	As in Dy_2O_3	6
Gd_2O_3†	10.812	As in Dy_2O_3	6
Ho_2O_3	10.606	As in Dy_2O_3	6
In_2O_3	10.117	As in Dy_2O_3	8
La_2O_3†	11.40	As in Dy_2O_3	9
Lu_2O_3	10.391	As in Dy_2O_3	6
β-Mn_2O_3	9.411	Mn: $x = 0.970$; O: $x = 0.385$, $y = 0.145$, $z = 0.380$	10
Nd_2O_3	11.076	As in Dy_2O_3	11
Pr_2O_3	11.04		12
Sc_2O_3	9.845	As in Dy_2O_3	13
Sm_2O_3	10.934	As in Dy_2O_3	6
Tb_2O_3	10.729		6
Tl_2O_3	10.543	As in Dy_2O_3	14
Tm_2O_3	10.487	As in Dy_2O_3	6, 7
Y_2O_3	10.602	As in Dy_2O_3	6
Yb_2O_3	10.433	As in Dy_2O_3	6
Y_2O_3:Nd^{3+}			15

† Not the stable form.

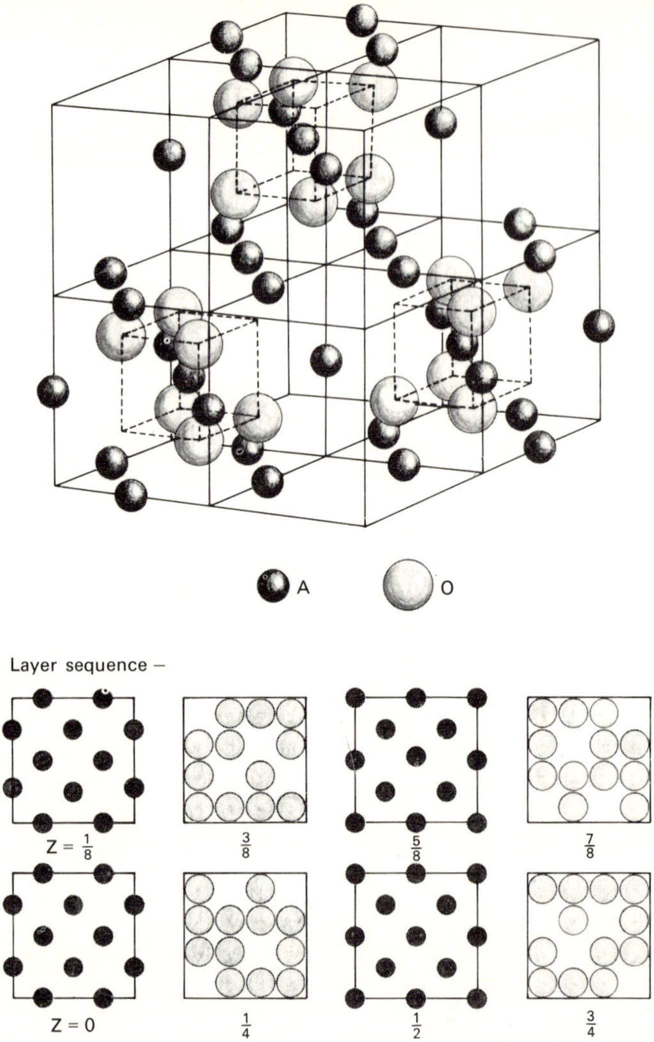

FIG. 5.7b. The C-rare earth structure (three filled octants of unit cell)

5.8. Pyrochlore Structure, $A_2B_2X_7$, *Fd3m*, Cubic

The pyrochlore structure is adopted by minerals with the general formula NaCa (Nb, Ta)$_2$O$_6$(OH, F) and a large number of other compounds (see Table 5.8). The structure can be regarded as an anion deficient fluorite structure which retains a face-centered cubic lattice. The A ions retain the eightfold coordination but the B ions are only in sixfold coordination. Thus, the A ion is larger than the B ion in compounds with this structure.

A unit cell of the pyrochlore structure is a cube with a cell edge of twice

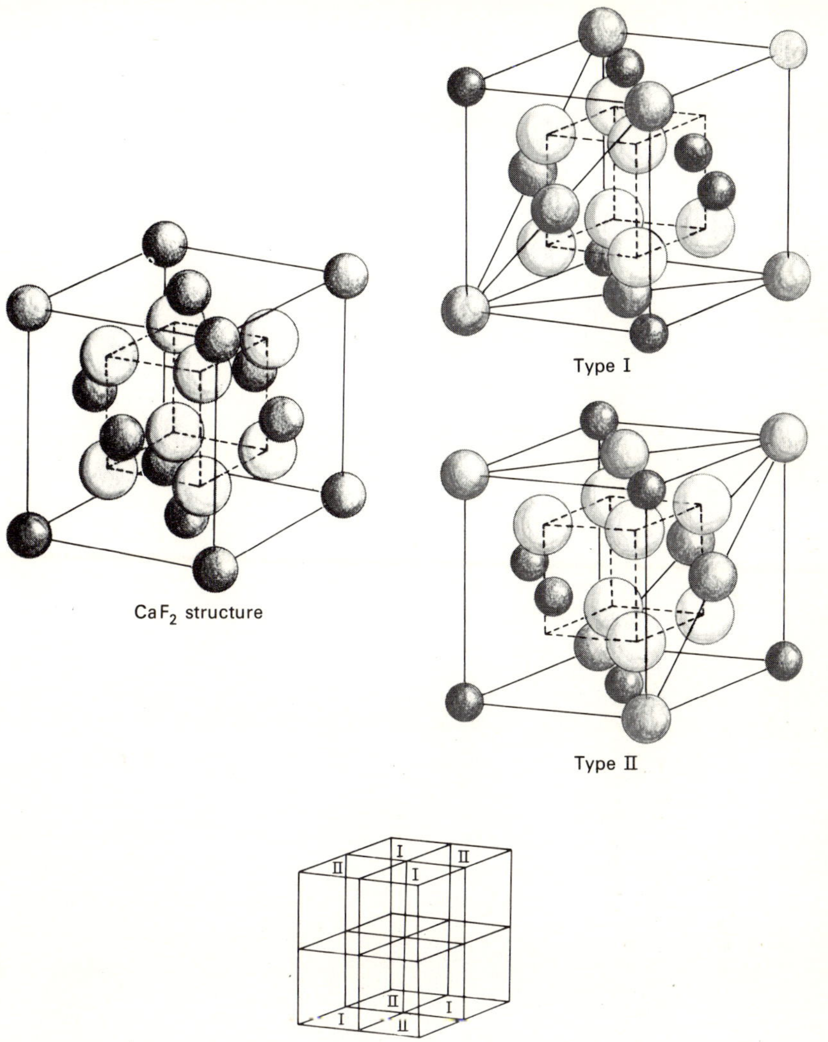

FIG. 5.8a. The pyrochlore structure (showing octants)

that of a unit cell of the CaF_2 structure. It can be described by dividing the large unit cell into eight smaller cubes, four of which are type I and four are type II. In the type I and II octants, the A and B ions are located at the corner and face-centered positions and the oxygen ions are in the tetrahedrally coordinated corners inside the cube. In the type I cubes, the A ions are located on face diagonals originating from an A ion at the lower left-hand corners and the oxygen opposite this corner is missing. In the type II cubes,

TABLE 5.8. *Phases with the Pyrochlore Structure*

Phase	Cell size, a_0(Å)	Refs.
Oxyfluorides		
$NaCaNb_2O_6F$	10.431	1
$NaLu_{0.67}Nb_2O_6F$	10.489	1
$BaSrNb_2O_6F$	10.525	1
$Ca_2Nb^{IV}Nb^{V}O_6F$	10.364	1
Oxides		
$Ca_2Sb_2O_7$	10.32	2
$Ca_2Ta_2O_7$	10.420	2
$Cd_2Nb_2O_7$	10.372	2
$Cd_2Sb_2O_7$	10.18	2
$Cd_2Ta_2O_7$	10.376	2
Dy_3NbO_7†	10.53	3
$Dy_2Ru_2O_7$	10.175	4
$Dy_2Sn_2O_7$	10.389	5
Dy_3TaO_7†	10.64	3
$Dy_2Tc_2O_7$	10.246	6
$Dy_2Ti_2O_7$	10.106	7
$Er_2Ru_2O_7$	10.120	4
$Er_2Sn_2O_7$	10.350	8
$Er_2Tc_2O_7$	10.194	6
$Er_2Ti_2O_7$	10.076	9
$Eu_2Ru_2O_7$	10.252	4
$Eu_2Sn_2O_7$	10.474	8
Gd_3NbO_7†	10.49	3
$Gd_2Ru_2O_7$	10.230	4
$Gd_2Sn_2O_7$	10.460	8
Gd_3TaO_7†	10.64	3
$Gd_2Ti_2O_7$	10.228	7
$Ho_2Ru_2O_7$	10.150	4
$Ho_2Sn_2O_7$	10.374	5
$La_2Hf_2O_7$	10.770	10
$La_2Sn_2O_7$	10.702	7
$La_2Zr_2O_7$	10.793	7
$Lu_2Ru_2O_7$	10.103	4
$Lu_2Sn_2O_7$	10.294	5
$Nd_2Hf_2O_7$	10.648	10
$Nd_2Ru_2O_7$	10.331	4
$Nd_2Sn_2O_7$	10.563	7
$Nd_2Zr_2O_7$	10.648	7
$Pr_2Ru_2O_7$	10.355	4
$Pr_2Sn_2O_7$	10.604	8
Sc_3NbO_7†	9.96	3
Sc_3TaO_7†	9.96	3
Sm_3NbO_7†	10.72	3
$Sm_2Ru_2O_7$	10.280	4
$Sm_2Sn_2O_7$	10.507	8
Sm_3TaO_7†	10.72	3
$Sm_2Tc_2O_7$	10.352	6
$Sn_2Ti_2O_7$	10.228	7
$Tb_2Ru_2O_7$	10.200	4

TABLE 5.8 *(cont.)*

Phase	Cell size, a_0(Å)	Refs.
$Tb_2Sn_2O_7$	10.428	5
$Tm_2Ru_2O_7$	10.096	4
$Tm_2Sn_2O_7$	10.330	5
Y_3NbO_7†	10.49	3
$Y_2Ru_2O_7$	10.144	4
$Y_2Sn_2O_7$	10.371	8
Y_3TaO_7†	10.49	3
$Y_2Ti_2O_7$	10.093	7
$Y_2Zr_2O_7$	10.402	7
$Yb_2Ru_2O_7$	10.087	4
$Yb_2Sn_2O_7$	10.304	8
$Yb_2Ti_2O_7$	10.030	7
$Zr_2Ce_2O_7$	10.699	11

† May have defect fluorite structures.

the A ions are located on face diagonals originating from the upper right-hand corner and the oxygen opposite this corner is missing (see Figs. 5.8a and 5.8b). These cubes are then brought together so the same type cubes are diagonally opposite one another but do not lie directly above one another. In this arrangement the A ions lie on face diagonals of the large unit cell starting from the A ion at the upper right-hand corner and lower left-hand corner of the cube. The layer sequence is shown in Fig. 5.8b. The ions are in the following positions:

$$(0, 0, 0; 0, \tfrac{1}{2}, \tfrac{1}{2}; \tfrac{1}{2}, 0, \tfrac{1}{2}; 0, \tfrac{1}{2}, \tfrac{1}{2})$$

$+ (\tfrac{1}{8}, \tfrac{1}{8}, \tfrac{1}{8}; \tfrac{1}{8}, \tfrac{3}{8}, \tfrac{3}{8}; \tfrac{3}{8}, \tfrac{1}{8}, \tfrac{3}{8}; \tfrac{3}{8}, \tfrac{3}{8}, \tfrac{1}{8})$ for 16 A ions,

$+ (\tfrac{5}{8}, \tfrac{5}{8}, \tfrac{5}{8}; \tfrac{5}{8}, \tfrac{7}{8}, \tfrac{7}{8}; \tfrac{7}{8}, \tfrac{5}{8}, \tfrac{7}{8}; \tfrac{7}{8}, \tfrac{7}{8}, \tfrac{5}{8})$ for 16 B ions,

$+ (\tfrac{1}{2}, \tfrac{1}{2}, \tfrac{1}{2}; \tfrac{3}{4}, \tfrac{3}{4}, \tfrac{3}{4})$ for 8 X ions,

$+ (x, 0, 0; x + \tfrac{1}{4}, \tfrac{1}{4}, \tfrac{1}{4}; \bar{x}, 0, 0; \tfrac{1}{4} - x, \tfrac{1}{4}, \tfrac{1}{4}$; and positions generated by rotating x, y and z) with $x \cong 0.20$ for the remaining X ions.

A number of compounds such as $Cd_2Nb_2O_7$, $Cd_2Ta_2O_7$ and $Ca_2Ta_2O_7$ which contain a large divalent cation and a smaller pentavalent cation form the pyrochlore structure. Many of the rare earth titanates also have this structure. In addition, a number of complex compounds with the pyrochlore structure have been produced by substituting some Nb^{4+}, Mo^{4+}, Ti^{4+}, Zr^{4+} and V^{4+} for the niobium and tantalum in NaCa (Nb, Ta)$_2$O$_6$(OH, F).

There has been considerable interest in compounds with this structure since the ferroelectric properties of $Cd_2Nb_2O_7$ were discovered in 1952. The main difference between this structure and the perovskite (Chapter 7) is in the relative orientation of the octahedra. In $Cd_2Nb_2O_7$ the octahedra are arranged so that the O-Nb-O chains lie on zig-zag lines approximately along the ⟨110⟩ directions.

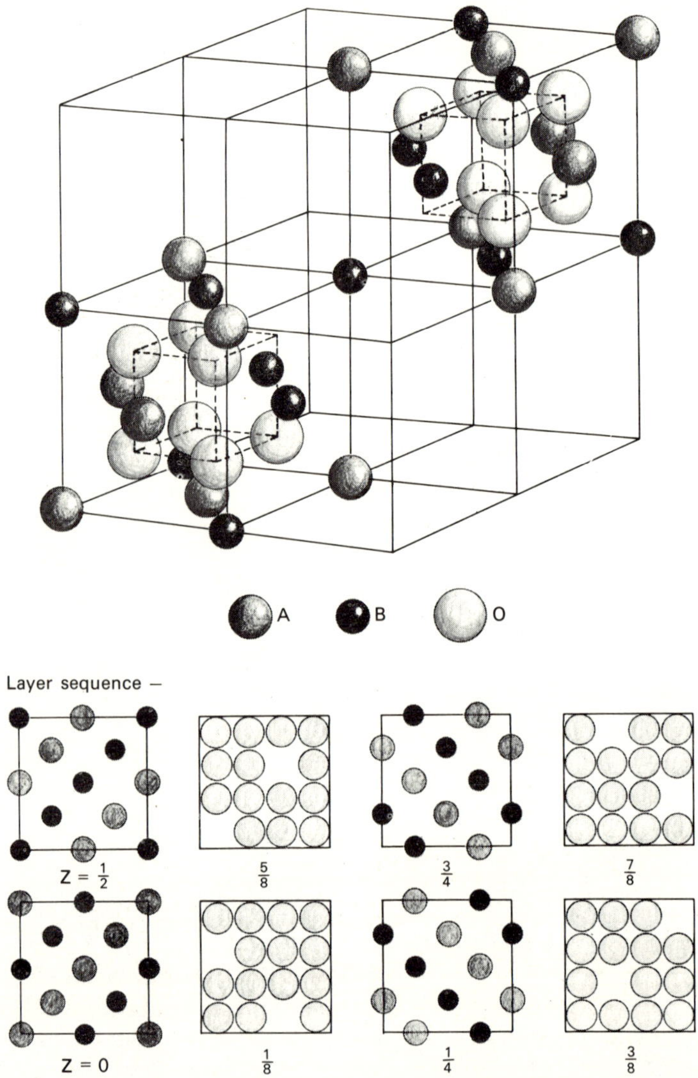

FIG. 5.8b. The pyrochlore structure (two filled octants of unit cell)

5.9. Scheelite Structure, $CaWO_4$ Type $I4_1/a$, Tetragonal

While it is not as obvious, the scheelite structure is related to the CaF_2 structure. Its unit cell can be visualized as two unit cells of the fluorite structure stacked on one another with calcium and tungsten atoms occupying the calcium atom positions (the corners and centers of the faces of the cubes) and so arranged that they alternate in each vertical row. The oxygen atoms replace the fluorine atoms (in the corners of the cubes) and are moved toward the tungsten atoms (see Fig. 5.9a). The scheelite unit cell and layer sequence are shown in Fig. 5.9b (see p. 110).

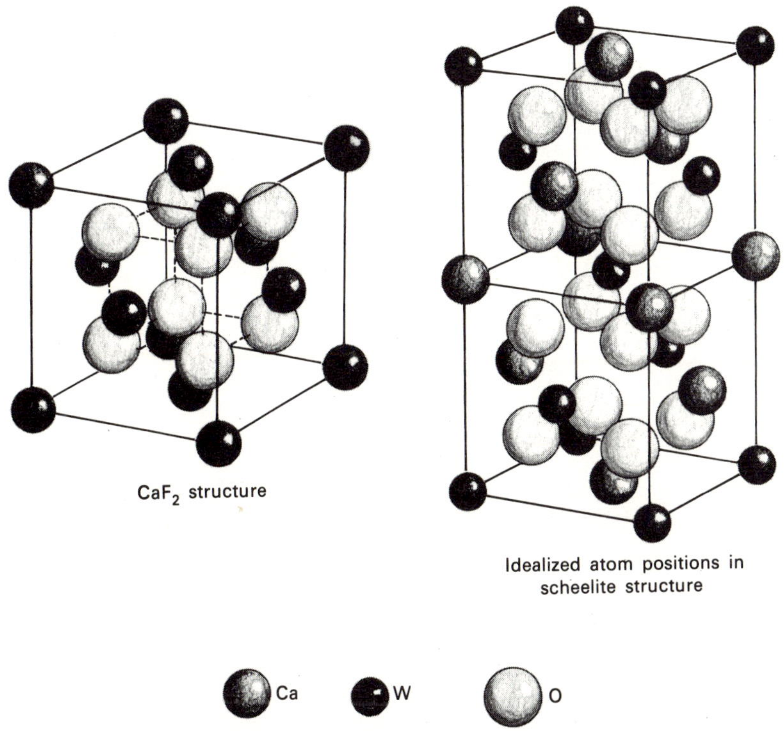

FIG. 5.9a. The scheelite structure (idealized)

The Ca atoms are at $0, 0, \frac{1}{2}; \frac{1}{2}, 0, \frac{1}{4}; \frac{1}{2}, \frac{1}{2}, 0; 0, \frac{1}{2}, \frac{3}{4}$; the W atoms are at $0, 0, 0; 0, \frac{1}{2}, \frac{1}{4}; \frac{1}{2}, \frac{1}{2}, \frac{1}{2}; \frac{1}{2}, 0, \frac{3}{4}$; and the O atoms are at $x, y, z; \bar{x}, \bar{y}, z; x, \frac{1}{2}+y, \frac{1}{4}-z; \bar{x}, \frac{1}{2}-y, \frac{1}{4}-z; \bar{y}, x, \bar{z}; y, \bar{x}, \bar{z}; \bar{y}, \frac{1}{2}+x, \frac{1}{4}+z; y, \frac{1}{2}-x, \frac{1}{4}+z$, B.C. $x = 0.241$, $y = 0.151$, $z = 0.081$.

A large number of compounds which form the scheelite structure are listed in Table 5.9a.

Properties

Because of the similarity of the pyrochlore and the CaF_2 structures, it is not surprising that compounds such as $CaWO_4$, $PbMoO_4$, $PbWO_4$ and $CaMoO_4$ also have been investigated in recent years as laser host materials

TABLE 5.9a. *Phases with the Scheelite Structure*

Phases	Cell size (Å)		x	y	z	Refs.
	a_0	c_0				
Halides, Oxyhalides						
$LiDyF_4$	5.188	10.83				1
$LiErF_4$	5.162	10.70				1
$LiEuF_4$	5.228	11.63				1
$LiGdF_4$	5.219	10.97				1
$LiHoF_4$	5.175	10.75				1
$LiLuF_4$	5.132	10.59				1
$LiTbF_4$	5.200	10.89				1
$LiTmF_4$	5.145	10.64				1
$LiYF_4$	5.175	10.74				1
$LiYbF_4$	5.132	10.59				1
$CsCrO_3F$	5.712	14.5				2
$KCrO_3F$	5.46	12.89				3
Oxides						
$AgIO_4$	5.37	12.10				4
$AgReO_4$	5.59	11.81				5
$BaMoO_4$	5.56	12.76	0.25	0.11	0.075	6
$BaWO_4$	5.64	12.70	0.25	0.11	0.075	6
$BiAsO_4$	5.08	11.70				7
$CaMoO_4$	5.226	11.430				1, 6, 8
$CaWO_4$	5.243	11.376	0.241	0.151	0.086	9
$CdMoO_4$	5.14	11.17				10
$CeGeO_4$	5.05	11.17				11
$HfGeO_4$	4.85	10.50				12
$KReO_4$	5.68	12.70				13
$KRuO_4$	5.61	12.99	0.244	0.117	0.073	10
NH_4IO_4	5.94	12.79				10
NH_4ReO_4	5.88	12.98				13
$NaIO_4$	5.32	11.93				14
$NaReO_4$	5.36	11.72	0.25	0.11	0.07	13
$NaTcO_4$	5.34	11.87				15
$PbMoO_4$	5.47	12.18	0.247	0.092	0.085	6
$PbWO_4$	5.44	12.01	0.25	0.13	0.075	6
$RbIO_4$	5.92	13.05				13
$RbReO_4$	5.805	13.17	0.25	0.11	0.07	13
$SrMoO_4$	5.36	11.94	0.25	0.14	0.075	6
$SrWO_4$	5.40	11.90	0.25	0.14	0.075	6
$ThGeO_4$	5.141	11.54				11
$TlReO_4$	5.761	13.33	0.25	0.11	0.07	13
$UGeO_4$	5.08	11.23	0.27	0.11	0.08	16
$YNbO_4$	5.16	10.91				17
$ZrGeO_4$	4.87	10.57				11
Complex Oxides						
$A'_{0.5}A''_{0.5}BO_4$						
$K_{0.5}Bi_{0.5}MoO_4$	5.380	11.92				18, 19, 20
$K_{0.5}Ce_{0.5}WO_4$	5.386	11.941				18, 19, 20
$K_{0.5}La_{0.5}WO_4$	5.443	12.03				18, 19, 20
$Li_{0.5}Bi_{0.5}MoO_4$	5.232	11.50				18, 19, 20
$Li_{0.5}La_{0.5}MoO_4$	5.307	11.67				18, 19, 20

TABLE 5.9a (*cont.*)

Phases	Cell size (Å)		x	y	z	Refs.
	a_0	c_0				
$Li_{0\cdot5}La_{0\cdot5}WO_4$	5.335	11.63				18, 19, 20
$Na_{0\cdot5}Bi_{0\cdot5}MoO_4$	5.267	11.55				18, 19, 20
$Na_{0\cdot5}Ce_{0.5}WO_4$	5.319	11.59				18, 19, 20
$Na_{0.5}La_{0.5}MoO_4$	5.328	11.70				18, 19, 20
$Na_{0\cdot5}La_{0\cdot5}WO_4$	5.345	11.63				18, 19, 20
$A(B'_{0.5}B''_{0.5})O_4$						
$Ce(Ti_{0\cdot5}Mo_{0\cdot5})O_4$	5.238	11.585				21
$Dy(Ti_{0\cdot5}Mo_{0\cdot5})O_4$	5.108	11.169				21
$Dy(Ti_{0\cdot5}W_{0\cdot5})O_4$	5.142	11.077				21
$Er(Ti_{0\cdot5}Mo_{0\cdot5})O_4$	5.071	11.059				21
$Er(Ti_{0\cdot5}W_{0\cdot5})O_4$	5.106	10.959				21
$Eu(Ti_{0\cdot5}Mo_{0\cdot5})O_4$	5.145	11.288				21
$Eu(Ti_{0\cdot5}W_{0\cdot5})O_4$	5.178	11.199				21
$Gd(Ti_{0\cdot5}Mo_{0\cdot5})O_4$	5.123	11.236				21
$Gd(Ti_{0\cdot5}W_{0\cdot5})O_4$	5.163	11.124				21
$Ho(Ti_{0.5}Mo_{0.5})O_4$	5.085	11.094				21
$Ho(Ti_{0\cdot5}W_{0\cdot5})O_4$	5.121	11.001				21
$Lu(Ti_{0\cdot5}Mo_{0\cdot5})O_4$	5.039	10.955				21
$Nd(Ti_{0\cdot5}Mo_{0\cdot5})O_4$	5.201	11.457				21
$Nd(Ti_{0\cdot5}W_{0\cdot5})O_4$	5.235	11.375				21
$Sm(Ti_{0\cdot5}Mo_{0\cdot5})O_4$	5.160	11.349				21
$Sm(Ti_{0\cdot5}W_{0\cdot5})O_4$	5.199	11.256				21
$Pr(Ti_{0\cdot5}Mo_{0\cdot5})O_4$	5.220	11.513				21
$Tb(Ti_{0\cdot5}Mo_{0\cdot5})O_4$	5.114	11.170				21
$Tb(Ti_{0\cdot5}W_{0\cdot5})O_4$	5.149	11.090				21
$Tm(Ti_{0\cdot5}Mo_{0\cdot5})O_4$	5.060	11.011				21
$Y(Ti_{0\cdot5}Mo_{0\cdot5})O_4$	5.083	11.060				21
$Yb(Ti_{0\cdot5}Mo_{0\cdot5})O_4$	5.051	11.983				21

TABLE 5.9b. *Lasers Which Use Compounds with the Scheelite Structure as Hosts*

Host	Doping	Temperature (°K)	Wavelength (μ)	Refs.
$CaMoO_4$	Nd^{3+}	295	1.067	1
$CaWO_4$	Er^{3+}	77	1.612	2
	Ho^{3+}	77	2.046	3
	Nd^{3+}	77	1.065	4
	Pr^{3+}	77	1.047	5
	Tm^{3+}	77	1.911	6
$PbMoO_4$	Nd^{3+}	295	1.059	7
$SrMoO_4$	Nd^{3+}	295	1.064	8
	Pr^{3+}		1.04	7
$SrWO_4$	Nd^{3+}	77	1.064	7

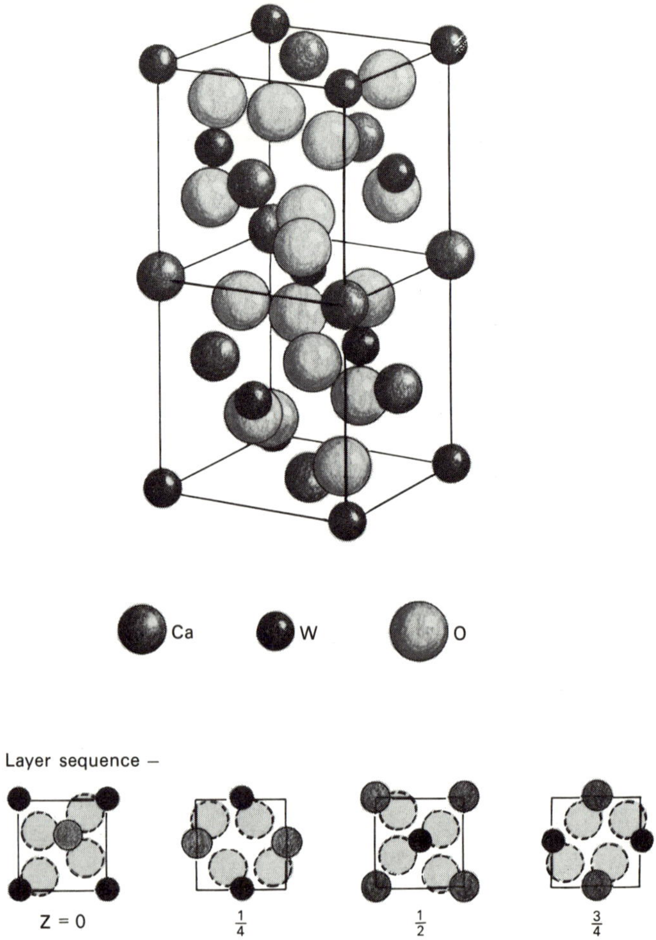

FIG. 5.9b. The scheelite structure

for rare earth ions. A list of lasers which employ these materials as hosts is given in Table 5.9b. Probably the most widely studied material is Nd^{3+} doped $CaWO_4$. Because of the difference of the valence states of Nd^{3+} and Ca^{2+}, the most satisfactory results have been obtained by using a compensating ion such as Na^+ in the preparations. Large single crystals of doped and pure scheelite compounds have been grown by the Czochralski technique.

5.10. Discussion

Structures (Relations)

In this chapter, the diamond structure was described as a cubic close-packed arrangement of atoms with half the tetrahedral holes filled. The zinc blende structure was considered an ordered diamond structure in whose unit cell the face-centers and corners were filled with atoms of one element and the tetrahedral holes with atoms of the other. In the high-temperature cristobalite structure these same tetrahedral holes are filled with SiO_4 tetrahedra. In the CaF_2 structure all the holes are filled with fluorine ions. Then, an ordered arrangement of these tetrahedrally coordinated positions results in the MgAgAs structure. In sort of an antifluorite structure $PtCl_6^{-2}$ octahedra are in the Ca ion positions of CaF_2 structure in K_2PtCl_6 with K ions in the fluorine atom positions. Eight CaF_2 unit cells of the CaF_2 structure were put together with anion vacancies in each cell to form the C-rare earth unit cell, and in the pyrochlore structure one anion was removed from each of eight unit cells of the CaF_2 structure stacked on one another with the cations ordered and anions displaced from the ideal positions.

Intermetallics

The elements and intermetallics discussed in this chapter are a very important group of materials. Germanium and silicon are basic materials in the semiconductor industry. When elements from groups III and V are combined, compounds with the zinc blende structure are formed. The energy gaps for these materials increase as elements higher in the groups are combined, just as the energy gaps of the group IV increase from grey Sn to diamond. For certain applications such as photoconducting cells, materials with small energy gaps are used. In transistors large energy gap materials are desired, so relatively large numbers of hole-electron pairs will not be generated at high temperatures. The mobility of the electrons also is quite different in these phases, being very high for InSb and InAs. Other semiconductor phases with the zinc blende structures also have been prepared by combining elements from group II and group VI. This large number of semiconductor phases with varied properties is being investigated for a number of semiconductor applications.

Among the AB_2 phases, those with R_A/R_B greater than 0.73 tend to form the CaF_2 structure in which the coordination number is 8 for the calcium ion. Sulfides, selenides and tellurides such as Li_2S, Li_2Se and Li_2Te fit into this category.

Halides

The AF_2 phases with R_A/R_F ratios between 0.41 and 0.73 form fluorides with the rutile structure, while the larger metal ions such as Ca^{+2}, Sr^{+2}, Ba^{+2}, Cd^{+2} and Hg^{+2} form CaF_2, SrF_2, BaF_2, CdF_2 and HgF_2 with the fluorite structure. The small beryllium ion forms a fluoride, BeF_2, with the β-cristobalite structure. At this point, we have the following picture of the AF and AF_2 compounds.

LiF							
NaF	MgF_2						
KF	CaF_2	MnF_2	FeF_2	CoF_2	NiF_2		ZnF_2
RbF	SrF_2					AgF	CdF_2
CsF	BaF_2						HgF_2

The compounds in the solid line box have the NaCl structure, the fluorides in the broken line box have the rutile structure and the others crystallize with the CaF_2 structure. Also included in this chapter were ternary halides with the K_2PtCl_6 structure. Many of the other halides as well as fluorides have interesting optical properties. The optical transmission range for some of them with the cesium chloride, sodium chloride, rutile and fluorite structure are shown in Fig. 5.10a. The refractive index, as a function of wavelength for some halides with the cesium chloride and sodium chloride structure is plotted in Fig. 5.10b.

Hydrides

A number of AH_2, rare earth hydrides, and YH_2, NbH_2 and ScH_2 have the CaF_2 structure.

Oxides

Of the AO oxides, BeO forms the zinc blende structure in which the coordination number is 4 for both ions. The alkali metal oxides Li_2O, Na_2O, K_2O, Rb_2O as well as the sulfides, selenides and oxides except for Rb_2Se and Rb_2Te form the fluorite structure in which the positions of cations and anions are interchanged. The dioxides ThO_2, HfO_2 and a number of rare earth and actinide metal oxides also have the fluorite structure, while zirconium dioxide has this structure only at high temperatures.

Silicon dioxide has a number of structures. The best known forms are quartz, tridymite and cristoballite. The β form of cristoballite can be related to the CaF_2 structure. Here again it is convenient to treat the SiO_4 tetrahedra as single units. These tetrahedra, which prevail in the silicates as large anions, are built up by joining SiO_4^{-4} in circles, chains, planes and three-dimensional structures.

A number of oxides with an A_2O_3 formula form the C-rare earth structure. Two oxygen atoms are removed from each of the eight unit cells of the

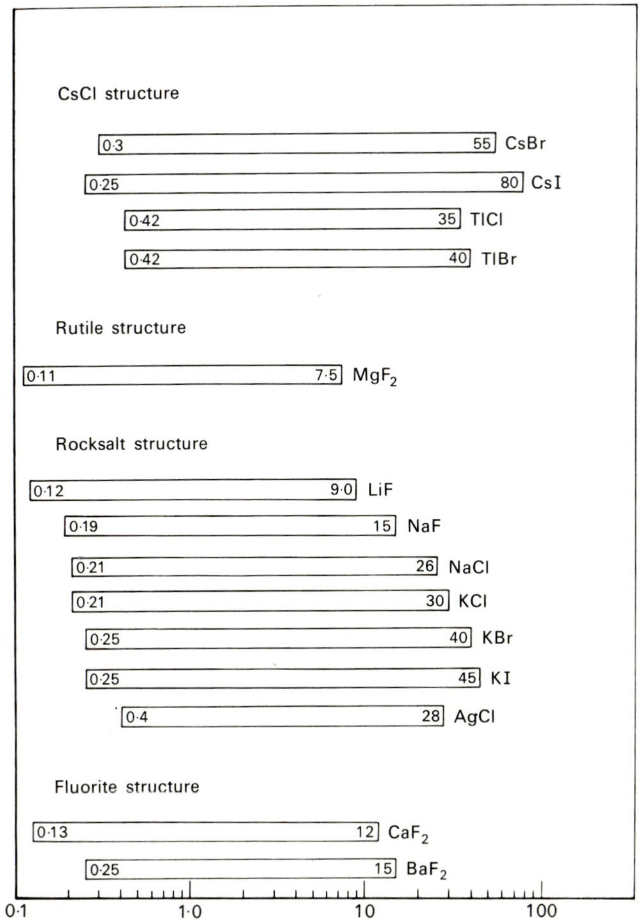

FIG. 5.10a. Range of transmittance for halide samples, 2 mm thick, 10% cut off

CaF_2 structure which make up the C-rare earth structure. Note that there are four CaF_2 molecules in each octant so that removal of two anions from each results in the A_2O_3 formula and a coordination of six instead of eight for the cation. Thus, Mn_2O_3, Sc_2O_3, Y_2O_3, Tl_2O_3 and the smaller rare earth metal oxides have this structure. Still smaller ions tend to form A_2O_3 oxides with the corundum structure and larger ones the A-rare earth structure.

When a larger ion is combined with a smaller ion to form an oxide, one of the possible structures it can have is the pyrochlore. The formula for these compounds is $A_2B_2O_7$. The A and B ions order in the Ca ion positions of

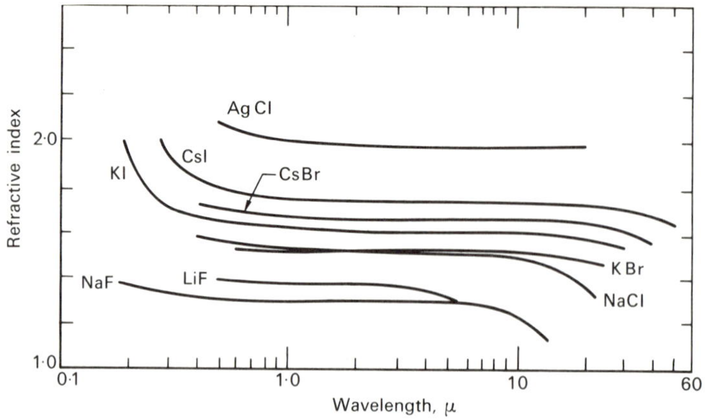

FIG. 5.10b. Refractive index, as a function of wavelength, for some halides

eight unit cells of the CaF_2 structure with one anion missing from each cell. The differences in size and charge of the A and B ions seem to be large enough to cause ordering, but not as large as they normally are in structures such as perovskites. Some of the compounds with this structure are $Ca_2Ta_2O_7$ and $Cd_2Ta_2O_7$.

The relation between the CaF_2 and $CaWO_4$ structures becomes more apparent if the formula is written as (Ca, W)O_4. The difference in charge in the A and B ions is large enough to cause ordering of these ions. In the $CaWO_4$ structure, the oxygen ions are also displaced toward the smaller tungsten ion as would be expected. Two unit cells of the CaF_2 structure stacked on one another are required to describe this structure. Compounds with the CaF_2 and $CaWO_4$ structures have been studied extensively as laser host materials for rare earth ions.

Table 5.1 *References*

1. Y. BASKIN and L. MEYER, *Phys. Rev.* **100**, 544 (1955).
2. S. GOTTLICHER and E. WOLFEL, *Z. Elektrochem.* **63**, 891 (1959).
3. R. DALVEN, *Infrared Phys.* **6**, 129 (1966).
4. P. BLUM and A. DURIF, *Acta Cryst.* **9**, 829 (1956).
5. M. E. STRAUMANIS, P. BORGEAUD and W. J. JAMES, *J. Appl. Phys.* **32**, 1382 (1961).
6. R. E. VOGEL and C. P. KEMPTER, U.S. At. Energy Comm. LA-2317 (1959).

Table 5.2a *References*

1. M. HOCH, *Quarterly Progress Report*, AF Contract AF 33(616)-6299 (June 15, 1960).
2. A. ADDAMIANO, *Acta Cryst.* **13**, 505 (1960).
3. A. ADDAMIANO, *J. Am. Chem. Soc.* **82**, 1537 (1960).
4. G. GIESECKE and H. PFISTER, *Acta Cryst.* **11**, 369 (1958).
5. E. D. PIERRON, D. L. PARKER and J. B. MCNEELY, *Acta Cryst.* **21**, 290 (1966).
6. G. OZOLINS, G. K. AVERKIEVA, N. A. GORYUNOVA and A. JEVINS, *Kristallografiya* **8**, 272 (1963).

7. J. A. Perri, S. LaPlaca and B. Post, *Acta Cryst.* **11**, 310 (1958).
8. S. Rundqvist, *Cong. Intern. Chim. Pure Appl.* 16, Paris (1957), Mem. Sect. Chim. Minerale, 539 (1958).
9. A. Taylor and R. M. Jones in J. R. O'Connor and J. Smittens (Eds.) *Silicon Carbide*, p. 147, Pergamon Press, New York (1960).
10. S. Hoshino, *J. Phys. Soc. Japan* **12**, 315 (1957).
11. S. Hoshino, *J. Phys. Soc. Japan* **7**, 560 (1952).
12. J. Krug and L. Sieg, *Z. Naturforsch.* **7A**, 369 (1952).
13. M. R. Lorenz and J. S. Prener, *Acta Cryst.* **9**, 538 (1956).
14. F. Ebert and H. Woitinek, *Z. Anorg. Chem.* **210**, 269 (1933).
15. R. H. Wentorf, Jr., *J. Chem. Phys.* **26**, 956 (1957).
16. W. H. Bragg and J. A. Darbyshire, *J. Metals* **6**, 238 (1954).
17. E. Staritzky, *Anal. Chem.* **28**, 915 (1956).
18. W. H. Zachariasen, *Z. Phys. Chem.* **119**, 201 (1926).
19. W. J. Müller and G. Löffler, *Z. Angew. Chem.* **46**, 538 (1933).
20. W. P. Lawson, S. Nielsen, E. H. Putley and A. S. Young, *Phys. Chem. Solids* **9**, 325 (1959).
21. J. Goodyear, W. F. Duffin and G. A. Steigmann, *Acta Cryst.* **14**, 1168 (1961).
22. H. Hahn and G. Frank, *Z. Anorg. Allgem. Chem.* **278**, 340 (1955).
23. H. Hahn and W. Klingler, *Z. Anorg. Chem.* **259**, 135 (1949).
24. J. C. Woolley and B. R. Pamplin, *J. Electrochem. Soc.* **108**, 874 (1961).
25. W. Hartwig, *Sitzgsber. Preuss. Akad. Wiss.* **10**, 79 (1926).
26. U. Zorll, *Z. Physik.* **138**, 167 (1954).
27. H. Schnaase, *Z. Phys. Chem.* **B20**, 89 (1933).
28. A. Baroni, *Z. Krist.* **99**, 336 (1938).
29. J. T. S. Van Aswegen and H. Verleger, *Naturwiss.* **47**, 131 (1960).
30. H. E. Swanson, R. K. Fuyat and G. M. Ugrinic, NBS Circular 539, **111**, 23 (1954).

Table 5.2b *References*

1. W. Paul, *J. Appl. Phys.* **32**, 2082 (1961).
2. S. D. Gromokov, Z. M. Latypov and P. S. Kirilyuk, *Zh. Fiz. Khim.* **40**, 1262 (1966).
3. H. L. Armstrong, *Electronics Industries* **18**, 90 (1959).
4. T. F. Kharakhornin, M. R. Poluboyarinova and U. G. Vinogradova, *Inorg. Materials* **2**, 25 (1966).
5. A. J. Cornish, *J. Electrochem. Soc.* **106**, 685 (1959).
6. G. Popovich, *Rev. Phys., Acad. Rep. Populaire Romaine* **8**, 269 (1963).
7. V. P. Schastlivyi and A. V. Vanyukov, *Izv. Akad. Nauk SSSR, Neorg. Materialy* **2**, 1378 (1966).

Table 5.3 *References*

1. V. A. Frank-Kamenetskii and I. E. Kamentsev, *Rost. Kristallow, Akad. Nauk SSSR, Inst. Krist.* **3**, 468 (1961).
2. R. E. Gibbs, *Proc. Roy. Soc. (London)* A **113**, 351 (1926).
3. T. Tokuda, *Nippon Kagaku Zasshi* **79**, 1063 (1958).
4. J. Shropshire, P. P. Keat and P. A. Vaughan, *Z. Krist.* **112**, 409 (1959).
5. N. A. Bendeliani and L. F. Vershchagin, *Dokl. Akad. Nauk SSSR* **158**, 819 (1964).

Table 5.4a *References*

1. D. C. Hamilton, C. J. Raub, B. T. Matthias, E. Corenzwit and G. W. Hull, Jr., *J. Phys. Chem. Solids* **26**, 665 (1965).
2. E. Zintl, A. Harder and W. Haucke, *Z. Phys. Chem.* **35B**, 354 (1937).
3. R. Juza and W. Schulz, *Z. Anorg. Chem.* **275**, 65 (1954).
4. H. Nowotny and K. Bachmayer, *Monatsh. Chem.* **80**, 734 (1949).
5. F. Bertraut and P. Blum, *Compt. Rend.* **231**, 626 (1950).
6. G. Brauer and J. Tiesler, *Z. Anorg. Chem.* **262**, 319 (1950).

7. M. Zumbusch, *Z. Anorg. Allgem. Chem.* **243**, 322 (1940).
8. O. Nial, *Svensk Kem. Tidskr.* **59**, 172 (1947).
9. J. B. Friauf, *J. Am. Chem. Soc.* **48**, 1906 (1926).
10. E. Zintl and H. Kaiser, *Z. Anorg. Chem.* **211**, 113 (1933).
11. F. W. Schönfield, E. M. Cramer, W. N. Miner, F. H. Ellinger and A. S. Coffinberry in *Progress in Nuclear Energy*, Ser. V, **2**, 579, Pergamon Press, N.Y. (1959).
12. G. Busch and U. Winkler, *Helv. Phys. Acta* **26**, 578 (1953).
13. R. F. Blunt, H. P. R. Frederikse and W. R. Hosler, *Phys. Rev.* **100**, 663 (1955).
14. S. Rundqvist and A. Hede, *Acta Chem. Scand.* **14**, 893 (1960).
15. C. J. Raub, W. H. Zachariasen, W. H. Geballe and B. T. Matthias, *J. Phys. Chem. Solids*, **24**, 1093 (1963).
16. K. Schubert and U. Rosler, *Z. Metallk.* **41**, 298 (1950).
17. L. Y. Markovskii, Y. D. Kondrashev and G. V. Kaputovskaya, *J. Gen. Chem. USSR* **25**, 1007 (1955).
18. E. Staritzky, *Anal. Chem.* **28**, 915 (1956).
19. W. H. Zachariasen, *Acta Cryst.* **4**, 231 (1951).
20. B. K. Vainshtein, *Tr. Inst. Krist., Akad. Nauk SSSR* **5**, 113 (1949).
21. A. Smakula and J. Kalmais, *Phys. Rev.* **99**, 1737 (1955).
22. H. M. Haendler and W. J. Bernard, *J. Am. Chem. Soc.* **73**, 5218 (1951).
23. W. Finkelnburg and A. Stein, *J. Chem. Phys.* **18**, 1296 (1950).
24. W. Doll and W. Klemm, *Z. Anorg. Allgem. Chem.* **241**, 239 (1939).
25. F. Ebert and H. Woitnek, *Z. Anorg. Chem.* **210**, 269 (1933).
26. A. Zalkin and D. H. Templeton, *J. Am. Chem. Soc.* **75**, 2453 (1953).
27. N. C. Baenziger, J. R. Holden, G. E. Knudson and A. I. Popov, *J. Am. Chem. Soc.* **76**, 4734 (1954).
28. I. I. Yamzin, L. Z. Nozik and N. V. Belov, *Dokl. Akad. Nauk SSSR* **138**, 110 (1961).
29. G. E. R. Schulze, *Z. Phys. Chem.* **32**, 430 (1936).
30. U. Croatto and M. Bruno, *Gazz. Chim. Ital.* **76**, 246 (1946).
31. E. Zintl and A. Udgard, *Z. Anorg. Allgem. Chem.* **240**, 150 (1939).
32. C. Ayphassorho, *Compt. Rend.* **247**, 1597 (1958).
33. A. Pebler and W. E. Wallace, *J. Phys. Chem.* **66**, 148 (1962).
34. D. L. Urich, *J. Chem. Phys.* **44**, 2202 (1966).
35. G. Brauer and H. Müller, *J. Inorg. Nucl. Chem.* **17**, 102 (1961).
36. J. C. McGuire and C. P. Kempter, *J. Chem. Phys.* **33**, 1584 (1960).
37. R. Juza, H. H. Weber and E. Meyer-Simon, *Z. Anorg. Chem.* **273**, 48 (1953).
38. R. E. Rundle, N. C. Baenziger, A. S. Wilson and R. A. McDonald, *J. Am. Chem. Soc.* **70**, 99 (1948).
39. L. B. Asprey, R. H. Ellinger, S. Fried and W. H. Zachariasen, *J. Am. Chem. Soc.* **77**, 1707 (1955).
40. G. Brauer and K. Gingerich, *J. Inorg. Nucl. Chem.* **16**, 87 (1960).
41. E. Zintl, A. Harder and B. Dauth, *Z. Elektrochem.* **40**, 588 (1934).
42. W. H. Zachariasen, *Acta Cryst.* **2**, 388 (1949).
43. A. W. Martin, *J. Phys. Chem.* **58**, 911 (1954).
44. K. W. Bagnall and R. W. M. D'Eye, *J. Chem. Soc.* **1954**, 4295 (1954).
45. A. Helms and W. Klemm, *Z. Anorg. Allgem. Chem.* **242**, 33 (1939).
46. N. C. Baenziger, H. A. Eick, H. S. Schuldt and L. Eyring, *J. Am. Chem. Soc.* **83**, 2219 (1961).
47. E. Slowinski and E. Norman, *Acta Cryst.* **5**, 768 (1952).
48. G. Brauer and H. Gradinger, *Z. Anorg. Allgem. Chem.* **276**, 209 (1954).
49. J. S. Anderson, I. O. Sawyer, H. W. Worner, G. M. Willis and M. J. Bannister, *Nature* **185**, 915 (1960).
50. C. D. West, *Z. Krist.* **88**, 97 (1934).
51. K. May, *Z. Krist.* **94A**, 412 (1936).

Table 5.4b *References*

1. G. V. SAMSONOV, *Silicides and Their Application in Technology,* AN Ukr. SSR, Kiev (1959).
2. G. S. MARKEVICH, Dissertation Abstract, Leningrad State Univ. (1961).
3. O. KUBASCHEWSKI and E. L. EVANS, *Metallurgical Thermochemistry,* Pergamon, London (1958).
4. N. K. HISTER, F. A. FERGUSON and N. FISHMAN, *Chem. Eng.* 237 (March 1957).
5. W. G. BRADSHAW and C. O. MATHEWS, *Properties of Refractory Materials,* LMSD-2466 (June 1958).
6. I. E. CAMPBELL, *High Temperature Technology,* Wiley, New York (1956).
7. O. HONIGSCHMID, *Monatsh.* **28**, 1017 (1907).
8. F. A. ROIGH and R. F. DICKERSON, *Nucleonics* **18**, 74 (1960).
9. E. M. SAVITSKU, *Mechanical Properties of Intermetallic Compounds*, Ed. J. H. WESTBROOK, Wiley, New York (1960).
10. G. R. FINLAY, *Chem. in Canada* **4**, 41 (1952).
11. S. M. LANG, *Properties of High Temperature Ceramics and Cermets,* N.B.S. Monograph No. 6 (March 1960).
12. J. M. WARDE, *Refractories for Nuclear Energy,* Techn. Bull. No. 94, Refractories Institute, Pittsburgh.
13. J. R. JOHNSON, *J. Metals,* **85**, 662 (1956).
14. O. H. KRIKORIAN, *Thermal Expansion of High Temperature Materials,* UCRL-6132 (September 1960).

Table 5.4c *References*

1. S. P. S. PORTO and A. YARIV, *Proc. IRE* **50**, 1542 (1962).
2. Z. J. KISS and R. C. DUNCAN, JR., *Proc. IRE* **50**, 1531 (1962).
3. E. CARNALL, JR., S. E. HATCH and W. F. PARSONS, *Mater. Sci. Res.* **3**, 165 (1966).
4. L. F. JOHNSON, *J. Appl. Phys.* **33**, 756, (1962).
5. W. KAISER, C. G. B. GARRETT and D. L. WOOD, *Phys. Rev.* **123**, 766 (1961).
6. Z. J. KISS and R. C. DUNCAN, JR., *Proc. IRE* **50**, 1532 (1962).
7. G. D. BOYD, R. J. COLLINS, S. P. S. PORTO, A. YARIV and W. A. HARGREVES, *Phys. Rev. Letters* **8**, 269 (1962).
8. D. L. WOOD and W. KAISER, *Phys. Rev.* **126**, 2079 (1962).
9. S. P. S. PORTO and A. YARIV, *Proc. IRE* **50**, 1543 (1962).

Table 5.5 *References*

1. H. NOWOTNY and W. SIBERT, *Z. Metallk.* **33**, 391 (1941).
2. H. NOWOTNY and B. GLATZL, *Monatsh. Chem.* **82**, 720 (1951).
3. G. LAVES, J. D'ANS and E. LAX, *FIAT Rev. Ger. Sci.* (1948).
4. H. NOWOTNY and B. GLATZ *Monatsh. Chem.* **83**, 237 (1952).
5. H. NOWOTNY, *Z. Metallk.* **34**, 237 (1942).
6. P. I. KRIPIAKEVICH, E. I. GLADYSHEVSKII and E. E. CHERKASHIN, *Dokl. Akad. Nauk SSSR* **75**, 205 (1950).
7. H. NOWOTNY and K. BACHMAYER, *Monatsh. Chem.* **81**, 488 (1950).
8. L. CASTELLIZ, *Monatsh. Chem.* **82**, 1059 (1951).
9. R. JUZA and F. HUND, *Naturwiss.* **33**, 121 (1946).

Table 5.6 *References*

1. W. KLEMM, W. BRANDT and R. HOPPE, *Z. Anorg. Allgem. Chem.* **308**, 179 (1961).
2. H. BODE and E. VOSS, *Z. Anorg. Allgem. Chem.* **286**, 136 (1956).
3. A. LAUBENGAYER, O. B. BILLINGS and A. E. NEWKIRK, *J. Am. Chem. Soc.* **62**, 546 (1940).
4. P. C. MOEWS, JR., *Inorg. Chem.* **5**, 5 (1966).
5. A. G. SHARPE, *J. Chem. Soc.* **1953**, 4177 (1953).
6. G. ENGEL, *Z. Krist.* **90A**, 341 (1935).
7. S. L. LAWTON and R. A. JACOBSON, *J. Am. Chem. Soc.* **88**, 616 (1966).

Table 5.7 *References*

1. M. V. Stackelberg and R. Paulus, *Z. Phys. Chem.* **22**, 305 (1933).
2. E. Zintl and E. Husemann, *Z. Phys. Chem.* **21**, 138 (1933).
3. H. Hartmann and H. J. Fröhlich, *Z. Anorg. Allgem. Chem.* **218**, 190 (1934).
4. R. Juza and H. Hahn, *Z. Anorg. Allgem. Chem.* **244**, 111 (1940).
5. R. E. Rundle, N. C. Baenzinger, A. S. Wilson and R. A. McDonald, *J. Am. Chem. Soc.* **70**, 99 (1948).
6. R. S. Roth and S. J. Schneider, *J. Res. NBS* **64A**, 309 (1960).
7. W. Hase, *Phys. Status Solids* **3**, K446 (1963).
8. M. Marezio, *Acta Cryst.* **20**, 723 (1966).
9. K. Löhberg, *Z. Phys. Chem.* **B28**, 402 (1935).
10. NBS Circular 539, **9** (1959).
11. L. Wolf, H. Schwab and S. Sieler, *J. Prakt. Chem.* **32**, 113 (1966).
12. I. G. Brauer and H. Gradinger, *Z. Anorg. Allgem. Chem.* **276**, 209 (1954).
13. H. E. Swanson, R. K. Fuyat and G. M. Ugrinic, NBS Circular 539, **3** (1953).
14. H. E. Swanson and R. K. Fuyat, NBS Circular 539, **2** (1953).
15. R. H. Hoskins and B. H. Soffer, *Appl. Phys. Letters* **4**, 22 (1964).

Table 5.8 *References*

1. R. Mazelsky and R. Ward, *J. Inorg. Nucl. Chem.* **20**, 39 (1961).
2. A. Bystrom, *Arkiv. Kemi, Mineral. Geol.* **18A**, 21 (1945).
3. H. P. Rooksby and E. A. White, *J. Am. Ceram. Soc.* **47**, 94 (1964).
4. F. Bertaut, F. Forrat and M. C. Montmory, *Compt. Rend.* **249**, 829 (1959).
5. C. G. Whinfrey and A. Tauber, *J. Am. Chem. Soc.* **83**, 755 (1961).
6. O. Müller, W. B. White and R. Roy, *J. Inorg. Nucl. Chem.* **26**, 2075 (1964).
7. R. S. Roth, *J. Res. Natl. Bur. Std.* **56**, 17 (1956).
8. C. G. Whinrey, D. W. Eckart and A. Tauber, *J. Am. Chem. Soc.* **82**, 2695 (1960).
9. O. Knop and F. Brisse, *Can. J. Chem.* **43**, 2812 (1965).
10. L. N. Komissarova, V. I. Spitsyn and K. S. Wang, *Dokl. Akad. Nauk SSSR* **150**, 816(1963).
11. J. J. Casey, L. Katz and W. C. Orr, *J. Am. Chem. Soc.* **77**, 2187 (1955).

Table 5.9a *References*

1. C. Keller and H. Schumtz, *J. Inorg. Nucl. Chem.* **27**, 900 (1965).
2. A. Ketelaar and E. Wegerif, *Rec. Trav. Chim.* **58**, 948 (1939).
3. J. A. A. Ketelaar and E. Wegerif, *Rec. Trav. Chim.* **57**, 1269 (1938).
4. L. Birkenback and F. Buschendorf, *Z. Phys. Chem.* **16B**, 102 (1932).
5. F. Buschendorf, *Z. Phys. Chem.* **20B**, 237 (1933).
6. L. Sillen and A. Nylander, *Arkiv. Kemi. Min. Geoll.* **17A**, No. 4 (1943).
7. R. C. L. Mooney, *Acta Cryst.* **1**, 163 (1948).
8. L. H. Brixner, *J. Electrochem. Soc.* **114**, 108 (1967).
9. M. I. Kay, B. C. Frazer and I. Almodovar, *J. Chem. Phys.* **40**, 504 (1964).
10. E. K. Brock, *Skrifter Oslo I Videnkaps-- Akad. Mat.—Naturv. Kl.* **8**, 61 (1929).Ni
11. F. Bertaut and A. Durif, *Compt. Rend.* **238**, 2173 (1954).
12. A. Durif, *Acta Cryst.* **14**, 312 (1961).
13. J. Beintema, *Z. Krist.* **97A**, 300 (1937).
14. E. A. Hazlewood, *Z. Krist.* **98A**, 439 (1938).
15. K. Schwochan, *Z. Naturforsch.* **17a**, 630 (1962).
16. A. Durif, *Acta Cryst.* **9**, 533 (1956).
17. A. I. Komkov, *Zap. Vses. Mineral Obshch.* **86**, 432 (1957).
18. L. G. Van Uitert, *J. Chem. Phys.* **37**, 98 (1962).
19. L. G. Van Uitert, *J. Electrochem. Soc.* **107**, 803 (1960).
20. L. G. Van Uitert, *J. Electrochem. Soc.* **110**, 46 (1963).
21. E. H. Brixner, *Inorg. Chem.* **3**, 600 (1964).

Table 5.9b *References*

1. R. C. Duncan, Jr., *J. Appl. Phys.* **36**, 874 (1965).
2. Z. J. Kiss and R. C. Duncan, Jr., *Proc. IRE* **50**, 1531 (1962).
3. L. F. Johnson and K. Nassau, *Proc. IRE* **49**, 1704 (1961).
5. A. Yariv, S. P. S. Porto and K. Nassau, *J. Appl. Phys.* **33**, 2519 (1962).
6. L. F. Johnson, G. D. Boyd and K. Nassau, *Proc. IRE* **50**, 86 (1962).
7. A. Yariv and J. P. Gordon, *Proc. IEEE* **51**, 4 (1963).
8. L. F. Johnson, G. D. Boyd, K. Nassau and R. R. Soden, *Phys. Rev.* **126**, 1406 (1962).

CHAPTER 6

NiAs TYPE AND RELATED STRUCTURES

THE nickel arsenide-type structures can be described by hexagonal unit cells with all of the atoms located on vertical lines which have x, y coordinates $0, 0; \frac{1}{3}, \frac{2}{3}$ and $\frac{2}{3}, \frac{1}{3}$ (see Fig. 6.0). The hexagonal close packed structure which is the simplest structure of this type, has a unit cell in which one atom is one-quarter of the cell edge up the $\frac{1}{3}, \frac{2}{3}, z$ line and another $\frac{3}{4}$ of the cell edge up the $\frac{2}{3}, \frac{1}{3}, z$ line.

A flow chart of the structures in this chapter are given in Table 6.0.

TABLE 6.0. NiAs *Type Structures*

- Hexagonal close-packed structure
  - 6.1 WC structure
  - 6.2 Wurtzite structure
  - 6.3 $Y(OH)_3$ structure
  - 6.4 MoS_2 structure
    - 6.5 Na_3As structure
  - 6.6 CdI_2 structure
    - 6.7 α-Rhomb. B structure
    - 6.8 AlB_2 structure
    - 6.9 A-Rare earth structure
    - 6.10 NiAS structure
      - 6.11 Ni_2In structure
        - 6.12 CuS structure
        - 6.13 $MgZn_2$ structure
        - 6.14 W_2B_5 structure
        - 6.15 γ′-MoC structure

6.1. WC Type, Bh, *P6m2*, Hexagonal

If a tungsten atom replaces an atom at the origin of the unit cell of a hexagonal close-packed structure and a carbon atom replaces the other metal atom, the unit cell becomes that of the WC structure. The tungsten atom is at 0, 0, 0 and the carbon atom is at the $\frac{2}{3}, \frac{1}{3}, \frac{1}{2}$ position (see Fig. 6.1).

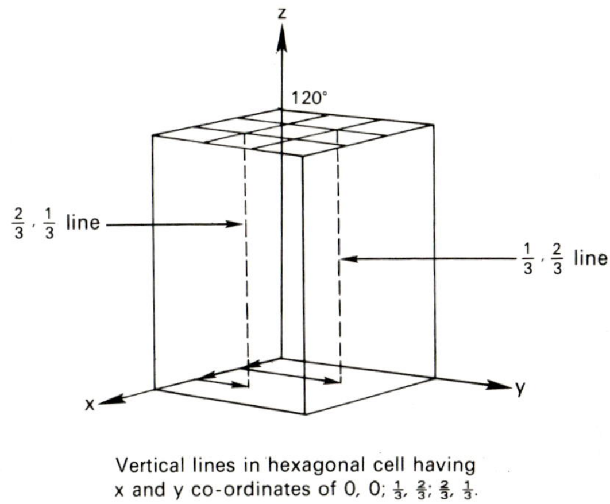

FIG. 6.0 Orientation of a hexagonal unit cell

Some of the compounds which form this structure are the carbides and nitrides presented in Table 6.1.

Properties

Tungsten carbide is extremely hard (microhardness 1780 kg/mm^2) and has a modulus of 99 × 10^6 psi. It has a melting point of 2720°C, thermal conductivity of 0.47 cal/cm-°C-sec, resistivity of 19.2 × 10^{-6} ohm-cm, and coefficient of thermal expansion of 3.84 × 10^{-6}/°C. Molybdenum phosphide

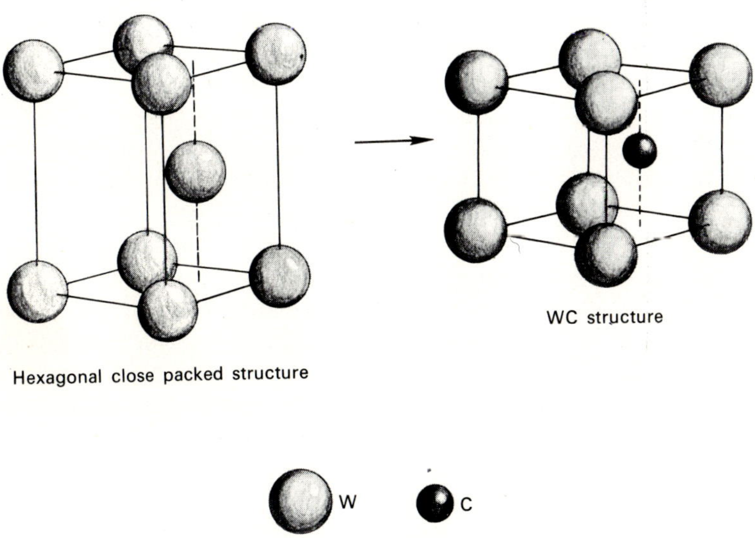

FIG. 6.1 The tungsten carbide structure

has a melting point which is lower than that of WC (1480°C) and an electrical resistivity which is higher, 1400×10^{-6} ohm-cm.

Tungsten carbide also has been produced in a cubic form by very rapid cooling of the melt. The resulting material has a superconducting transition temperature of 10°K.

TABLE 6.1. *Phases with the* WC *Structure*

Phases	Cell size (Å)		Refs.
	a_0	c_0	
Intermetallics			
MoP	3.23	3.207	1, 2
$NbS_{<1}$	3.32	3.23	3
Carbides			
RuC			4
WC	2.9065	2.8366	5, 6†
Nitrides			
γ-$NbN_{(0.8-0.9)}$ (not common form)	2.950–2.958	2.772–2.779	7
$(Ni_{0.3}Ti_{0.7})$ N	2.94	2.89	8
δ-$TaN_{0.9}$ (not common form)	2.938	2.883	9

†Has been produced as cubic form by rapid cooling of the melt.

6.2. Wurtzite Structure, Zincite Structure, ZnO, B4, *C6mc*, Hexagonal

The ZnO structure can be visualized by first superimposing atoms of zinc and oxygen at the atomic positions in the unit cell of a hexagonal close-packed structure, and then moving the oxygen atom array up $\frac{3}{8}$ of a vertical unit cell edge (see Fig. 6.2). In this cell the zinc atoms are at the corners of a hexagonal unit cell, and midway on the $\frac{1}{3}, \frac{2}{3}, z$-line and oxygen atoms are at 0.875 on the $\frac{1}{3}, \frac{2}{3}, z$-line and at 0.375 on the vertical edges. More formally, the zinc atoms are at 0, 0, 0; $\frac{1}{3}, \frac{2}{3}, \frac{1}{2}$ and O atoms at 0, 0, z; $\frac{1}{3}, \frac{2}{3}, z + \frac{1}{2}$ where $z = 0.375$. Each atom has a tetrahedron of atoms of the other element around it as is found in the zinc blende structure.

The ZnO structure appears to be formed by compounds which are intermediate in covalency between NaCl type compounds and the compounds with the zinc blende structure. In general, then, materials with small radius ratio ($R_A/R_B = 0.41$) tend to form this structure, except for the more covalent compounds with the same ratios, which form the zinc blende structure. Many AB compounds such as ZnS and SiC will form both structures under different conditions.

A list of the compounds with this structure is given in Table 6.2a.

Properties

BeO is one of the most important oxides with the wurtzite structure. It

TABLE 6.2a. *Phases with the Wurtzite Structure*

Phase	Cell size (Å)		Atomic parameter, z	Refs.
	a_0	c_0		
Carbides				
α-SiC	3.076	5.048		1
Halides				
AgI	4.592	7.51		2
CuBr	4.06	6.66		3
CuCl	3.91	6.42		4
CuI	4.31	7.09		3
NH_4F	4.39	7.02		5
Hydrides				
CuH	2.893	4.614		6
Nitrides				
AlN	3.08	4.93	0.385	7, 8
GaN	3.160	5.125	0.375	9
InN	3.537	5.70	0.375	10
NbN (not common form)	3.017	5.580		11
TaN (not common form)	3.05	4.94		12
Oxides				
BeO	2.698	4.379	0.375	13
ZnO	3.24961	5.20653	0.345	14
Sulfides, Selenides, Tellurides				
$AgInS_2$ (H.T.)	4.129	6.687		15
Al_2Se_3	3.890	6.30	0.375	16
CdS	4.136	6.714	0.375	17
CdSe	4.309	7.021	0.39	18
β-Ga_2S_3 (H.T.)	3.685	6.028	0.375	19
MgTe	4.54	7.39		20
MnS (red)	3.986	6.44	0.375	21
γ-MnS_2	4.13	6.73		22
MnTe	4.087	6.701		23
ZnS	3.820	6.260	0.375	24
ZnSe	4.01	6.54		25
ZnTe	4.27	6.99		26

TABLE 6.2b. *Properties of Materials with the* ZnO *Structure*

	Melting point (°C)	Resistivity (ohm-cm)	Thermal conductivity (cal/cm-°C-sec)	Coefficient of thermal expansion (10^{-6}/°C)
AlN	>2400 (1)	10^{10} (4)	0.072 (6)	4.0 (6)
BeO	2530 (2)	18 (1000°) (5)	0.52 (7)	9.03 (8)
ZnO	1975 (3)			6.77 (8, 9)

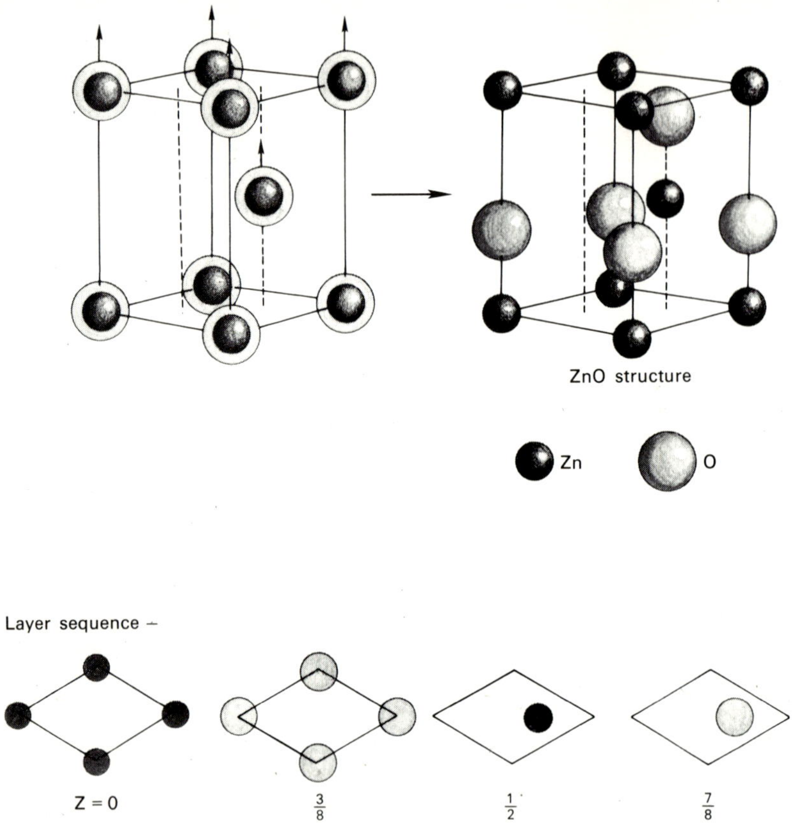

FIG. 6.2 The wurtzite structure

exhibits very high thermal conductivity and very low electrical conductivity. It has been used in crucibles, in high temperature insulators, and as nose-caps for certain types of re-entry vehicles. Only its toxicity prevents it from being used more extensively.

Zinc oxide, like Cu_2O, is a semiconductor. However, while Cu_2O conduction depends on an oxygen excess, zinc oxide conductivity depends on a zinc atom excess.

Some properties of materials with the ZnO structure are listed in Table 6.2b.

6.3. $Y(OH)_3$ Type, $P6_3/m$, Hexagonal

A number of halides and hydroxides form the $Y(OH)_3$ structure. The centers of planar triangles of $(OH)_3$ units are in the same positions as the atoms in the unit cells of the close packed hexagonal structure and these

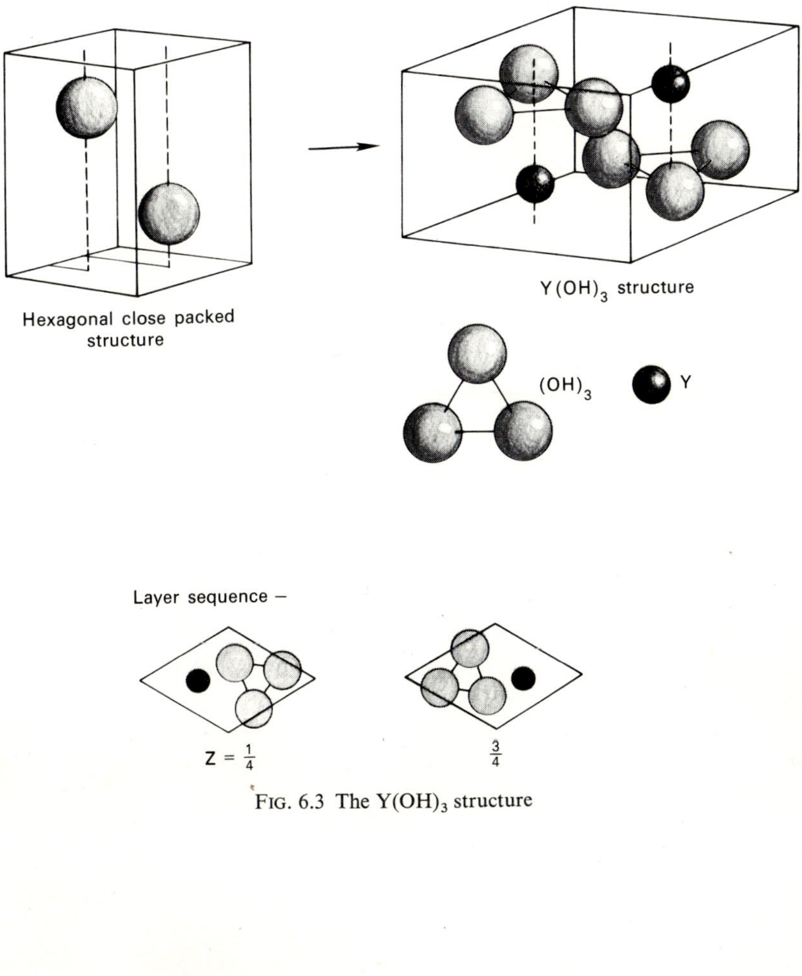

FIG. 6.3 The $Y(OH)_3$ structure

triangles are parallel to the base (see Fig. 6.3). Opposite these triangles are yttrium atoms on the $\frac{1}{3}, \frac{2}{3}, z$ and $\frac{2}{3}, \frac{1}{3}, z$-chains. Thus, the yttrium atoms are at $\frac{2}{3}, \frac{1}{3}, \frac{1}{4}; \frac{1}{3}, \frac{2}{3}, \frac{3}{4}$ and OH molecules are at $\pm\,(x, y, \frac{1}{4}; \bar{y}, x - y, \frac{1}{4}; y - x, \bar{x}, \frac{1}{4})$ with $x = 0.287$ and $y = 0.382$. Phases with this structure are listed in Table 6.3.

TABLE 6.3. *Compounds with the* $Y(OH)_3$ *Structure*

Compound	Cell size (Å)		Refs.
	a_0	c_0	
Halides			
$CfCl_3$	7.393	4.090	1
$CeBr_3$	7.95	4.44	2
$CeCl_3$	7.45	4.31	3
$EuCl_3$	7.37	4.13	3
$GdCl_3$	7.36	4.11	3
$LaBr_3$	7.97	4.51	2
$LaCl_3$	7.48	4.38	2
$NdCl_3$	7.40	4.23	2
$PrBr_3$	7.93	4.39	2
$PrCl_3$	7.42	4.28	3
$SmCl_3$	7.38	4.17	3
UBr_3	7.94	4.44	2
UCl_3	7.44	4.32	2
Others			
$Gd(OH)_3$	6.27	3.54	4
$La(OH)_3$	6.52	3.86	5
$Nd(OH)_3$	6.42	3.74	5
$Pr(OH)_3$	6.48	3.77	2
$Sm(OH)_3$	6.31	3.59	5
$Y(OH)_3$	6.24	3.53	6
$Yb(OH)_3$	6.22	3.50	6

6.4. MoS_2 Type, C7, $P6_3/mmc$

Another structure which has an atomic arrangement which can be visualized by starting with metal atoms in the same position as they are in the unit cell of a hexagonal close packed structure is that of MoS_2. The S–S units are centered on the $\frac{1}{3}, \frac{2}{3}, z$ and $\frac{2}{3}, \frac{1}{3}, z$-chains directly opposite the molybdenum atoms.

Two molybdenum atoms are $\frac{1}{4}$ of the unit cell height up on the $\frac{1}{3}, \frac{2}{3}, z$-line and $\frac{3}{4}$ of the height up the $\frac{2}{3}, \frac{1}{3}, z$-line ($\frac{1}{3}, \frac{2}{3}, \frac{1}{4}$ and $\frac{2}{3}, \frac{1}{3}, \frac{3}{4}$) and four sulfur atoms are at $\frac{1}{3}, \frac{2}{3}, z$; $\frac{2}{3}, \frac{1}{3}, \bar{z}$; $\frac{2}{3}, \frac{1}{3}, \frac{1}{2} + z$; and $\frac{1}{3}, \frac{2}{3}, \frac{1}{2} - z$ with $z = 0.629$. The structure is shown in Fig. 6.4.

Sulfides, selenides and tellurides with the MoS_2 structure are listed in Table 6.4.

The coordination number of the metal atom in the MoS_2 structure is 6 as it is in the TiO_2 and CdI_2 structures. Thus it would be expected that phases with radius ratios (R_A/R_B) between 0.41 and 0.73 would form any of these structures. However, the more ionic compounds form the rutile structure, the covalent compounds have the CdI_2 structure, and those which are intermediate adopt the MoS_2 structure.

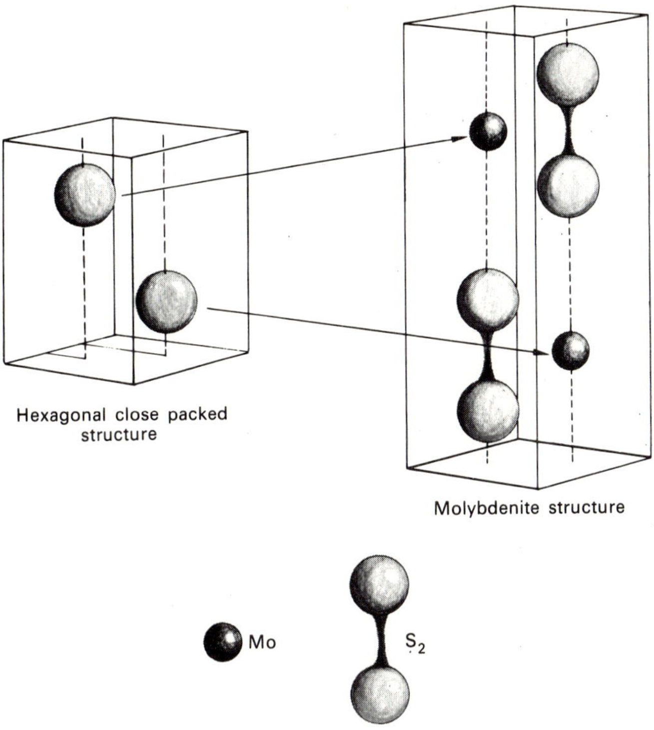

FIG. 6.4 The molybdenite structure.

Properties

MoS_2 is probably the most interesting phase in Table 6.4. It is an excellent lubricant. Its melting point is greater than 1820°C. The selenide, WSe_2, has a resistivity of 719×10^{-3} ohm-cm and its Seebeck coefficient is 700 μV/°C. It has been combined with $TaSe_2$ to form thermoelectric materials.

TABLE 6.4. *Compounds with the Molybdenite Structure*

Compound	Cell size (Å)		Refs.
	a_0	c_0	
Sulfides, Selenides, Tellurides			
MoS_2	3.1604	12.295	1, 2
$MoSe_2$	3.288	12.931	3
$MoTe_2$	3.5182	13.974	4
WS_2	3.145	12.25	5
WSe_2	3.29	12.97	5

6.5. Na_3As Type, LaF_3, Tysonite Structure, $D0_{18}$, $P6_3/mmc$, Hexagonal

The Na_3As is a more complex structure than those discussed previously in this chapter. However, it can be easily visualized as a unit cell of the MoS_2 structure with Na–Na units replacing the S–S molecules, the As atoms replacing the molybdenum atoms and two additional sodium atoms placed $\frac{1}{4}$ and $\frac{3}{4}$ of the way up the vertical edges of the cell (see Fig. 6.5). The center of the Na–Na units are $\frac{3}{4}$ of the unit cell height up the $\frac{1}{3}, \frac{2}{3}, z$-line and $\frac{1}{4}$ of the way up the $\frac{2}{3}, \frac{1}{3}, z$-line, respectively, with arsenic atoms situated directly opposite these centers on the same lines and also on the vertical edges. Two sodium atoms are at $0, 0, \frac{1}{4}$ and $0, 0, \frac{3}{4}$; four others are at $\frac{1}{3}, \frac{2}{3}, z$;

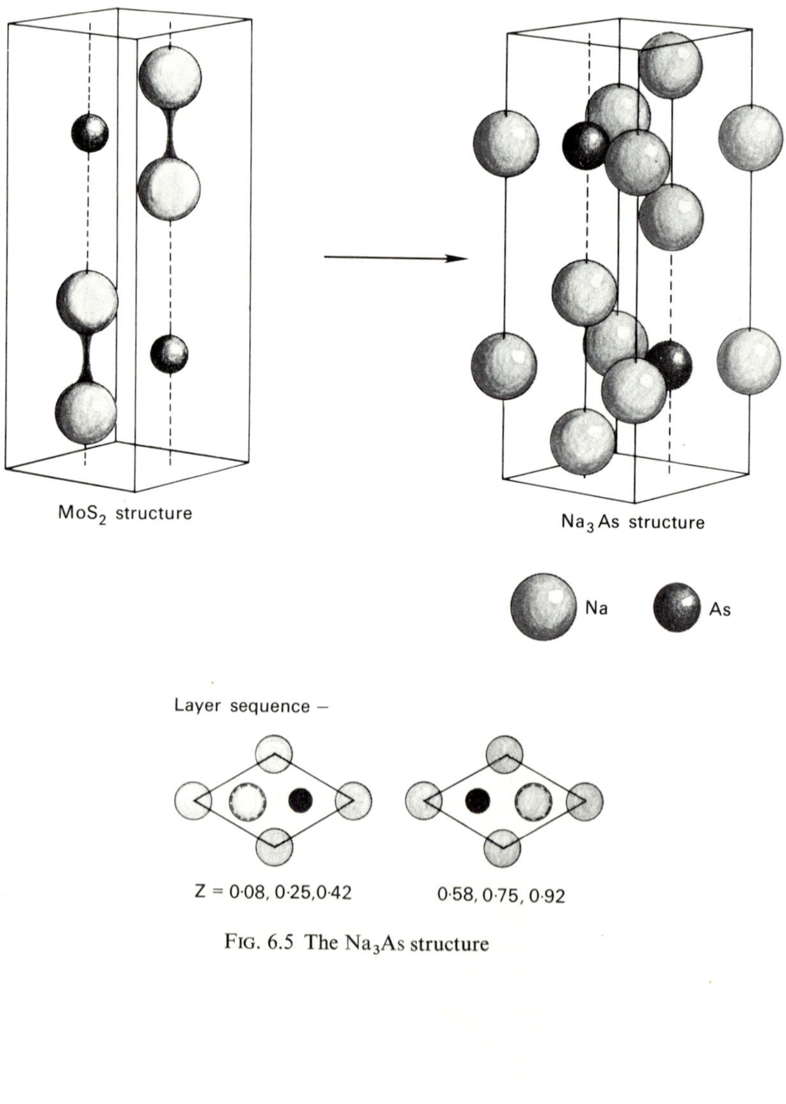

FIG. 6.5 The Na_3As structure

$\frac{2}{3}, \frac{1}{3}, \bar{z}$; $\frac{2}{3}, \frac{1}{3}, \frac{1}{2} + z$; and $\frac{1}{3}, \frac{2}{3}, \frac{1}{2} - z$ with $z = 0.6$ and As atoms at $\frac{1}{3}, \frac{2}{3}, \frac{1}{4}$ and $\frac{2}{3}, \frac{1}{3}, \frac{3}{4}$.

There are a number of phosphides and other arsenides which have the Na_3As structure. These are listed in Table 6.5.

In addition, binary halides such as LaF_3, other rare earth fluorides, and complex fluorides such as $BaUF_6$ and $SrThF_6$, and the hydrides of rare earth elements Dy, Er, Gd, Ho, Lu, Sm, Tb, and Tm crystallize with this structure.

Properties

Several intermetallic phases listed in Table 6.5 have been studied. The melting point of Na_3Sb, Na_3Bi, K_3Sb, K_3Bi and Rb_3Bi were found to be 856, 775, 812, 671 and 640°C in that order. The fluoride, LaF_3, fluoresces at 1.06 μ when it contains Nd^{3+} and therefore has been considered to be a possible laser host material.

TABLE 6.5. *Phases with the* Na_3As *Structure*

Phase	Cell size (Å)		Refs.
	a_0	c_0	
Intermetallics			
AsK_3	5.794	10.243	1
$AsLi_3$	4.377	7.801	2
$AsNa_3$	5.098	9.000	1
$AsRb_3$	6.057	10.73	3
$AuMg_3$	4.64	8.46	4
α-BiK_3	6.191	10.955	1
$BiNa_3$	5.459	9.674	1
$BiRb_3$	6.42	11.46	5
$HgMg_3$	4.868	8.656	6
K_3P	5.691	10.05	3
K_3Sb	6.037	10.717	1
Li_3P	4.273	7.594	1
α-Li_3Sb	4.710	8.326	1
Na_3P	4.990	8.814	1
Na_3Sb	5.366	9.515	1
Rb_3Sb	6.28	—	3
Halides			
LaF_3	7.190	7.367	7

6.6 CdI_2 Type, C6, *P*3*m*1, Trigonal, Hexagonal

The CdI_2 structure also can be derived from the hexagonal close-packed structure by replacing the metal atoms in the unit cell by iodine atoms and by adding cadmium atoms at the corners of the unit cell (see Fig. 6.6).

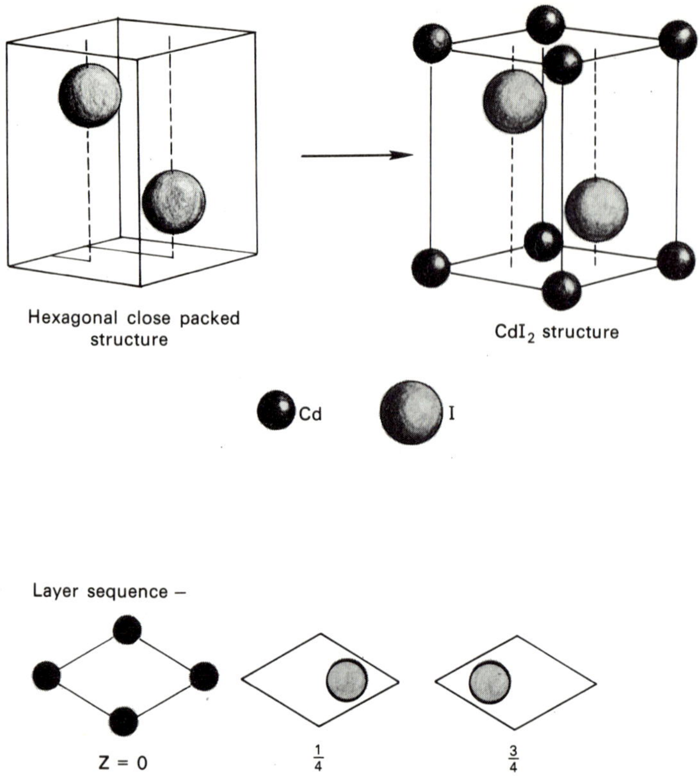

FIG. 6.6 The CdI_2 structure

Thus the iodine atoms are in close-packed layers with cadmium atoms between them. The cadmium atoms are at the corners (0, 0, 0) and the iodine atoms are at $\frac{1}{3}, \frac{2}{3}, z$ and $\frac{2}{3}, \frac{1}{3}, \bar{z}$ with $z = \frac{1}{4}$.

Other iodides as well as sulfides, selenides, tellurides, bromides and chlorides such as the ones shown in Table 6.6, also form this structure.

The more covalent AB_2 phases tend to form the CdI_2 structure. Thus, the larger polarizable iodides and bromides form this structure with highly polarizing cations while the fluorides favor the rutile structure.

Properties

The nitride Mn_2N is ferromagnetic with a Curie point of 753°K. The carbide W_2C has a melting point of approximately 2780°C and ZnS_2 a melting point of 1550°C.

TABLE 6.6. *Phases with the* CdI_2 *Structure*

Phase	Cell size (Å)		Atomic parameter, z	Refs.
	a_0	c_0		
Intermetallics				
Cd_2Ce	5.073	3.450	Cd: 0.42	1
Cd_2La	5.075	3.458		1
Cd_2Pr	5.035	3.466		1
Carbides				
W_2C	2.994	4.724		2
Halides				
Ag_2F	2.989	5.710	0.3	3
BiTeBr	4.23	6.47		4
BiTeI	4.31	6.83		4
CaI_2	4.48	6.96	0.25	5
CdI_2	4.24	6.84		6
$CoBr_2$	3.68	6.12	0.25	7
CoI_2	3.96	6.65	0.25	7
$FeBr_2$	3.74	6.17	0.25	7
FeI_2	4.04	6.75	0.25	7
GeI_2	4.13	6.79	0.25	8
$MgBr_2$	3.81	6.26	0.25	7
MgI_2	4.14	6.88	0.25	5
$MnBr_2$	3.82	6.19	0.25	7
MnI_2	4.16	6.82	0.25	7
PbI_2	4.555	6.977	0.265	9
ThI_2	4.13	7.02		10
$TiBr_2$	3.629	6.492		11
$TiCl_2$	3.561	5.875	0.25	12
TiI_2	4.110	6.820		13
TmI_2	4.520	6.967		14
VBr_2	3.768	6.180		13
VCl_2	3.601	5.835		15
VI_2	4.000	6.670		13
YbI_2	4.503	6.972		14
ZnI_2	4.25	6.54	0.25	16
Nitride				
Mn_2N	2.834	4.514		17
Sulfides, Selenides, Tellurides				
$CoTe_2$ (H.T.)	3.792	5.414		18
HfS_2	3.635	5.837		19
$HfSe_2$	3.748	6.159		19
$IrTe_2$	3.930	5.393		20
$NiTe_2$	3.869	5.308	0.25	20
$PdTe_2$	4.0365	5.1262	0.25	21
PtS_2	3.5432	5.0388	0.25	22
$PtSe_2$	3.7278	5.0813	0.38	22
$PtTe_2$	4.0259	5.2209	0.25	22
$RhTe_2$ (H.T.)	3.92	5.41	0.25	23
$SiTe_2$	4.28	6.71	0.265	24
SnS_2	3.646	5.880	0.25	25
$SnSe_2$	3.811	6.137		26

TABLE 6.6 *(cont.)*

Phase	Cell size (Å)		Atomic parameter, z	Refs.
	a_0	c_0		
α-TaS_2	3.319	6.275	0.25	27
TiS_2	3.4080	5.7014	0.25	19
$TiSe_2$	3.5356	6.0041	0.25	19, 27
$TiTe_2$	3.764	6.526		29
VSe_2	3.355	6.134		30
ZrS_2	3.662	5.813	0.25	19
$ZrSe_2$	3.771	6.148	0.25	31
$ZrTe_2$	3.950	6.630		31
Others				
$Ca(OH)_2$	3.5844	4.8962	0.233	32
$Cd(OH)_2$	3.48	4.67		33
$Co(OH)_2$	3.73	4.640	0.22	34
$Fe(OH)_2$	3.258	4.605		35
$Mg(OH)_2$	3.147	4.769		33
$Mn(OH)_2$	3.34	4.68		33
$Ni(OH)_2$	3.117	4.595		36

6.7. α-Rhombohedral Boron–Rhombohedral, Hexagonal

While there are many atoms in a unit cell of the α-rhombohedral structure, if a boron icosahedron is treated as a single atom, the similarity between this structure and that of CdI_2 can be seen (see Figs. 6.6 and 6.7a). Icosahedra are located at the corners, three-quarters of the vertical height up the $\frac{1}{3}$, $\frac{2}{3}$, z-line, and one-quarter of the distance up the $\frac{2}{3}$, $\frac{1}{3}$, z-line. The layer sequence is shown in Fig. 6.7b.

Other structures adopted by boron are given in Table 6.7.

Either the α-rhombohedral form of boron or amorphous boron is prepared at temperatures below 1300°C. At higher temperatures a tetragonal type of boron is formed which can be described as a flattened Cu_2O structure with boron atoms replacing the oxygen atoms and boron icosahedra replacing the copper atoms. Thus the boron atoms are at the corners and center of the unit cell, and the boron icosahedra are in diagonally opposite corners in the top and bottom of the cell, but not in the corners directly above one another. The β-rhombohedral structure is much more complex and is made up of units consisting of boron icosahedra sharing faces of a central icosahedron.

When molten boron is cooled from the melt the β-rhombohedral form normally results. However, by very rapid cooling of the melt an amorphous form of boron can be obtained.

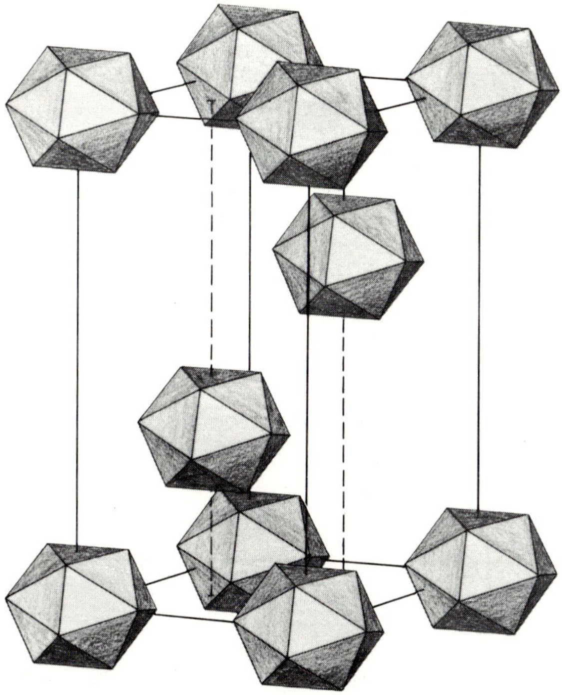

FIG. 6.7a The α-rhombohedral boron

TABLE 6.7. *Structures Adopted by Boron*

Type	Space group	Resistivity (ohm-cm)	Cell size (Å)	Refs.
α-rhomb. B	$R3m$		R (a_0 = 5.057, α = 58.06°)	1
			H (a_0 = 4.908, c_0 = 12.567)	
Tetrag. B	$P4_2/nnm$		a_0 = 8.75, c_0 = 5.06	2
β-rhomb. B	$R3m$	7×10^5 (300°K)	R (a_0 = 10.14, α = 65.05°)	2, 3
		1 (750°K)	H (a_0 = 10.94, c_0 = 23.8)	4, 5†

† An amorphous or highly faulted form of boron is obtained by pyrolytic deposition of boron on a hot tungsten wire at temperatures below 1300°C, or by very rapid cooling of molten boron.

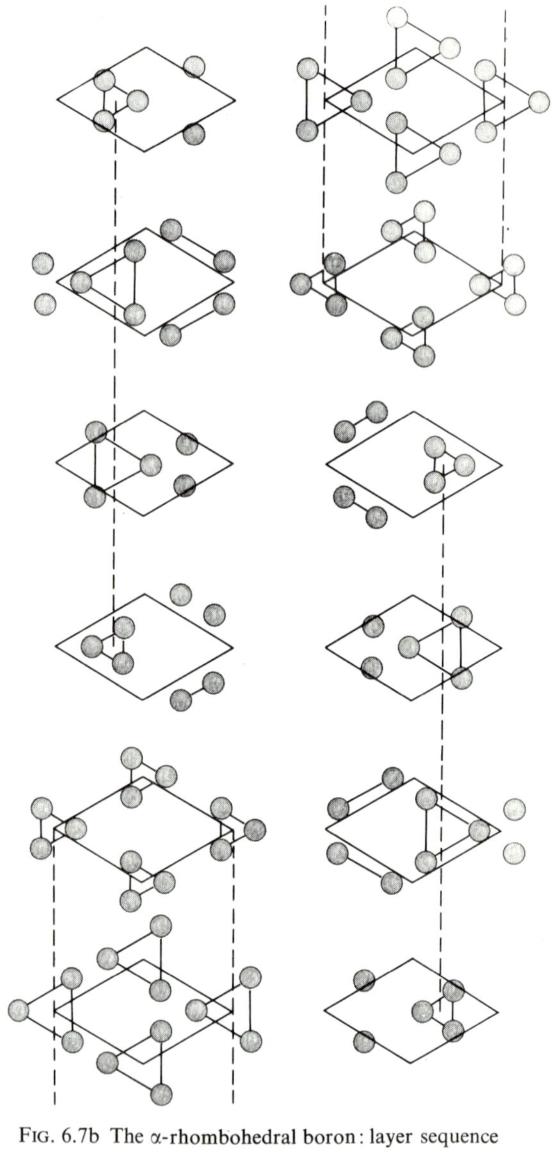

FIG. 6.7b The α-rhombohedral boron: layer sequence

6.8. AlB_2 Type, C32, *P6/mmm*, Hexagonal

The AlB_2 structure also can be derived from the CdI_2 structure by replacing the cadmium atoms with aluminum at the corners and the iodine atoms with boron on the $\frac{2}{3}, \frac{1}{3}, z$- and $\frac{1}{3}, \frac{2}{3}, z$-lines but moving each boron atom to one-half the cell height (see Fig. 6.8). The aluminum atoms are at

(0, 0, 0) and the boron atoms atoms are at $\frac{1}{3}, \frac{2}{3}, \frac{1}{2}$ and $\frac{2}{3}, \frac{1}{3}, \frac{1}{2}$. The aluminum atoms are surrounded by twelve boron atoms; the boron atoms are in sixfold coordination with aluminum. The boron atoms are arranged in layers with layers of aluminum atoms between them. The arrangement of the boron atoms in the layers is the same as that in graphite. There are a number of other borides as well as intermetallic compounds which form this structure. A table of these phases and the cell parameters are presented below in Table 6.8a.

Properties

Some of their properties are listed in Table 6.8b. Note that a number of borides are good conductors and have high melting temperatures.

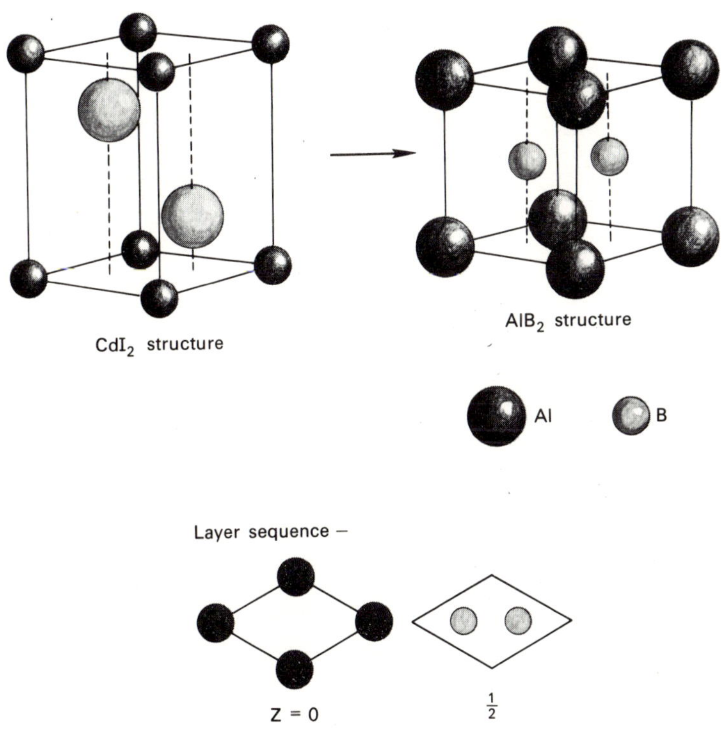

FIG. 6.8 The AlB_2 structure

TABLE 6.8a. *Phases with the Hexagonal AlB_2 Structure*

Compound	Cell size (Å)		Refs.
	a_0	c_0	
Al_2Th	4.393	4.162	1
$BaGa_2$	4.432	5.064	2
Be_2Zr	3.82	3.24	3
$CaGa_2$	4.323	4.323	4
$CaHg_2$	4.887	3.573	6
$CeGa_2$	4.32	4.34	5
Cu_2Th	4.387	3.472	7
$EuHg_2$	4.970	3.705	6
Ga_2Eu	4.345	4.520	8
Ga_2La	4.320	4.416	5
Ga_2Pr	4.272	4.298	5
Ga_2Sr	4.344	4.732	2
Hg_2La	4.958	3.640	9
Hg_2Sr	4.929	3.869	6
Hg_2V	4.976	3.218	10
Ni_2Th	3.960	3.844	1
β-Si_2U	4.028	3.952	11
Si_2Th	4.136	4.126	12
Si_2U	4.028	3.852	12
TiV_2	4.828	2.847	13
Borides			
AgB_2	3.00	3.24	14
AuB_2	3.14	3.52	14
θ-CrB_2	2.969	3.066	15
HfB_2	3.142	3.477	16
LuB_2	3.246	3.704	17
MgB_2	3.0834	3.5213	18
MnB_2	3.007	3.037	19
MoB_2 (H.T.)	3.05	3.113	20
NbB_2	3.089	3.303	21. 22
OsB_2	2.8761	2.8709	23
PuB_2	3.18	3.90	24
RuB_2	2.852	2.855	23
ScB_2	3.146	3.517	25
TaB_2	3.078	3.265	26
TiB_2	3.0245	3.2326	27
UB_2	3.13	3.99	28
VB_2	3.006	3.056	29
ZrB_2	3.167	3.529	30

TABLE 6.8b. *Properties of Phases with the* AlB_2 *Structure*

Compound	Melting point (°C)	Electrical resistivity (μohm-cm)	Thermal e,m.f. (μV/o)	Micro-hardness (kg/mm²)	Thermal conductivity (cal/cm-deg.-sec)	Coeff. of thermal expansion (10^6/°C)
$BaSi_2$	1850 [1]	38 × 10^4 [10]	600 [1]	930 [1]	0.0037 [1]	8.2 [25]
$DySi_2$	1550 [2]	3020 [11]				
USi_2	1700 [2]					
CrB_2	2200 [3]	84 [12]	−0.05 [17]	2100 [17]	0.0534 [21]	11 [13]
HfB_2	3250 [4]	9 [13]		2900 [13]		5.7 [13]
MoB_2	2100 [5]	45 [14]		1200 [14]		
NbB_2			−1.4 [12]	2600 [18]		
PuB_2					0.040 [22]	
ScB_2		7–15 [15]		1780 [15]		
TaB_2	3100 [6]	37 [12]	−3.1 [12]	2500 [19]	0.026 [23]	11.4 [27]
TiB_2	2980 [7]	14 [12]	−5.1 [12]	3370 [20]	0.010 [24]	5.5 [28]
UB_2	2385 [8]					8.5 [29]
VB_2	2400 [9]	3.5 [16]	9.2 [12]	2800 [20]		5.3 [19]
ZrB_2		17 [12]	1.2 [12]	2252 [20]	0.058 [22]	6.8 [26]

6.9. A-Rare Earth Structure, La_2O_3 Type, $D5_2$, *P3m*1, Trigonal Hexagonal

The similarity between the A-rare earth structure (La_2O_3 type) and the CdI_2 structure can be seen in Fig. 6.9. The iodine atoms are replaced by lanthanum atoms, cadmium atoms by oxygen atoms and extra oxygen atoms are added. These atoms are added in the unit cell two-fifths of the vertical edge of the unit cell edge directly above one lanthanum atom and another two-fifths of the vertical distance of the unit cell edge directly below the other lanthanum atom. The two lanthanum atoms are then at $\frac{1}{3}, \frac{2}{3}, z; \frac{2}{3}, \frac{1}{3}, \bar{z}$, $z = 0.25$, one oxygen atom is at 0, 0, 0 and two others are at $\frac{1}{3}, \frac{2}{3}, z; \frac{2}{3}, \frac{1}{3}, \bar{z}$ where $z = 0.63$ for La_2O_3.

The larger rare earth metal oxides form the A-rare earth structure (see Table 6.9).

A striking feature of the structure is that there is an unusual 7-coordination of metal atoms. Since the metal atoms in the C-rare earth structure are co-ordinated with six atoms, the A_2O_3 compounds with small A atoms prefer to form the C-rare earth structure.

Properties

The nitride U_2N_3 exists in two modifications, the Mn_2O_3 type being the more stable. The β-form, which has the A-rare earth structure, is ferro-magnetic with a Curie point of 186°K.

The conductivity of Mg_3Sb_2 has been studied. Its thermal conductivity is 0.01 cal/cm-deg-sec and it is transparent to visible light. It is a photo-conductor when it is not stoichiometric.

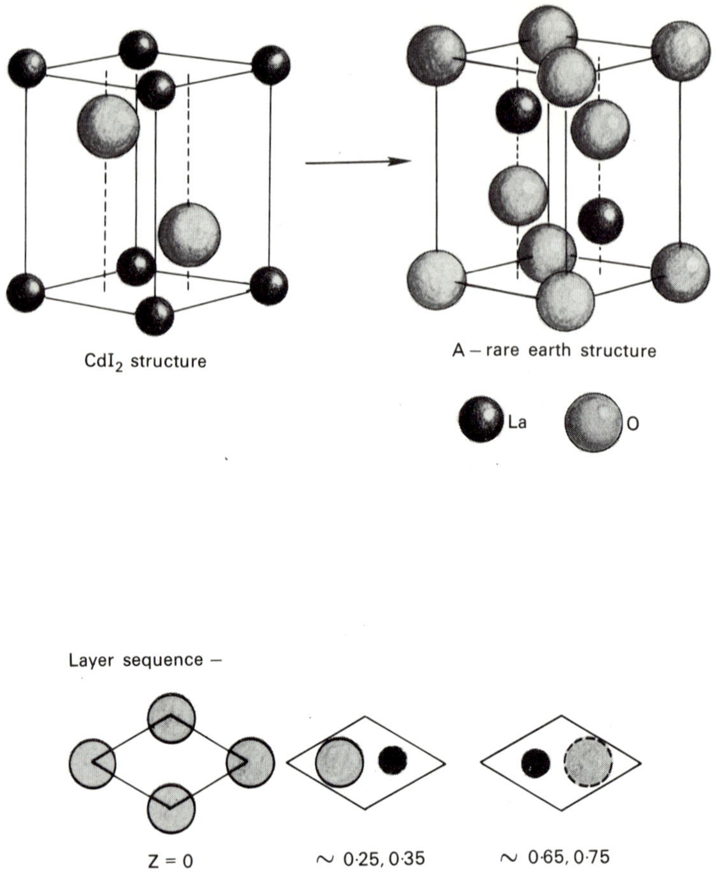

FIG. 6.9 The A- rare earth structure

Two oxides, La_2O_3 and Gd_2O_3, have been used as laser host materials for Nd^{3+} ions. Lanthanum oxide has a melting point of 2300°C and a coefficient of thermal expansion of 11.9×10^{-6}/°C from room temperature to 780°C.

TABLE 6.9. *Phases with the A-Rare Earth Structure*

Phase	Cell size (Å)		Atomic parameter, z	Refs.
	a_0	c_0		
Intermetallics				
α-Bi_2Mg_3 (L.T.)	4.675	7.416	Bi: $z = 0.235$; Mg: $z = 0.63$	1
α-Mg_3Sb_2	4.582	7.243	Mg: $z = 0.63$; Sb: $z = 0.235$	2
Nitrides				
Th_2N_3	3.883	6.187	Th: $z = 0.235$; N: $z = 0.63$	3
U_2N_3	3.69	5.83		4
Oxides				
Ac_2O_3	4.08	6.30	Ac: $z = 0.235$; O: $z = 0.63$	3
Gd_2O_3	3.76	5.89	As Ac_2O_3	5
La_2O_3	3.937	6.130	As Ac_2O_3	6
Nd_2O_3	3.82	5.98	As Ac_2O_3	5
Pr_2O_3	3.802	5.954		6
Gd_2O_3: Nd^{3+}				7
La_2O_3: Nd^{3+}				8

6.10. NiAs Type, $B8_1$, $P6_3/mmc$, Hexagonal

If, in the unit cell of the CdI_2 structure, arsenic atoms are placed at the iodine atom positions, nickel atoms at the cadmium atom positions and additional nickel atoms at the center of the vertical edges the unit cell for the NiAs structure is obtained. The metal atoms are located at the corners (0, 0, 0) and center of the vertical edges (0, 0, $\frac{1}{2}$) and the arsenic atoms at $\frac{1}{3}, \frac{2}{3}, \frac{1}{4}; \frac{2}{3}, \frac{1}{3}, \frac{3}{4}$. There are six nickel atoms surrounding the arsenic atoms and six arsenic atoms around each nickel atom (see Fig. 6.10).

Compounds of Cr, Mn, Pd, Co, Ni, Pt and Fe with S, Se, Te, Sb, As and Bi tend to form this structure (see Table 6.10a). Since the coordination number in this structure is six, AB compounds with radius ratios R_A/R_B between 0.73–0.41 should adopt the nickel arsenide structure, except for the more ionic phases which tend to form the sodium chloride structure. Phases with lower R_A/R_B values tend to form the ZnO and ZnS structures.

Properties

A number of interesting magnetic materials have this structure, such as NiS, CoS, FeS, VS, FeSe, CrSe, VSe, CoTe, FeTe and MnTe which are antiferromagnetic and CrTe, MnAs, MnBi and MnSb which are ferromagnetic. In the structure of each of the ferromagnetic manganese phases the Mn–Mn

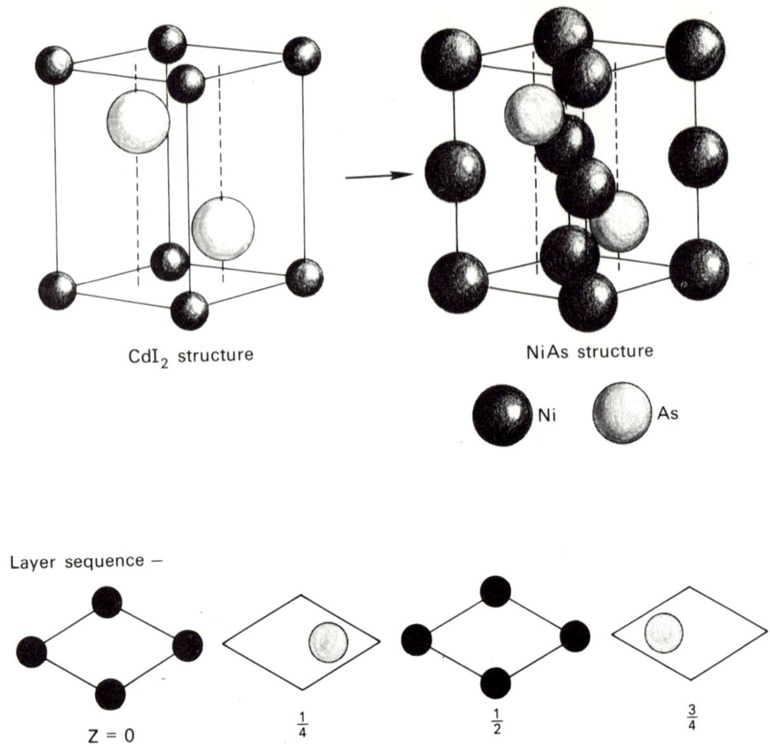

FIG. 6.10 The NiAs structure

TABLE 6.10a. *Phases with the* NiAs *Structure*

Phase	Cell size (Å)		Refs.
	a_0	c_0	
Intermetallics			
AsMn	3.724	5.706	1
AsNi	3.619	5.034	2
β-AsTi	3.64	6.15	3
AuSn	4.323	5.523	4
BiMn	4.27	6.15	5
BiNi	4.070	5.35	6
BiPt	4.315	5.490	7
BiRh	4.075	5.669	8
CuSb	3.874	5.193	9
CrSb	4.13	5.51	10
CuSn	4.198	5.096	11
FeSb	4.072	5.140	12

TABLE 6.10a *(cont.)*

Phase	Cell size (Å)		Refs.
	a_0	c_0	
IrPb	3.993	5.566	13
IrSb	3.978	5.521	14
IrSn	3.988	5.567	15
MnSb	4.22, 4.15	5.95, 5.78	5, 10
NiSb	3.942	5.155	16
NiSn	4.048	5.123	17
PV	3.18	6.22	18
PbPt	4.258	5.467	15
PdSb	4.078	5.593	19
PdSn	4.11	5.44	4
PtSb	4.13	5.483	20
PtSn	4.111	5.439	21
RhSn	4.340	5.553	15
SbTi	4.070	6.306	22
Borides			
PtB (anti-NiAs)	3.358	4.058	23
Nitrides			
δ′-NbN (not common form)	2.968	5.535	24
Sulfides, Selenides, Tellurides			
CoS	3.374	5.187	25
CoSe	3.6294	5.3006	26
	3.894	5.371	12
CrS (53.2 at. % 5)	3.464	5.763	27
CrSe	3.71	6.03	10
CrTe	3.93	6.15	28
β-FeS (ferromagnetic)	3.450	5.882	29
FeSe	3.617	5.88	30
FeTe (60 at. % Te)	3.816	5.654	31
MnTe	4.087	6.701	32
NbS	3.32	6.46	33
β-NiS (H.T.)	3.4392	5.3484	34
β-NiSe	3.6613	5.3562	35
NiTe	3.98	5.38	36
PdTe	4.152	5.672	37
RhTe	3.99	5.66	38
ScTe	4.120	6.748	39
TiS	3.299	6.380	40
TiSe	3.5722	6.205	41
VS	3.33	5.82	42
VSe	3.66	5.95	42
VTe	3.942	6.126	43
ZrTe	3.953	6.647	44

bond distance is different. The variation in their Curie temperatures (45°C, MnNi; 313°C, MnSb; 360°C, MnBi) with Mn–Mn bond distance showed that a maximum exchange energy existed between the manganese atoms in MnBi. This compound has been studied as a permanent magnet material because it has an extremely large coercive force.

Other materials with various structures which have been considered for permanent magnet application are listed for comparison in Table 6.10b. Next to CoPt, which is extremely expensive, and rare earth cobalt phases such as Co_5Sm (see Table 8.1b p. 216), MnBi has the largest H_c value

TABLE 6.10b. *Permanent Magnet Materials*

Phase	(T°K)	H_c (oersteds)	Refs.
$Al_{0.89}Mn_{1.11}$	653	1760 (6000)†	1, 2
AlMnGe	518	(7200)†	3
CoPt	813	3925 (9000)‡	4
CrTe	334		5
FeP (worked)	215		6
Fe_2P	273	2500	4
Fe_3P	716	107	6
FePd	738	260 (3000)†	4
FePt	743	1300	4
$Fe_{1.7}Co_{0.3}P$	423	3100	4
$Fe_{1.9}Ni_{0.1}P$	328	250	4
Co_2B	429		4
Fe_2B	1013	50	4
MnB	582	183	7
Fe_5P_3	635	367	4
Fe_2C	627	800	8
$Fe_{20}C_9$	528		9
Fe_3C	483		9
Fe_4N	761	150	4
Fe_3N	548	150	4
MnBi	633	3650	4
Alnico 8 (Al–Co–Ni)		1810	10
$BaFe_{12}O_{19}$ (oriented)	725	2600	10
$SrFe_{12}O_{19}$		3200	10

† Under mechanical loading.
‡ Very fine particles.

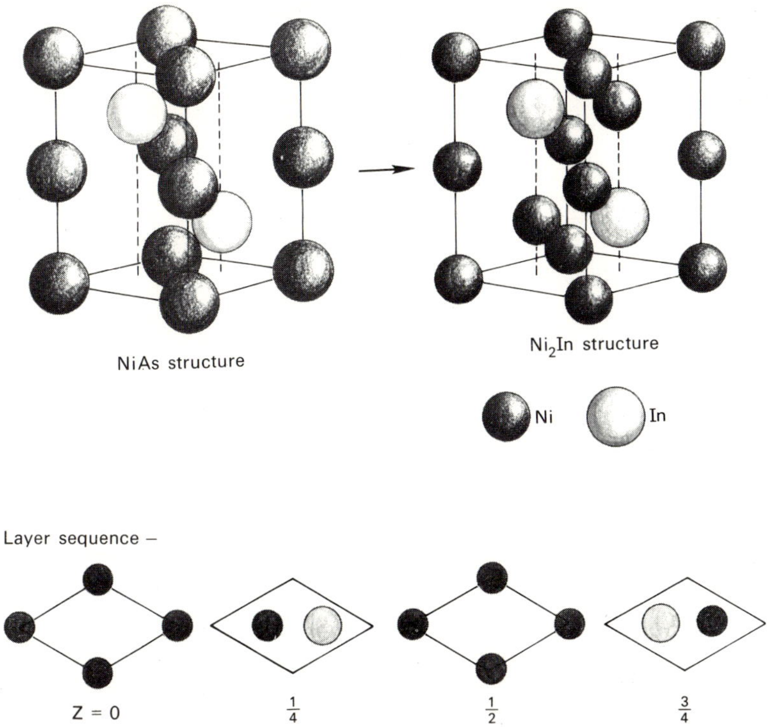

FIG. 6.11 The Ni_2In structure

6.11. Ni_2In (filled up) Type, $B8_2$, $P6_3/mmc$, Hexagonal

The Ni_2In structure is called a filled NiAs structure because it contains additional nickel atoms filling holes above one of the arsenic sites and below the other arsenic site in the unit cell. The additional atoms are at the same level as the arsenic atom positions. Two atoms are located at 0, 0, 0; 0, 0, $\frac{1}{2}$ and two others at $\frac{1}{3}, \frac{2}{3}, \frac{3}{4}$ and $\frac{2}{3}, \frac{1}{3}, \frac{1}{4}$ and 2 indium atoms at $\frac{1}{3}, \frac{2}{3}, \frac{1}{4}$; $\frac{2}{3}, \frac{1}{3}, \frac{3}{4}$ (see Fig. 6.11).

The phases with this structure and the cell sizes are given in Table 6.11.

TABLE 6.11. *Phases with the* Ni_2In *Structure*

Phase Intermetallics	Cell size (Å)		Refs.
	a_0	c_0	
BeSiZr	3.71	7.19	1
β-Bi_3Pd_5 (H.T.)	4.51	5.82	2
CoFeGe	3.978	5.029	3
CoGeMn	4.042	5.252	3
CoNiSb	3.995	5.168	3
CoNiSn	4.095	5.208	3
Co_3Sn_2 (H.T.)	4.111	5.179	4
Cu_2In	4.278	5.249	5
$Fe_{1.7}Ge$	4.017	5.005	3
FeGeMn	4.104	5.224	3
FeGeNi	4.016	5.082	3
ε-Fe–Sb	4.108	5.147	6
γ-Ga_2Ni_3 (H.T.)	4.000	4.983	7
GeMnNi	4.066	5.392	3
$GeNi_{1.70}$	3.920	5.046	3
β-$InNi_2$	4.179	5.131	8
Mn_2Sn	4.379	5.486	9
θ-Ni_2Si (H.T.)	3.805	4.890	10
Ni_3Sn_2	4.145	5.214	11
Pb_2Pd_3	4.4650	5.7040	12
Pd_5Sb_3 (660°C)	4.45	5.83	2
Pd_3Sn_2	4.399	5.666	13
Rh_3Sn_2	4.340	5.553	2
$SnTi_2$	4.653	5.700	14

6.12. CuS Type, Covelite Type, B18, $P6_3/mmc$, Hexagonal

The structure of CuS appears to be complex. However, the unit cell can be constructed from a CdI_2 unit cell by replacing nickel atoms on the edge with S–S units, replacing nickel atoms with copper atoms on the $\frac{1}{3}, \frac{2}{3}, z$ and $\frac{2}{3}, \frac{1}{3}, z$-lines, and substituting Cu–S–Cu units for the indium atoms (see Fig. 6.12). Other phases with the CuS structure are listed in Table 6.12.

Two copper atoms are at $\frac{1}{3}, \frac{2}{3}, \frac{3}{4}$ and $\frac{2}{3}, \frac{1}{3}, \frac{1}{4}$, while four other copper atoms are at $\frac{1}{3}, \frac{2}{3}, z$; $\frac{2}{3}, \frac{1}{3}, \bar{z}$; $\frac{2}{3}, \frac{1}{3}, \frac{1}{2} + z$ and $\frac{1}{3}, \frac{2}{3}, \frac{1}{2} - z$ with $z = 0.107$. Two sulfur atoms are at $\frac{1}{3}, \frac{2}{3}, \frac{1}{4}$ and $\frac{2}{3}, \frac{1}{3}, \frac{3}{4}$ and four more sulfur atoms are at $0, 0, z$; $0, 0, \bar{z}$; $0, 0, \frac{1}{2} + z$ and $0, 0, \frac{1}{2} - z$ with $z = 0.064$.

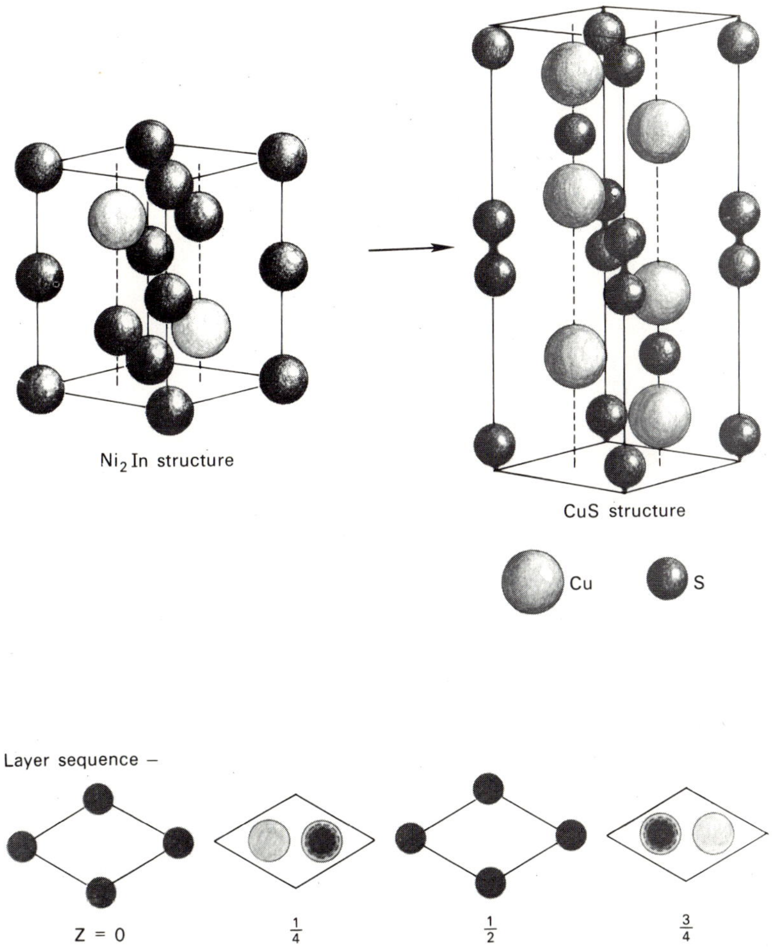

FIG. 6.12 The CuS structure

TABLE 6.12. *Compounds with the* CuS *Structure*

Compound	Cell size (Å)		Refs.
	a_0	c_0	
Sulfides and Selenides			
CuS	3.794	16.332	1
CuSe	3.94	17.25	2

6.13. $MgZn_2$ Type, C14, $P6_3/mmc$, Hexagonal

The unit cell of $MgZn_2$ also is similar to that of Ni_2In in many respects. It can be described by a unit cell of the Ni_2In structure in which zinc atoms replace the nickel atoms at the edges and corners, triangles of zinc atoms replace the nickel atoms on the $\frac{1}{3}, \frac{2}{3}, z$ and $\frac{2}{3}, \frac{1}{3}, z$, chains and the Mg–Mg

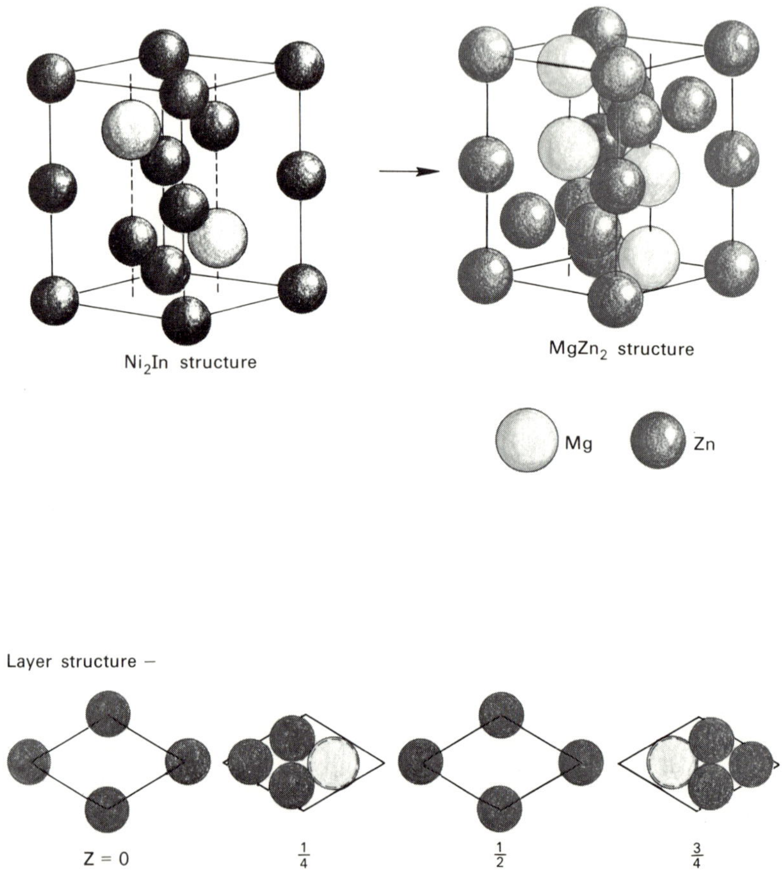

FIG. 6.13 The $MgZn_2$ structure

dumb-bells substitute for the indium atoms (see Fig. 6.13). The centers of these Mg–Mg units are not exactly at the same positions as the indium atoms. The Mg atoms are at $\frac{1}{3}, \frac{2}{3}, z; \frac{2}{3}, \frac{1}{3}, \bar{z}; \frac{2}{3}, \frac{1}{3}, \frac{1}{2} + z; \frac{1}{3}, \frac{2}{3}, \frac{1}{2} - z$ with $z = 0.062$, two zinc atoms are at 0, 0, 0 and 0, 0, $\frac{1}{2}$ and six more zinc atoms are at $x, 2x, \frac{1}{4}; 2\bar{x}, \bar{x}, \frac{1}{4}; x, \bar{x}, \frac{1}{4}; \bar{x}, 2\bar{x}, \frac{3}{4}; 2x, x, \frac{3}{4}$ and $\bar{x}, x, \frac{3}{4}$ with $x = 0.830$.

Phases with this structure, called Laves phases, are listed in Table 6.13.

TABLE 6.13. *Phases with the* $MgZn_2$ *Structure*

Phase	Cell size (Å)		Refs.
	a_0	c_0	
Al_2Zr	5.282	8.748	1
$BaMg_2$	6.663	10.557	2
Be_2Cr	4.27	6.92	3
β-Be_2Fe	4.212	6.853	4
Be_2Mn	4.240	6.924	5
Be_2Mo	4.433	7.341	6
Be_2Re	4.354	7.101	5
Be_2V	4.394	7.144	5
Be_2W	4.446	7.289	5
$CaCd_2$	5.993	9.654	7
$CaLi_2$	6.260	10.25	8
$CaMg_2$	6.23	10.12	9
(Co, Cr) Ta	4.856	7.952	10
CoTaTi	4.892	8.140	10
CoTaV	4.893	7.909	10
CrNiTa	4.855	7.888	10
Cr_2Ta (H.T.)	4.925	8.062	11
Cr_2Ti (H.T.)	4.932	7.961	12
Cr_2Zr (L.T.)	5.102	8.230	12
Cu_3Mg_2Si (L.T.)	5.014	7.889	13
Fe_2Nb	4.830	7.882	14
Fe_2Ta	4.81	7.86	14
Fe_2Ti (37 at.% Ti)	4.81	7.85	15, 16
Fe_2W	4.745	7.722	17
$HfOs_2$	5.184	8.468	18
$HfRe_2$	5.239	9.584	18
KNa_2	7.50	12.30	19
KPb_2	6.66	10.76	20
$LuOs_2$	5.254	8.661	18
$LuRu_2$	5.204	8.725	18
Mg_2Sr	6.439	10.494	7
$MgZn_2$	5.22	8.56	21
Mn_2Nb	4.891	7.969	22
Mn_2Sc	5.03	8.19	23
Mn_2Ta	4.852	7.911	24
Mn_2Th	5.476	8.931	25
Mn_2Ti	4.826	7.924	25
Mn_2Zr	5.039	8.279	25
NiTaV	4.848	7.909	9
Ni_2V	4.966	8.252	26
Os_2Sc	5.307	8.786	27
Os_2Y	5.219	8.538	28
Os_2Zr	5.219	8.538	29
Re_2Zr	5.2701	8.6349	30
Ru_2Zr	5.144	8.504	29
$TiZn_2$	5.064	8.210	26

Properties

The phase $FeBe_2$ is ferromagnetic with a Curie point of 520°C. It has a melting point of 1480°C. Other beryllides, VBe_2 and $MoBe_2$, have melting points of >1650°C and >1870°C. The melting points of intermetallics, Fe_2Ta, Fe_2Ti, Fe_2Nb, Cr_2Ta and Mn_2Nb, are 1780°, 1520°, 1630°, 1980° and 1480°C respectively.

6.14. W_2B_5 Structure, D8h, *P*6_3/*mmc*, Hexagonal

The W_2B_5 structure is another complex structure which can be visualized easily if it is treated as a derivative of the Ni_2In structure. The unit cell can

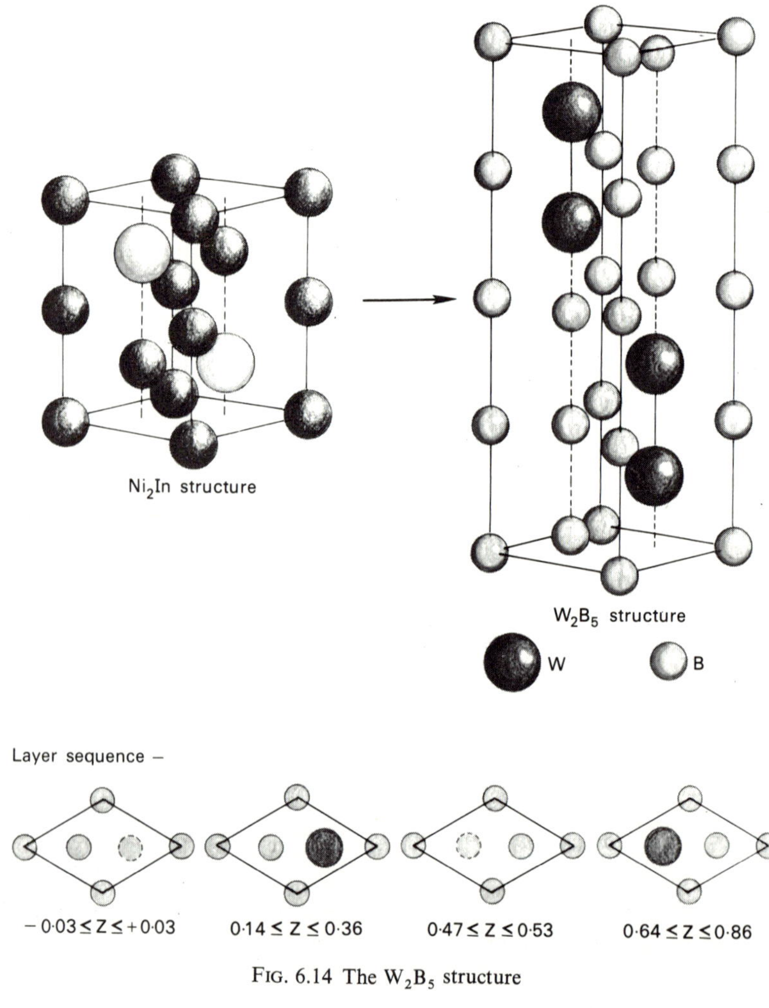

FIG. 6.14 The W_2B_5 structure

be constructed by replacing a nickel atom at the corner of a unit cell of the Ni_2In structure with a boron atom, replacing a nickel atom at the edge with a B–B–B unit aligned vertically, replacing the nickel atoms on the $\frac{1}{3}, \frac{2}{3}, z$ and $\frac{2}{3}, \frac{1}{3}, z$-lines with B–B–B units and substituting a W–W unit for the indium atoms (see Fig. 6.14).

These tungsten borides have been studied extensively recently because of the interest in boron fibers which are produced on a tungsten wire substrate. Boron trichloride and hydrogen are passed over a moving hot tungsten wire forming boron on a boride core. Various tungsten borides have been identified in the core of the fiber.

Other borides with this structure are listed in Table 6.14a and other borides of tungsten are listed in Table 6.14b.

The four W atoms are at $\pm(\frac{1}{3}, \frac{2}{3}, z; \frac{2}{3}, \frac{1}{3}, z + \frac{1}{2})$ with $z = 0.139$. Two B atoms are at $\pm(0, 0, \frac{1}{4})$, two others are at $\pm(\frac{2}{3}, \frac{1}{3}, \frac{1}{4})$, four more are at $\pm(\frac{1}{3}, \frac{2}{3}, z; \frac{2}{3}, \frac{1}{3}, z + \frac{1}{2})$ with $z = 0.028$ and two are at 0, 0, 0; 0, 0, $\frac{1}{2}$.

TABLE 6.14a. *Compounds with the* W_2B_5 *Structure*

Compound	Cell size (Å)		Refs.
	a_0	c_0	
Os_2B_5	2.91	12.91	1
Ru_2B_5	2.89	12.81	1
Ti_2B_5	2.98	13.98	2
W_2B_5	2.982	13.87	3, 4

TABLE 6.14b. *Tungsten–Boron Compounds*

Compound	Type	Cell size (Å)	Refs.
W_2B	tetr. $CuAl_2$	$a = 5.564$, $c = 4.740$	1
α-WB	tetr. MoB	$a = 3.115$, $c = 16.93$	1
WB	orthor. NiB	$a = 3.19$, $b = 8.40$, $c = 3.07$	2
W_2B_5	defect. Hex.	$a = 2.982$, $c = 13.87$	1
WB_2	tetr. AlB_2	$a = 3.02$, $c = 3.05$	3
WB_4	hex.	$a = 5.189$, $c = 6.332$	4

6.15. γ′-MoC Structure, Bi, $P6_3/mmc$, Hexagonal

The γ′-MoC structure also can be derived from the Ni_2In structure by placing carbon atoms in the nickel atom positions and linear Mo–Mo units at the indium atom positions with their axis lying along the vertical direction of the unit cell (see Fig. 6.15).

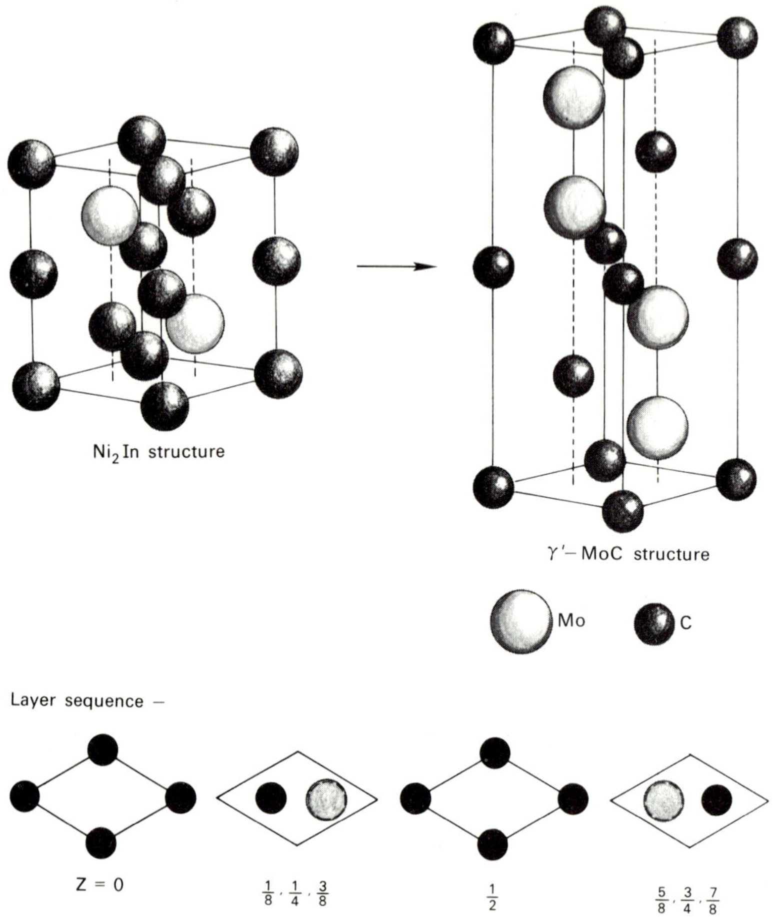

FIG. 6.15 The γ'-MoC structure

Four molybdenum atoms are at $\frac{1}{3}, \frac{2}{3}, z$; $\frac{2}{3}, \frac{1}{3}, \bar{z}$; $\frac{2}{3}, \frac{1}{3}, \frac{1}{2} + z$ and $\frac{1}{3}, \frac{2}{3}, \frac{1}{2} - z$. Two carbon atoms are at 0, 0, 0 and 0, 0, $\frac{1}{2}$, and two more are at $\frac{1}{3}, \frac{2}{3}, \frac{3}{4}$ and $\frac{2}{3}, \frac{1}{3}, \frac{1}{4}$.

Phases with this structure are listed in Table 6.15.

Properties

Molybdenum carbide has a melting point of 2700°C, a modulus of 28.6×10^6 psi and a coefficient of thermal expansion of 5.67×10^{-6}/°C. It has been prepared as a cubic phase by rapid quenching of the melt. The resulting material is a superconductor with a transition temperature of 13°K.

TABLE 6.15. *Phases with the γ-MoC Structure*

Phase	Cell size (Å)		Atomic parameter, z	Refs.
	a_0	c_0		
Intermetallics				
α-AsTi (at 45 at.% Ti)	3.65	12.29	0.125	1
PHf	3.65	12.37		2
PTi	3.487	11.65	0.125	3
β-PZr	3.677	12.53	0.125	3
Carbides				
γ-MoC	2.932	10.97	0.125	4, 5†
Nitrides				
ε-$NbN_{1.00}$ (not common form)	2.952	11.25	0.125	3

† Has been produced in cubic form by very rapid cooling of melt.

6.16. Discussion

Structure Relations

Because of the large number of compounds in this chapter, they probably can best be summarized in flow sheet form as the one shown in Fig. 6.0. The WC can be considered to be an ordered hexagonal close-packed structure and the zincite structure can be visualized as interpenetrating hexagonal close-packed arrangements of oxygen and zinc atoms. The $Y(OH)_3$ structure and molybdenite structure have $(OH)_3$ triangles and S_2 units respectively in place of some of the metal atoms in the hexagonal close-packed arrangement and the CdI_2 structure can be described by a hexagonal close-packed arrangement of I ions with Cd ions at the corners of a hexagonal unit cell. The Na_3As structure is derived from the molybdenite structure. The α-rhombohedral boron, the AlB_2, the A-rare earth and NiAs structures are derived from the CdI_2 structure. The Ni_2In structure is formed by filling the holes in the CdI_2 structure, and the CuS, $MgZn_2$, W_2B_5 and γ′-MoC structure can be derived from the Ni_2In structure.

Intermetallics

Many ionic type intermetallics which contain alkali metals and group V metals of the B_2 elements form the Na_3As structure. Another large group of intermetallics are the Laves phases with the $MgZn_2$ structure. Covalent AB intermetallic phases with R_A/R_B equal to 0.73—0.41 tend to form the nickel arsenide structure where the coordination number of each atom is six. The magnetic phases MnNi, MnSb and MnBi have drawn a great deal of attention to this structure. When the holes in this structure are filled the Ni_2In structure is formed.

Borides

In Chapter 3 it was pointed out that the UB_4, CaB_6 and UB_{12} structures contain B_6 units in the first two structures and B_{12} units in the latter. In the AlB_2 structure there are layers of B atoms with layers of aluminum atoms in between. The tungsten borides are listed in Table 6.14b. In the structure of W_2B, there are isolated B atoms, the WB structure contains boron in zigzag chains, the W_2B_5 structure has continuous and short straight chains of B atoms, the WB_2 structure contains layers of B atoms and the structure of WB_4 has not been completely determined.

The structure of boron itself contains large B structural units. It has a low density and high modulus. However, it is brittle, so that for structural applications, it must be used in fibrous form in matrices. Some of the other materials which have been considered for use as fibers for composite structures are also listed in Table 6.16. The first seven materials are all characterized as having low densities, high moduli and high melting points.

TABLE 6.16. *Data for Possible Fiber Materials*

Materials	Form	Modulus (10^6 psi)	Ultimate tensile strength (10^3 psi)	Thermal expansion coefficient (10^{-6}/°C)	Melting point (°C)	Density (g/cm³)	Refs.
Al_2O_3	Whisker	60	160–2600	8.8	2000	3.98	1
B	Fiber	55	350–500	4.3 (8 bulk)	2200	2.35	2
B_4C	Whisker	65	200–900	4.5	2450	2.51	3
Be	Wire	42	220	13–18	1285	1.85	4
BeO	Whisker	60	2000–2800	9	2550	3.01	5
C	Fiber	30	180		3700	1.86	6
SiC	Fiber	55	300	5.9	2700	3.22	7
Steel	Wire	29	100–500	11.76	1537	7.87	8
TiB_2	Bulk	54–77	19	9	2940	4.5	9
ZrB_2	Bulk	64	29	7.5	3060	6.1	10
ZrC	Bulk	50–60	28	9	3400	6.56	9
WC	Bulk	74–102	50	5	2600	18.7	9
SiO_2 (glass)	Fiber	8–10	100–900	0.5		2.54	11

Carbides

The most important carbides discussed in this chapter are SiC with the zincite structure and tungsten carbide. The zincite form of SiC is not as common as the zinc blende form, but it also has a low density and high modulus. Tungsten carbide has been used in cutting tools and as coatings. It is extremely hard and has a melting point of 2867°C.

Halides

In Chapter 5 it was noted that except for BeF_2 the metal fluorides MF_2 have either the rutile or fluorite structure. The smaller metal ions form fluorides with the rutile structure and those with larger metal ions form fluorides with the fluorite structure. In addition, $CaCl_2$, $CaBr_2$ and $SrCl_2$ form a distorted rutile structure.

Calcium iodide and PbI_2 have the CdI_2 structure, while the other AX_2 halides of calcium, strontium, barium and lead adopt either the $PbCl_2$ or $SrBr_2$ structures. In the $PbCl_2$ structure each lead atom is surrounded by nine chlorine atoms and the $SrBr_2$ structure is related.

Trichlorides, tribromides and hydroxides of the trivalent rare earth and actinide elements form the $Y(OH)_3$ structure. Each metal is coordinated with nine anions. In the next chapter, the trifluorides with the ReO_3 structure containing the metal atom in sixfold coordination will be described. The trifluorides with the Na_3As structure contain the metal atoms in fivefold coordination. Most of the rare earth and actinide fluorides have this structure.

Hydrides

The most interesting hydrides in this chapter are the trihydrides of the rare earth metals with the Na_3As structure. They are usually nonstoichiometric and represent transitions between ionic hydrides such as LiH, CaH_2, etc., and metallic hydrides. The metals are electropositive like the metals in the ionic hydrides, but the hydrides are less dense than the parent metals like the metallic hydrides.

Nitrides

Of the group II metal nitrides with the A_3N_2 formula, Be_3N_2, Mg_3N_2 and Ca_3N_2 form the C-rare earth structure. In the next group, boron forms BN with the layer structure and Al, Ga and In form AlN, GaN and InN with the wurtzite structure. The nitrides Th_2N_3 and U_2N_3 adopt the A-rare earth structure.

Oxides

Zinc oxide and BeO are the only AO oxides with the wurtzite structure. The metal ions are surrounded with four oxygen ions. The larger rare earth metals form A_2O_3 type oxides with the C-rare earth structure. The sevenfold coordination about the cation makes this structure more suitable for the larger trivalent ions than the C-rare earth structure in which the metal ions are octahedrally coordinated.

Table 6.1 *References*

1. K. Bachmayer, H. Nowotny and A. Kohl, *Monatsh. Chem.* **86**, 39 (1955).
2. N. Schönberg, *Acta Chem. Scand.* **8**, 226 (1954).
3. G. Hagg and N. Schönberg, *Arkiv. Kemi.* **7**, 371 (1954).
4. C. P. Kempter, *J. Chem. Phys.* **41**, 1515 (1964).
5. J. Leciejewicz, *Acta Cryst.* **14**, 200 (1961).
6. R. H. Willens and E. Buchler, *Appl. Phys. Letters* **7**, 25 (1965).
7. N. Schönberg, *Acta Chem. Scand.* **8**, 208 (1954).
8. N. Schönberg, *Acta Met.* **2**, 427 (1954).
9. N. Schönberg, *Acta Chem. Scand.* **8**, 199 (1954).

Table 6.2*a References*

1. R. F. Adamsky and K. M. Merz, *Z. Krist.* **111**, 350 (1959).
2. *Natl. Bur. Stds. Circular* 539, **8** (1958).
3. J. Krug and L. Sieg, *Z. Naturforsch.* **7A**, 369 (1952).
4. M. R. Lorenz and J. S. Prener, *Acta Cryst.* **9**, 538 (1956).
5. W. H. Zachariasen, *Z. Phys. Chem. (Leipzig)* **127**, 218 (1927).
6. J. A. Goedkoop and A. F. Anderson, *Acta Cryst.* **8**, 118 (1955).
7. K. M. Taylor and C. Lenie, *J. Electrochem. Soc.* **107**, 308 (1960).
8. J. Pastinak and L. Roskovcova. *Phys. Status Solidi* **14**, K5 (1966).
9. J. V. Lirman and G. S. Zhdanov, *Acta Physicochim. URSS* **6**, 306 (1937).
10. R. Juza and H. Hahn, *Z. Anorg. Allgem. Chem.* **239**, 282 (1938).
11. G. Brauer and R. Esselborn, *Z. Anorg. Allgem. Chem.* **309**, 151 (1961).
12. G. Brauer and K. H. Zapp, *Z. Anorg. Allgem. Chem.* **277**, 129 (1954).
13. G. A. Jeffrey, G. S. Parry and R. L. Mozzi, *J. Chem. Phys.* **25**, 1024 (1956).
14. T. B. Rymer and G. D. Archard, *Research* **5**, 292 (1952).
15. H. Hahn, G. Frank, W. Klinger, A. D. Meyer and G. Storger, *Z. Anorg. Chem.* **271**, 153 (1953).
16. A. Schneider and G. Gattow, *Z. Anorg. Chem.* **277**, 49 (1954).
17. S. Ibuki, *J. Phys. Soc. Japan* **14**, 1181 (1959).
18. N. A. Goryunova, V. A. Kotovich and V. Frank-Kamenetskii, *Zh. Tekh. Fiz.* **25**, 2419 (1955).
19. H. Hahn and W. Klingler, *Z. Anorg. Chem.* **259**, 135 (1949).
20. W. Klemm and K. Wahl, *Phys. Rev.* **83**, 1270 (1951).
21. H. Schnaase, *Z. Phys. Chem.* **B20**, 89 (1933).
22. A. Baroni, *Z. Krist.* **99A**, 336 (1938).
23. N. P. Grazhdankina and D. I. Gurfel, *Sov. Phys.-JETP* **8**, 631 (1959).
24. H. E. Swanson and R. K. Fuyat, *Natl. Bur. Stds. (U.S.) Circ.* 539, II, 14, 16 (1953).
25. A. S. Pashinkin, G. N. Tishchenko, I. V. Korneeva and B. N. Ryzhenko, *Sov. Phys.-Cryst.* **5**, 243 (1960).
26. Y. D. Chistyakov and E. Cruceaunu, *Rev. Phys., Acad. Rep. Populaire Roumaine* **6**, 211 (1961).

Table 6.2*b References*

1. I. Wolf, *Z. Anorg. Chem.* **87**, 120 (1914).
2. O. J. Whittemore, Jr., *J. Can. Ceram. Soc.* **28**, 43 (1949).
3. I. E. Cambell, *High Temperature Technology*, Wiley, New York (1956).
4. T. Renner, *Z. Anorg. Chem.* **298**, 22 (1959).
5. G. R. Finlay, *Chem. in Can.* **4**, 41 (1952).
6. K. Taylor and C. Lenie, *J. Electrochem. Soc.* **107**, 308 (1960).
7. W. G. Bradshaw and C. O. Mathews, *Properties of Refractory Materials, Collected Data and References*, LMSD-2466 (June 24, 1958).

8. O. H. Krikorian, *Thermal Expansion of High Temperature Materials*, UCRL-6132 (September 1960).
9. C. Heiland and H. Iback, *Solid State Commun.* **4**, 353 (1966).

Table 6.3 *References*

1. J. L. Green, UCRL-16516, Univ. of Calif., Berkeley (1966).
2. W. H. Zachariasen, *Acta Cryst.* **1**, 265 (1948).
3. D. H. Templeton and C. H. Dauben, *J. Am. Chem. Soc.* **76**, 5237 (1954).
4. R. Fricke and A. Seitz, *Z. Anorg. Allgem. Chem.* **254**, 107 (1947).
5. R. Roy and H. A. McKinstry, *Acta Cryst.* **6**, 365 (1953).
6. R. Fricke and W. Dürrwächter, *Z. Anorg. Allgem. Chem.* **259**, 305 (1949).

Table 6.4 *References*

1. J. Jellinek, G. Brauer and H. Mueller, *Nature* **185**, 376 (1960).
2. R. E. Bell and R. E. Herfert, *J. Am. Chem. Soc.* **79**, 3351 (1957).
3. P. B. James and M. T. Lavik, *Acta Cryst.* **16**, 1183 (1963).
4. O. Knop and R. D. MacDonald, *Can. J. Chem.* **39**, 897 (1961).
5. S. M. Samoilov and A. M. Rubinshtein, *Bull. Acad. Sci. USSR, Div. Chem. Sci.* **1959**, 1819 (1959).

Table 6.5 *References*

1. G. Brauer and E. Zintl, *Z. Phys. Chem.* **B37**, 323 (1937).
2. R. E. Tate and F. W. Schonfeld, *Trans. AIME* **215**, 296 (1959).
3. G. Gnutzmann, F. W. Dorn and W. Klemm, *Z. Anorg. Allgem. Chem.* **309**, 210 (1961).
4. K. Schubert and K. Anderko, *Z. Metallk.* **42**, 321 (1951).
5. W. W. Zhuravlev, V. A. Smirnov and T. A. Mingazin, *Sov. Phys.-Cryst.* **5**, 134 (1960).
6. G. Brauer and R. Rudolph, *Z. Anorg. Chem.* **248**, 405 (1941).
7. M. Wansmann, *Z. Anorg. Allgem. Chem.* **331**, 98 (1964).

Table 6.6 *References*

1. A. Iandelli and R. Ferro, *Gazz. Chim. Ital.* **84**, 463 (1954).
2. L. N. Butorina and Z. G. Pinsker, *Sov. Phys.-Cryst.* **5**, 560 (1961).
3. H. Terrey and H. Diamond, *J. Chem. Soc.* **1928**, 2820 (1928).
4. E. Donges, *Z. Anorg. Allgem. Chem.* **265**, 56 (1951).
5. H. Blum, *Z. Phys. Chem.* **22B**, 298 (1933).
6. R. S. Mitchell, *Z. Krist.* **108**, 296 (1956).
7. A. Ferrari and F. Giorgi, *Rend. Accad. Lincei* **10**, 522 (1929).
8. H. M. Powell and F. M. Brewer, *J. Chem. Soc.* **1938**, 197 (1938).
9. Z. G. Pinsker, L. I. Tatarinova and V. A. Norikova, *Acta Physicochim. USSR* **18**, 378 (1943).
10. J. S. Anderson and R. W. M. D'Eye, *J. Chem. Soc.* **1949**, S244 (1949).
11. P. Ehrlich, W. Gutsche and H. J. Seifert, *Z. Anorg. Allgem. Chem.* **312**, 80 (1961).
12. N. C. Baenziger and R. E. Rundle, *Acta Cryst.* **1**, 274 (1948).
13. W. Klemm and L. Grimm, *Z. Anorg. Allgem. Chem.* **249**, 198 (1941).
14. L. B. Asprey and F. H. Kruse, *J. Inorg. Nucl. Chem.* **13**, 32 (1960).
15. P. Ehrlich and H. J. Seifert, *Z. Anorg. Allgem. Chem.* **301**, 282 (1959).
16. H. R. Oswald, *Helv. Chim. Acta* **43**, 77 (1960).
17. A. E. Vol, *Structure and Properties of Binary Metal Systems*, Vol. I, Fizmatgiz., Moscow (1959).
18. S. Tengner, *Z. Anorg. Allgem. Chem.* **239**, 126 (1938).
19. F. K. McTaggart and A. D. Wadsley, *Australian J. Chem.* **11**, 445 (1958).

20. E. F. HOCKINGS and J. G. WHITE, *J. Phys. Chem.* **64**, 1042 (1960).
21. C. J. RAUB, V. B. COMPTON, T. H. GEBALLE, B. T. MATTHIAS, J. P. MAITA and G. W. HULL, JR., *J. Phys. Chem. Solids*, **26**, 2051 (1965).
22. A. KJEKSHUS, *Acta Chem. Scand.* **15**, 159 (1961).
23. S. GELLER, *J. Am. Chem. Soc.* **77**, 2641 (1955).
24. A. WEISS and A. WEISS, *Z. Anorg. Chem.* **273**, 124 (1953).
25. I. OFTEDAL, *Z. Phys. Chem.* **134**, 301 (1928).
26. G. BUSCH, G. FROHLISH, C. HULLINGER and E. STEIGMEIER, *Helv. Phys. Acta* **34**, 359 (1961).
27. G. HAGG and N. SCHÖNBERG, *Arkiv. Kemi.* **7**, 371 (1954).
28. B. BERNUSSET, *Rev. Chim. Minerale* **3**, 135 (1966).
29. P. EHRICH, *Z. Anorg. Allgem. Chem.* **260**, 1 (1949).
30. E. HOSCHEK and W. KLEMM, *Z. Anorg. Chem.* **242**, 49 (1939).
31. H. HAHN and P. NESS, *Z. Anorg. Allgem. Chem.* **302**, 37 (1959).
32. H. E. PETCH, *Acta Cryst.* **14**, 950 (1961).
33. G. NATTA, *Gazz. Chim. Ital.* **58**, 344 (1928).
34. W. FEITKNECHT, *Helv. Chim. Acta* **21**, 766 (1938).
35. A. L. MACKAY, *Croat. Che. Acta* **31**, 67 (1959).
36. W. LOTMAR and W. FEITKNECHT, *Z. Krist.* **93A**, 368 (1936).

Table 6.7 *References*

1. B. F. DECKER and J. S. KASPER, *Acta Cryst.* **12**, 503 (1959).
2. J. L. HOARD and A. E. NEWKIRK, *J. Am. Chem. Soc.* **82**, 70 (1960).
3. R. A. BRUNGS, *Boron, Int. Symp. Paris,* 119 (1964) (pub. 1965).
4. J. L. HOARD and R. E. HUGHES, *Boron, Int. Symp. Paris* 81 (1964) (pub. 1965).
5. F. GALASSO, R. VASLET and J. PINTO, *Appl. Phys. Letters* **8**, 331 (1966).

Table 6.8*a References*

1. A. BROWN, *Acta Cryst.* **14**, 860 (1961).
2. A. IANDELLI, *Atti. Accad. Nazl. Lincei, Rend. Classe Sci. Fis. Mat. Nat.* **19**, 39 (1955).
3. J. W. NIELSEN and N. C. BAENZIGER, *Acta Cryst.* **7**, 132 (1954).
4. F. LAVES, *Naturwiss.* **31**, 145 (1943).
5. S. E. HASZKO, *Trans. AIME* **221**, 201 (1961).
6. A. IANDELLI and A. PALENZONA, *Atti. Accad. Nazl. Lincei, Rend. Classe Sci. Fis.* **37**, 165 (1964).
7. B. T. MATTHIAS, V. B. COMPTON and E. CORENZWIT, *Phys. Chem. Solids* **19**, 130 (1961).
8. D. E. DZYANA, V. YAMARKIV and E. I. GLADISHEVS'KII, *Dopvidi. Akad. Nauk uhr. RSR* **9**, 1177 (1964).
9. A. IANDELLI, *Atti. Accad. Nazl. Lincei, Rend. Classe Sci. Fis. Mat. Nat.* **29**, 62 (1960).
10. R. E. RUNDLE and A. S. WILSON, *Acta Cryst.* **2**, 148 (1949).
11. A. BROWN and J. J. NORREYS, *Nature* **191**, 61 (1961).
12. G. E. CO., A. BROWN, Brit. 997,077, June 30 (1965).
13. C. J. RAUB, V. B. COMPTON, T. H. GEBALLE, T. B. MATTHIAS, J. P. MAITA and G. W. HULL, JR., *J. Phys. Chem. Solids* **26**, 2051 (1965).
14. W. OBROWSKI, *Naturwiss.* **48**, 428 (1961).
15. R. KEISSLING, *Acta Chem. Scand.* **3**, 595 (1949).
16. E. RUDY and F. BENESOVSKY, *Monatsh. Chem.* **92**, 415 (1961).
17. M. PRZYBYLSKA, A. REDDOCH and G. RITTER, *J. Am. Chem. Soc.* **85**, 407 (1963).
18. M. E. JONES and R. E. MARSH, *J. Am. Chem. Soc.* **76**, 1434 (1954).
19. I. BINDER and B. POST, *Acta Cryst.* **13**, 356 (1960).
20. F. BERTAUT and P. BLUM, *Acta Cryst.* **4**, 72 (1951).
21. L. H. ANDERSSON and R. KIESSLING, *Acta Chem. Scand.* **4**, 160 (1950).

22. B. Post, R. W. Glaser and D. Moskowitz, *Acta Met.* **2**, 20 (1954).
23. C. P. Kempter and R. J. Fries, *J. Chem. Phys.* **34**, 1994 (1961).
24. B. J. McDonald and W. L. Stuart, *Acta Cryst.* **13**, 447 (1960).
25. N. N. Zhuravler and A. A. Stepanova, *Sov. Phys.-Cryst.* **3**, 76 (1958).
26. R. Kiessling, *Acta Chem. Scand.* **3**, 603 (1949).
27. I. Higashi and T. Atoda, *Sci. Papers Inst. Phys. Chem. Res.* **60**, 32 (1966).
28. B. W. Howlett, *J. Inst. Metals* **88**, 91 (1959–1960).
29. G. A. Meerson and G. V. Samsonov, *Zh. Prikl. Khim.* **27**, 1115 (1954).
30. H. Nowotny, E. Rudy and F. Benesovsky, *Monatsh. Chem.* **91**, 963 (1960).

Table 6.8*b References*

1. V. S. Neshpor and V. L. Yupko, *Zhur. Priklad. Khim.* **36**, 1139 (1963).
2. I. Binder, *J. Am. Ceram. Soc.* **43**, 287 (1960).
3. V. S. Neshpor and P. S. Kislyi, *Agneupory* **23**, 231 (1959).
4. F. Glaser, D. Moskowitz and B. Post, *J. Metals* **5**, 1119 (1953).
5. P. Gilles and B. Polock, *J. Metals* **5**, 1537 (1943).
6. C. Agte and K. Moers, *Z. Anorg. Chem.* **198**, 233 (1931).
7. A. Polty, H. Margolin and J. Nilsen, *Trans. ASM* **46**, 312 (1954).
8. B. Howlett, *J. Inst. Metals* **88**, 467 (1959–60).
9. H. Nowotny, F. Benesovsky and R. Kieffer, *Z. Metallk.* **50**, 258 (1959).
10. V. S. Neshpor and G. V. Samsonov, *Fiz. Tverd. Tela.* **2**, 2101 (1960).
11. J. Binder and R. Steinitz, *Planseeber. Pulvermet.* **7**, 18 (1959).
12. S. N. L'vov, V. F. Nemchenko and G. V. Samsonov, *Dokl. Akad. Nauk SSSR* **135**, 577 (1960).
13. Yu. B. Paderno, T. I. Serebryakova and G. V. Samsonov, *Tsvetnye Metal* **11**, 48 (1959).
14. R. Steinitz, J. Binder and D. Moskowitz, *J. Metals* **4**, 148 (1952).
15. G. V. Samsonov, *Dokl Akad. Nauk SSSR* **133**, 1344 (1960).
16. K. Moers, *Z. Anorg. Chem.* **198**, 243 (1931).
17. P. S. Kislyi, S. N. L'vov, V. F. Memchenko and G. V. Samsonov, *Izv. Akad. Nauk SSSR, Otdel. Tekh. Nauk, Ser. Metallurgiya i Toplivo* No. 6 (1962).
18. H. Nowotny, F. Benesovsky and R. Kieffer, *Z. Metallk* **50**, 258 (1959).
19. G. V. Samsonov and V. S. Neshpor, Problems of Powder Metallurgy and the Strength of Materials. No. 5, *Izd., An. UKr. SSR, Kiev.* **3** (1958).
20. G. V. Samsonov, *Dokl. Akad. Nauk SSSR* **86**, 319 (1952).
21. S. N. L'vov, V. F. Nemchenko and G. V. Samsonov, *Poroshkovaya Metallurgiya* **1**, 68 (1961).
22. S. Sindeband and P. Schwarzkopf, *Powder Met. Bull.* **5**, 42 (1950).
23. P. Schwarzkopf and S. Si debond, *Proc. Electrochem. Soc. Clev.* (1950).
24. *Industrial Heating* **28**, 137 (1961).
25. V. S. Neshpor and M. I. Reznichenko, *Agneuproy* **3**, 134 (1963).
26. V. S. Neshpor and G. V. Samsonov, *Fiz. Metal i Metalloved.* **4**, 181 (1957).
27. H. Holleck, F. Benesovsky, E. Laube and H. Nowotny, *Montsh. Chem.* **93**, 1075 (1962).
28. A. M. Belikov and Ya S. Umanskii, *Kristallografiya* **4**, 684 (1959).
29. G. Beckmann and R. Kiessling, *Nature* **178**, 1341 (1956).

Table 6.9 *References*

1. E. Zintl and E. Husemann, *Z. Phys. Chem.* **B 21**, 138 (1933).
2. E. Zintl and E. Husemann, *Z. Phys. Chem.* **B 21**, 148 (1933).
3. W. H. Zachariasen, *Acta Cryst.* **2**, 388 (1949).
4. D. A. Vaughan, *J. Metals* **8**, 78 (1956).
5. R. S. Roth and S. J. Schneider, *J. Res. NBS* **64A**, 309 (1960).
6. V. B. Glushkova and E. K. Keler, *Dokl. Akad. Nauk SSSR* **152**, 611 (1963).
7. H. Mueller-Buschbaum, *Z. Anorg. Allgem. Chem.* **343**, 6 (1966).
8. B. H. Soffer and R. H. Hoskins, *Appl. Phys. Letters* **4**, 113 (1964).
9. R. H. Hoskins and B. N. Soffer, *J. Appl. Phys.* **4**, 323 (1964).

Table 6.10*a References*

1. Z. S. Basinski, R. O. Kornelsen and W. B. Pearson, *Trans. Indian Inst. Metals* **13**, 141 (1960).
2. R. D. Heyding and L. D. Calvert, *Can. J. Chem.* **35**, 1205 (1957).
3. K. Bachmayer, H. Nowotny and A. Kohl, *Monatsh. Chem.* **86**, 39 (1955).
4. D. C. Hamilton, C. J. Raub, B. T. Matthias, E. Corenzwit and G. W. Hull, Jr., *J. Phys. Chem. Solids* **26**, 665 (1965).
5. S. A. Shchukarev, M. P. Morozova and T. A. Stolyarova, *J. Gen. Chem., USSR* **31**, 1657 (1961).
6. N. N. Zhuravlev, N. E. Alekseevskii and G. S. Zhdonov, *Vestniv Moskov Univ., Ser. Mat. Mekh., Astron. Fiz. i Khim.* **14**, 1117 (1959).
7. N. N. Zhuravlev, A. A. Stepanova and N. I. Zyuzin, *Sov. Phys. JETP* **7**, 627 (1959).
8. V. P. Glagolenva and G. S. Zhdanov, *Zh. Eksperim. Teoret. Fiz.* **25**, 248 (1953).
9. M. Furst and F. Halla, *Z. Phys. Chem.* **408**, 285 (1938).
10. F. K. Lotgering and E. W. Gorter, *Phys. Chem. Solids* **3**, 238 (1957).
11. A. Westgren and G. Phragmen, *Z. Anorg. Chem.* **175**, 80 (1928).
12. I. Oftedal, *Z. Phys. Chem. (Leipzig)* **128**, 135 (1927).
13. H. Pfisterev and K. Schubert, *Z. Metallk.* **41**, 358 (1950).
14. R. H. Kuzmin, *Sov. Phys.-Cryst.* **3**, 367 (1958).
15. H. Nowotny, K. Schubert and U. Dettinger, *Z. Metallk.* **37**, 137 (1946).
16. D. F. Hewitt, *Econ. Geol.* **43**, 408 (1948).
17. O. Nial, *Svensk. Kem. Tidskr.* **59**, 172 (1947).
18. N. Schönberg, *Acta Chem. Scand.* **8**, 226 (1954).
19. L. Thomassen, *Z. Phys. Chem. (Leipzig)* **135**, 383 (1928).
20. L. Thomassen, *Z. Phys. Chem.* **4B**, 277 (1929).
21. I. Offedal, *Z. Phys. Chem. (Leipzig)* **132**, 208 (1928).
22. H. Nowotny and J. Pesi, *Monatsh. Chem.* **82**, 336 (1951).
23. B. Aronsson, E. Stenberg and J. Aselius, *Acta Chem. Scand.* **14**, 733 (1960).
24. N. Schönberg, *Acta Chem. Scand.* **8**, 208 (1954).
25. D. Lundquist and A. Westgren, *Z. Anorg. Allgem. Chem.* **239**, 85 (1938).
26. F. Bohm, F. Gronvold, H. Haraldsen and H. Prydz, *Acta Chem. Scand.* **9**, 1510 (1955).
27. F. Jellinek, *Acta Cryst.* **10**, 620 (1957).
28. H. A. Wilhelm, A. S. Newton, A. H. Daane and C. Neher, U.S. Atomic Energy Comm. CT-3714 (1946).
29. H. Haraldsen, *Z. Anorg. Chem.* **246**, 169 (1941).
30. K. Hirakawa, *J. Phys. Soc. Japan* **12**, 929 (1957).
31. F. Gronvold, H. Haraldsen and J. Vihovde, *Acta Chem. Scand.* **8**, 1927 (1954).
32. N. P. Grazhdankina and D. I. Gurfel, *Soviet Phys.-JETP* **8**, 631 (1959).
33. G. Hägg and N. Schönberg, *Arkiv. Kemi.* **7**, 371 (1954).
34. M. Laffitte, *Bull. Soc. Chim. France* **1959**, 1211.
35. J. E. Heller and W. Wegener, *Neues. Jahrb. Mineral. Abhandl.* **94**, 1147 (1960).
36. W. Klemm and N. Fratini, *Z. Anorg. Allgem. Chem.* **251**, 222 (1943).
37. F. Gronvold and E. Rost, *Acta Chem. Scand.* **10**, 1620 (1956).
38. S. Geller, *J. Am. Chem. Soc.* **77**, 2641 (1955).
39. A. A. Menkov, L. N. Komissarova, Yu. P. Simanov and V. I. Spitsyn, *Proc. Acad. Sci., USSR, Chem. Sect.* **141**, 1137 (1961).
40. S. F. Bartram, *Dissertation Abstr.* **19**, 1216 (1958).
41. P. Bernusset, *Rev. Chim. Minerale* **3**, 135 (1966).
42. I. Tsubokawa, *J. Phys. Soc. Japan* **14**, 196 (1959).
43. F. Gronvold, O. Hagberg and H. Haraldsen, *Acta Chem. Scand.* **12**, 971 (1958).
44. H. Hahn and P. Ness, *Z. Anorg. Allgem. Chem.* **302**, 136 (1959).

Table 6.10*b References*

1. A. H. Zijlstra and H. B. Haanstra, *J. of Appl. Phys.* **37**, 2853 (1966).

2. A. J. J. Kock, P. Hokkeling, M. G. V. D. Steeg and K. J. De Vos, *J. Appl. Phys. Suppl.* **31**, 755 (1960).
3. J. Wernick, S. E. Hoszko and W. J. Romanow, *J. Appl. Phys.* **32**, 2495 (1961).
4. W. A. J. J. Velge and K. J. De Vos, *Z. Angew. Phys.* **21**, 115 (1966).
5. M. Suchet, *Compt. Rend.* **257**, 1756 (1963).
6. M. C. Cadeville and A. J. P. Meyer, *Compt. Rend.* **251**, 1621 (1960).
7. N. Miryasov and A. P. Parsanov, *Izv. Zuest. Akad. Nauk SSSR* **23**, 285 (1959).
8. W. D. Johnston, R. R. Heikes and J. Petrolo, *J. Phys. Chem.* **64**, 1720 (1960).
9. Ph. Duenner and S. Mueller, *Z. Naturforsch* **18a**, 1012 (1963).
10. J. H. Cochrane, *Machine Design*, 194 (September 15, 1966).

Table 6.11 *References*

1. J. W. Nielsen and N. C. Boenziger, *Acta Cryst.* **7**, 132 (1954).
2. K. Schubert *et al., Naturwiss* **40**, 269 (1953).
3. L. Castelliz, *Monatsh. Chem.* **84**, 765 (1953).
4. O. Nial, *Z. Anorg. Chem.* **238**, 287 (1938).
5. F. Laves and H. J. Wallbaum, *Z. Angew. Min.* **4**, 17 (1941).
6. N. V. Ageev and E. S. Makarov, *Izv. Akad. Nauk SSSR (Khim)* 87 (1943).
7. E. Hellner, *Z. Metallk.* **41**, 480 (1950).
8. E. S. Makarov, *Izv. Akad. Nauk SSSR (Khim)* 114 (1944).
9. H. Nowotny and K. Schubert, *Z. Metallk.* **37**, 17 (1946).
10. K. Toman, *Acta Cryst.* **5**, 329 (1952).
11. O. Nial, *Svensk. Kem. Tid.* **59**, 165 (1937).
12. H. Nowotny, K. Schubert and W. Dettinger, *Z. Metallk.* **37**, 137 (1946).
13. C. J. Raub, W. H. Zachariasen, T. H. Geballe and B. T. Matthias, *J. Phys. Chem. Solids* **24**, 1093 (1963).
14. P. Pietrokowsky, *J. Metals.* **4**, 211 (1952).

Table 6.12 *References*

1. S. Djurle, *Acta Chem. Scand.* **12**, 1415 (1958).
2. C. A. Taylor and F. A. Underwood, *Acta Cryst.* **13**, 361 (1960).

Table 6.13 *References*

1. B. T. Matthias, T. H. Geballe and V. B. Compton, *Rev. Modern Phys.* **35**, 1 (1963).
2. W. E. Zeek, *Dissertation Abstr.* **17**, 1209 (1957).
3. A. R. Edwards and S. T. M. Johnstone, *J. Inst. Metals* **84**, 313 (1956).
4. R. J. Teitel and M. Cohen, *Trans. AIME* **185**, 285 (1949).
5. L. Misch, *Metallwirtschaft* **15**, 163 (1936).
6. S. G. Gorden, J. A. McGurty, G. E. Klein and W. J. Koshuba, *Trans. AIME* **191**, 637 (1951).
7. H. Nowotny, *Z. Metallk.* **37**, 31 (1946).
8. E. Hellner and F. Laves, *Z. Krist.* **A105**, 134 (1943).
9. H. Witte, *Naturwiss.* **25**, 795 (1937).
10. K. Kuo, *Acta Metal.* **1**, 720 (1953).
11. P. Duwez and H. Martens, *J. Metals* **4**, 72 (1952).
12. R. P. Elliott, Armour Research Foundation, Tech. Rept. 10SR Tech. Note OSR-TN-247, (Aug. 1954).

13. H. WITTE, *Metallwirt.* **18**, 459 (1939).
14. H. J. WALLBAUM, *Z. Krist.* **A103**, 391 (1941).
15. Y. MURAKAMI, H. KIMURA and Y. NISHIMURA, *Trans. Natl. Res. Inst. Metals (Tokyo)* **1**, 7 (1959).
16. H. ARNFELT, *Iron Steel Inst. (London), Carnegie Schol. Mem.* **17**, 1 (1928).
17. R. P. ELLIOT, TR. No. 1, Contract No. AF 18(600)-642 (1954).
18. V. B. COMPTON and B. T. MATTHIAS, *Acta Cryst.* **12**, 651 (1959).
19. F. LAVES and H. J. WALLBAUM, *Z. Anorg. Chem.* **250**, 110 (1942).
20. D. GILDE, *Z. Anorg. Chem.* **284**, 142 (1956).
21. W. DÖRING, *Metallwirt.* **14**, 918 (1935).
22. E. M. SAVITSKII and CH. V. KOPETSKII, *Russ. J. Inorg. Chem.* **5**, 1363 (1960).
23. P. T. KRYS'YAKEVICH, V. S. PROTASOV and YU B. KUZ'MA, *Dopvidi. Abstr. Nauk uhr. RSR* **2**, 212 (1964).
24. E. M. SAVITSKII and CH. V. KIPETSKII, *Russ. J. Inorg. Chem.* **5**, 1274 (1960).
25. A. E. DWIGHT, U.S. At. Energy Comm. ANL-6330, 156 (1960).
26. N. C. BAENZIGER, R. E. RUNDLE, A. I. SNOW and A. S. WILSON, *Acta Cryst.* **3**, 34 (1950).
27. B. T. MATTHIAS, *J. Appl. Phys.* **31**, 325 (1960).
28. B. T. MATTHIAS, V. B. COMPTON and E. CORENZWIT, *J. Phys. Chem. Solids* **19**, 130 (1961).
29. B. T. MATTHIAS, V. B. COMPTON and E. CORENZWIT, *Phys. Chem. Solids* **19**, 130 (1961).
30. N. H. KRIKORIAN, W. G. WITTEMAN and M. G. BOWMAN, *J. Phys. Chem.* **64**, 1517 (1960).

Table 6.14*a References*

1. C. P. KEMPTER and R. J. FRIES, *J. Chem. Phys.* **34**, 1994 (1961).
2. B. POST and F. W. GLASER, *J. Chem. Phys.* **20**, 1050 (1952).
3. R. KIESSLING, *Acta Chem. Scand.* **1**, 893 (1947).
4. YU B. KUZ'MA, V. I. LAKH, B. I. STADNYK and YU V. VORSHILOV, *Poroshkovaya Met. Akad. Nauk Ukr. SSR* **6**, 73 (1966).

Table 6.14*b References*

1. R. KEISSLING, *Acta Chem. Scand.* **1**, 893 (1947).
2. B. POST and F. W. GLASER, *J. Chem. Phys.* **20**, 1050 (1952).
3. H. P. WOODS, F. E. WAWNER, JR. and G. B. FOX, *Science* **151**, 75 (1966).
4. F. GALASSO and A. PATON, *Trans. AIME* **236**, 115 (1966).

Table 6.15 *References*

1. K. BACHMAYER, H. NOWOTNY and A. KOHL, *Monatsh. Chem.* **86**, 39 (1955).
2. W. JEITSCHKO and H. NOWOTNY, *Monatsh. Chem.* **93**, 1197 (1962).
3. N. SCHÖNBERG, *Acta Chem. Scand.* **8**, 226 (1954).
4. H. NOWOTNY, E. PARTHÉ, R. KIEFFER and F. BENESOVSKY, *Monatsh. Chem.* **85**, 255 (1954).
5. L. E. TOTH, E. RUDY, J. JOHNSTON and E. R. PARKER, *J. Phys. Chem. Solids* **26**, 517 (1965).

Table 6.16 *References*

1. S. S. BRENNER, *J. Metals* **14**, 809 (1962).
2. F. GALASSO, M. SALKIND, D. KUEHL and V. PATARINI, *Trans. AIME* **236**, 117 (1966).
3. W. H. SUTTON *et al.*, Contract Now60-0465d, *First Quarterly Report,* May 1960 through 16*th Quarterly Report,* June 1964.
4. *Berylco, Bull.* **2125**, 8 (1964).
5. D. K. SMITH and R. W. NEWKIRK, Lawrence Radiation Laboratory, UCRL 7245, 20 (1963).
6. *Union Carbide, Tech. Info. Bull.* No. 116HD Rev. No. 4.
7. F. GALASSO, M. BASCHE and D. KUEHL, *Appl. Phys. Letters* **9**, 37 (1966).

8. J. J. GILMAN, *Am. Ceram. Soc. Educ. Symp.* (April 1962).
9. BATTELLE, *Ceramics for Aerospace*, *Am. Ceram. Soc.* (1964).
10. A. GATTI, R. CREE, E. FEINGOLD and R. MEHAN, G. E. Company, Contract NASW-670, Final Report (July 10, 1964).
11. T. D. CALLINAN, *Modern Material,* Vol. 1, p. 150, Academic Press Inc., New York, 150 (1958).

CHAPTER 7

PEROVSKITE TYPE AND RELATED STRUCTURES

THE perovskite-type structures are formed by ABX_3-type compounds where the A atoms replace some of the X atoms in close-packed cubic layers and the B atoms fit in the octahedrally coordinated sites. In the ordered Cu_3Au structure there are no B atoms; in the ReO_3 structure the A atoms are missing so that there are holes in the close-packed X atom layers. These close-packed layers are perpendicular to the ⟨111⟩ directions (body diagonals) in the cubic unit cell. Related structures can be built up as close-packed layers of X atoms or by stacking cubic unit cells. The flow diagram showing the relationships between these structures is presented in Table 7.0.

TABLE 7.0.

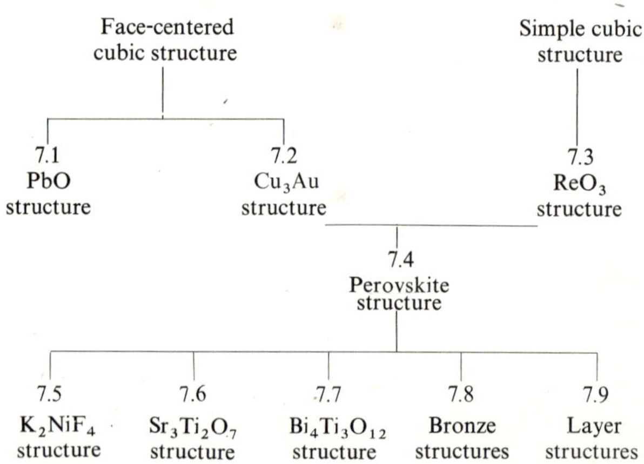

7.1. Red Lead Oxide, PbO, B10, *P*4/p*mm*, Tetragonal

This lead oxide structure can be visualized by starting with an ordered cubic close-packed structure. Lead atoms are placed in the centered positions on the vertical faces of a unit cell and the oxygen atoms in the centered position on the top and bottom faces as well as at the cell corners. The cubic cell is elongated in the c direction; the lead atoms are displaced vertically downwards in the front and back faces and upwards in the side faces. This arrangement corresponds to placement of the atoms in the following special positions in space group $P4/nmm$:

$$2\text{Pb at (2c)}: 0, \tfrac{1}{2}, z; \tfrac{1}{2}, 0, \bar{z} \text{ with } z = 0.2385;$$
$$2\text{O at (2a)}: 0, 0, 0; \tfrac{1}{2}, \tfrac{1}{2}, 0.$$

In this structure both the lead and oxygen atoms are in fourfold coordination with atoms of the other type. The oxygen atoms are in a tetrahedron of lead atoms, while the lead atoms are at the vertex of a square pyramid with oxygen atoms at the base. In the horizontal oxygen sheets, the atoms are in square planar coordination or roughly cubic packing. This structure is illustrated in Fig. 7.1. Other compounds which adopt this structure are listed in Table 7.1.

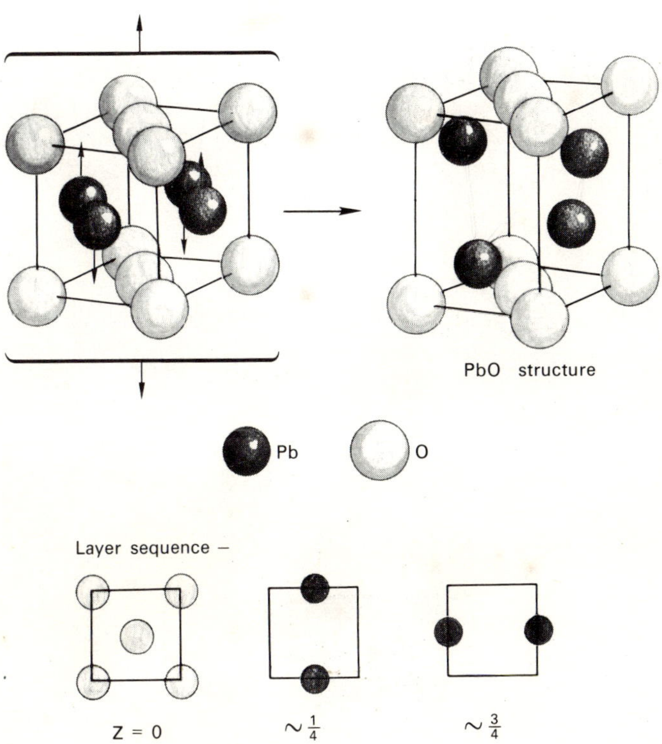

FIG. 7.1 Lead oxide structure

TABLE 7.1. *Phases with the Red Lead Oxide Structure*

Phase	Cell size (Å)		Atomic parameter, z	Refs.
	a_0	c_0		
FeSe (L.T.)	3.773	5.529	Se : 0.26	1
FeTe	3.9230–3·8198	6.2767–6.2805	Te : 0.285	2
InBi (anti-PbO)	5.000	4.773	Bi : 0.393	3
LiOH	3.546	4.334	OH : 0.18–0.22	4
PbO	3.947	4.988	Pb : 0.2385	5, 6, 7
	3.975	5.023		
SnO	3.796	4.816	Sn : 0.2356	5, 8
	3.802	4.836		

7.2. Ordered Cu_3Au Structure, LI_2, *Pm3m*, Cubic

When the atoms at the face-centered positions of a cubic unit cell are different from the atoms at the corners, the lattice type is simple cubic. In the ordered Cu_3Au structure the gold atoms are located at 0, 0, 0 and the copper atoms are at $\frac{1}{2}, \frac{1}{2}, 0; \frac{1}{2}, 0, \frac{1}{2}$ and $0, \frac{1}{2}, \frac{1}{2}$ (see Fig. 7.2). The layer sequence is shown in this figure.

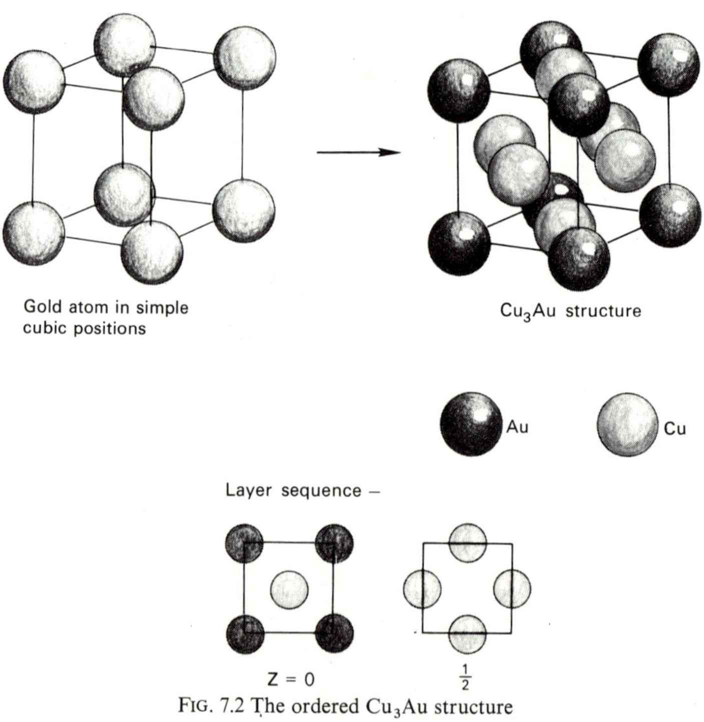

FIG. 7.2 The ordered Cu_3Au structure

Intermetallics, Cu_3Au, Cu_3Pd, Cu_3Pt and Ni_3Fe have a random cubic close-packed structure at high temperatures, but if they are cooled very slowly they adopt the ordered structure shown in Fig. 7.2. Some of the phases which have this structure are listed in Table 7.2.

TABLE 7.2. *Phases with the Ordered Cu_3Au Structure*

Phase	Cell size, a_0 (Å)	Refs.
Intermetallics		
α''-Ag_3Pt (L.T.)	3.887	1
γ-$AgPt_3$ (85 at.% Pt)	3.885	1
α'-$AlCo_3$ (H.T.)	3.658	2
Al_3Er	4.214	3
Al_3Ho	4.248	3
α'-$AlNi_3$	3.567	4
Al_3Np	4.262	5
Al_3U	4.254	6
Al_3Yb	4.2036	7
$AlZr_3$	4.372	8
$AuCu_3$	3.748	9
α'-Au_3Pt	3.926	10
$CaPb_3$	4.901	11
$CaSn_3$	4.742	11
$CaTl_3$	4.804	11
Cd_3Nb	4.215	12
α'-$CdPt_3$	3.977	13
$CeIn_3$	4.68	14
Ce_3In	5.0006	7
$CePb_3$	4.875	11
$CePt_3$	4.162	7
$CeSn_3$	4.722	11
$CoPt_3$	3.831	15
α'-$CrIr_3$	3.801	16
Cr_3Pt (62 at.% Cr)	3.775	17
α'-Cu_3Pd	3.662	18
Cu_3Pt	3.682	19
$ErIn_3$	4.563	3
$ErPt_3$	4.050	3
$ErTl_3$	4.659	3
$FeNi_3$	3.5523	20
$FePd_3$	3.851	21
$GaNi_3$	3.5823	22
Ga_3U	4.2475	23
$GeNi_3$	3.57	24
Ge_3Pu	4.223	25
Ge_3U	4.205	26
δ-$HgTi_3$ (H.T.)	4.165	27
Hg_3Zr	4.365	23
$HoIn_3$	4.570	3
$HoPt_3$	4.058	3
$HoTl_3$	4.667	3

TABLE 7.2 *(cont.)*

Phase	Cell size, a_0 (Å)	Refs.
$InLa_3$	5.075	7
$InNd_3$	4.9296	7
$InPr_3$	4.9636	7
In_3Pr	4.670	28
$InPu_3$	4.702	29
In_3Pu	4.607	29
In_3Sc	4.477	30
In_3Tb	4.5896	7
In_3Tm	4.561	7
In_3U	4.6013	23
In_3Yb	4.6164	7
γ'-$IrMn_3$	3.794	31
$LaPb_3$	4.903	32
$LaPd_3$	4.233	33
$LaPt_3$	4.0745	7
$LaSn_3$	4.782	32
$MnPt_3$	3.900	31
γ'-Mn_3Pt	3.834	31
γ'-Mn_3Rh	3.812	31
β-$NaPb_3$	4.884	34
Nb_3Si	4.211	35
$NdPt_3$	4.0590	7
$NdSn_3$	4.709	11
β-Ni_3Si	3.5040	36
$PbPd_3$	4.0216	36
Pb_3Pr	4.867	32
$PbPu_3$	4.053	37
Pb_3U	4.7834	23
Pd_3Sn	3.978	37
Pd_3Y	4.076	33
$PrPt_3$	4.0656	7
$PrSn_3$	4.725	11
Pt_3Sc	3.958	33
Pt_3Sm	4.0633	7
Pt_3Sn	3.993	39
Pt_3Tb	4.0839	7
Pt_3Ti	3.898	39
Pt_3Tm	4.0423	7
Pt_3Y	4.075	33
Pt_3Yb	4.0455	7
Pt_3Zn	3.893	13
$PuSn_3$	4.630	25
Rh_3Sc	3.898	33
Ru_3U	3.988	40
Si_3U	4.0353	23
Sn_3U	4.626	23
$TbTl_3$	4.679	7
$TiZn_3$	3.9322	27
Tl_3Tm	4.6554	7
Tl_3U	4.675	41

The ordered Cu_3Au structure has been used in many studies as a typical example of an ordered structure. The compound Cu_3Au has been studied extensively to determine the effect of ordering on various properties.

7.3. ReO_3 Structure, $D0_9$ Type, ABX_3, $E2_1$, *Pm3m*

The CsCl structure is derived from the one atom per unit cell cubic structure by adding an atom of a different element at the center of each cube of atoms, the ordered Cu_3Au structure by adding atoms at the faces and the ReO_3 structure by adding atoms at the edges (see Fig. 7.3). The rhenium atoms are at 0, 0, 0 and the oxygen atoms are located at 0, $\frac{1}{2}$, 0; $\frac{1}{2}$, 0, 0 and 0, 0, $\frac{1}{2}$. Other phases with this structure are listed in Table 7.3.

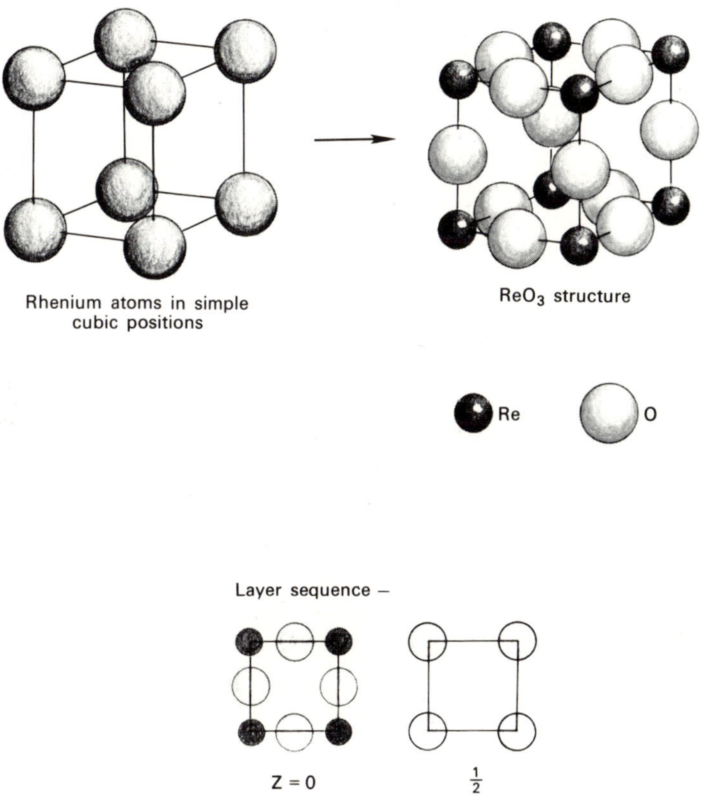

FIG. 7.3 The ReO_3 structure

TABLE 7.3. *Phases with the* ReO_3 *Structure*

Phase	Cell size, a_0 (Å)	Refs.
Halides, Oxyhalides		
FeF_3†	3.734	1
MoF_3	3.8985	2
NbF_3	3.903	3
NbO_2F	3.902	4
TaF_3	3.9012	3
TaO_2F	3.896	4
TiO_2F	3.798	5
Nitrides		
Cu_3N	3.814	6
Oxides		
ReO_3	3.7510	7
UO_3 (unstable)	4.156	8
Others		
$CaSn(OH)_6$		9
$CoSn(OH)_6$		9
$CuSn(OH)_6$		9
$FeSn(OH)_6$		9
$MgSn(OH)_6$		9
$NiSn(OH)_6$		9
$ZnSn(OH)_6$		9

† Distorted.

7.4. Perovskite Structure, $CaTiO_3$ Type, ABX_3, $E2_1$, *Pm3m*

The perovskite structure which is adopted by many compounds with the formula ABX_3 can be described by a cubic unit cell which contains an A atom at the center of a cube, B atoms at the corners and X atoms at the center of the cell edges (A unit), where X is most often O or F atoms. When the origin of the unit cell is taken at the A atom, the perovskite structure can also be described by a cubic unit cell with the A atom at the corner, B atom in the center and X atoms in the faces (B unit). There are several ways of deriving this structure.

Starting with AB in a CsCl arrangement of atoms, X atoms can be added to the center of the edges of the unit cell, or, starting with B and X atoms in an ReO_3 arrangement of the atoms, the A atoms can be added to the center of the cubic unit cell. The addition of a B atom in the center of the unit cell of an ordered Cu_3Au arrangement of AX_3 also produces the perovskite structure (see Fig. 7.4a). These two types of unit cells are called A and B, respectively. The relationship between them is shown in the layer sequence in Fig. 7.4b.

In the B-type unit cell, the A atoms are located at 0, 0, 0; the B atoms are at $\frac{1}{2}, \frac{1}{2}, \frac{1}{2}$ and three X atoms are at $0, \frac{1}{2}, \frac{1}{2}; \frac{1}{2}, 0, \frac{1}{2}; \frac{1}{2}, \frac{1}{2}, 0$. The A atoms are surrounded by twelve X atoms and the B atoms are coordinated with six X atoms. Ideally, since the A and X atoms close pack, they should be about the same size and the B atoms should be smaller.

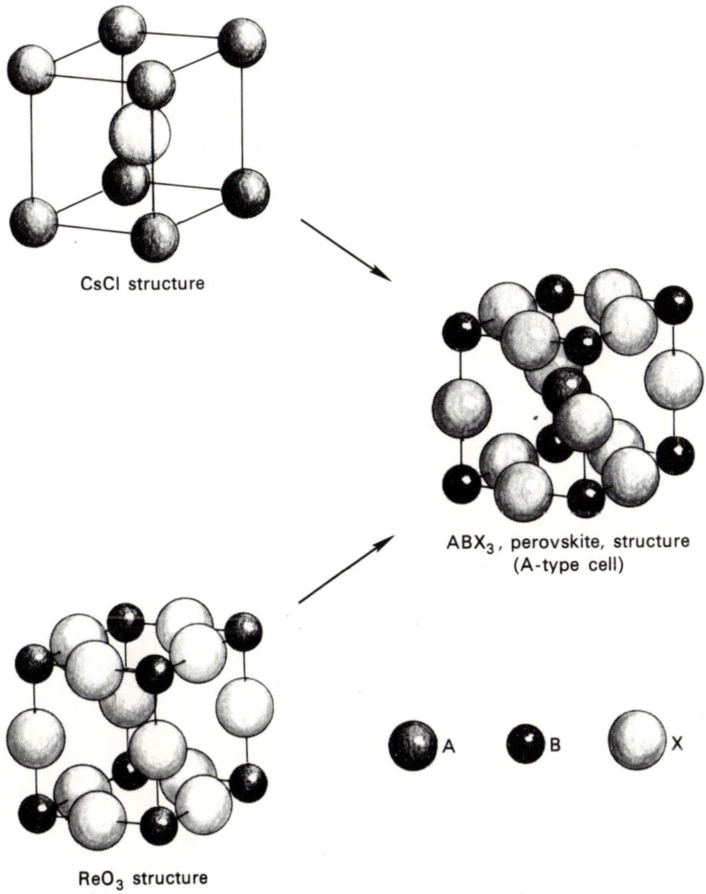

FIG. 7.4a The perovskite structure

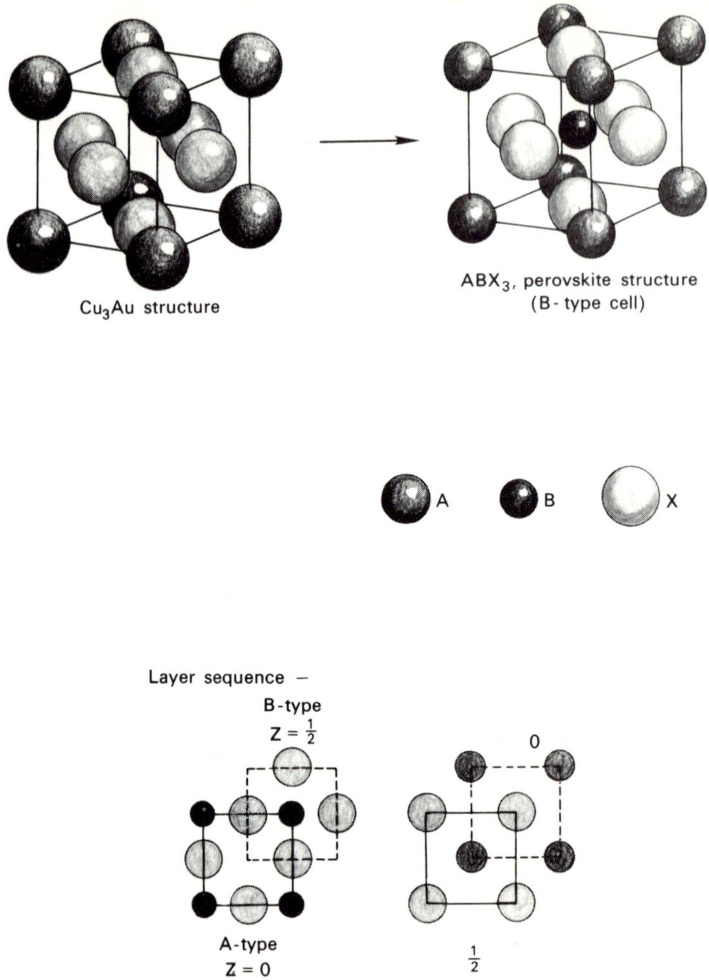

FIG. 7.4b The perovskite structure

Properties

Actually, there are very few perovskite-type compounds which form the ideal cubic structure. In a way this is fortunate since the ferroelectric properties exhibited by many of these compounds are due to dipoles which are caused by distortions in the cells. The slight tetragonal distortion of the

$BaTiO_3$ lattice can be seen in Fig. 7.4c. When the distortion is eliminated by heating barium titanate to 120°C it no longer exhibits ferroelectric properties. Figure 7.4d shows how the dielectric constant of barium titanate changes with temperature. Note that the dielectric constant rises rapidly with increasing temperature near the Curie point and then drops after the structure acquires a center of symmetry. This is typical of ferroelectric and antiferroelectric materials. Some of the simple and complex perovskites

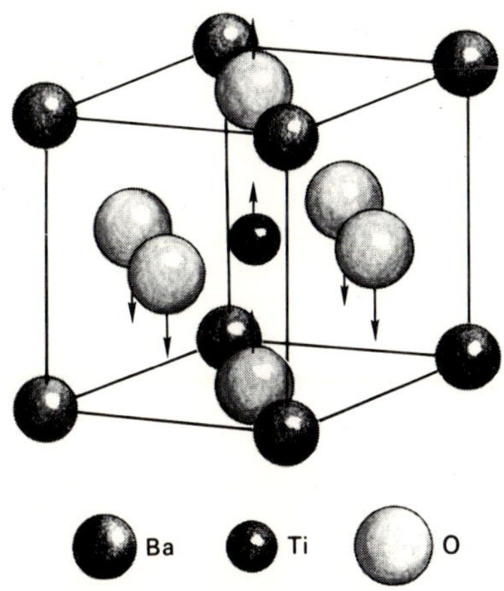

FIG. 7.4c The barium titanate structure

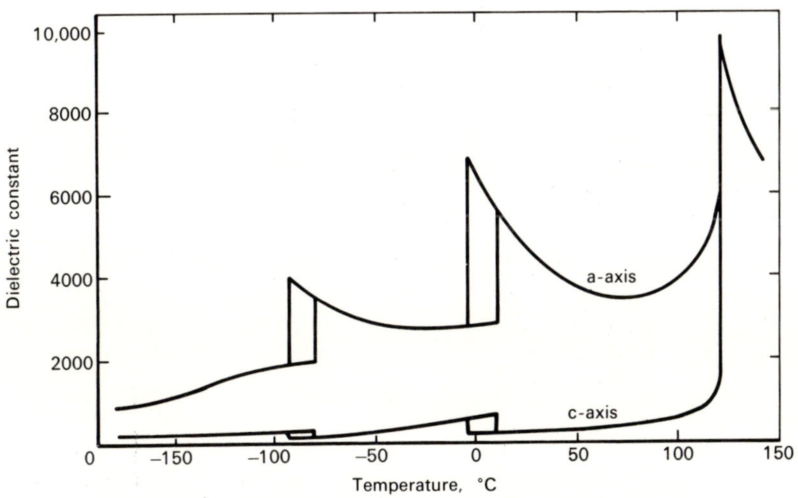

FIG. 7.4d Dielectric constant of $BaTiO_3$ as a function of temperature (after W. J. Merz, *Phys. Rev.* **75**, 687 (1949))

which have been shown to be ferroelectric and their transition temperatures are listed in Table 7.4a.

Other compounds such as the rare earth ferrites have a monoclinic distortion and have to be indexed on an orthorhombic unit cell of approximately $\sqrt{2}a$ for the a- and b-axis and $2a$ for the c-axis where a is the volume of the small perovskite unit cell. The structure of one of these, $GdFeO_3$, is shown in Fig. 7.4e. In this figure, only atoms contained in one unit cell are shown while the layer sequence includes atoms from neighboring cells. The dotted lines represent the outlines of simple perovskite unit cells.

Some compounds containing two different B ions with different valence states such as $Ba(Fe_{0.5}Ta_{0.5})O_3$ also have been shown to have the perovskite structure.

TABLE 7.4a. *Ferroelectric Data*

	T (°C)	P_s (at T°C) (10^{-6}c/cm^2)	Refs.
$BaTiO_3$	120	26.0 (23)	1
$PbTiO_3$	490	750.0 (23)	2
$KNbO_3$	435	0.9 (23) 30 (250)	3
$KTaO_3$	−260		3
$NaNbO_3$	−200		4
$CdTiO_3$	−218		5
$Pb(Cd_{0.5}W_{0.5})O_3$			6
$Pb(Sc_{0.5}Nb_{0.5})O_3$	90	3.6 (18)	7
$Pb(Sc_{0.5}Ta_{0.5})O_3$	26		7
$Pb(Fe_{0.67}W_{0.33})O_3$	−75		8
$Pb(Fe_{0.5}Nb_{0.5})O_3$	112		9
$Pb(Fe_{0.5}Ta_{0.5})O_3$	−30		10
$Pb(Mg_{0.33}Nb_{0.67})O_3$	−12	14 (−130)	11
$Pb(Ni_{0.33}Nb_{0.67})O_3$	−120		11
$Pb(Ni_{0.33}Ta_{0.67})O_3$	−180		11
$Pb(Mg_{0.33}Ta_{0.67})O_3$	−98		11
$Pb(Co_{0.33}Nb_{0.67})O_3$	−70		11
$Pb(Co_{0.33}Ta_{0.67})O_3$	−140		11
$Pb(Zn_{0.33}Nb_{0.67})O_3$	−140		11

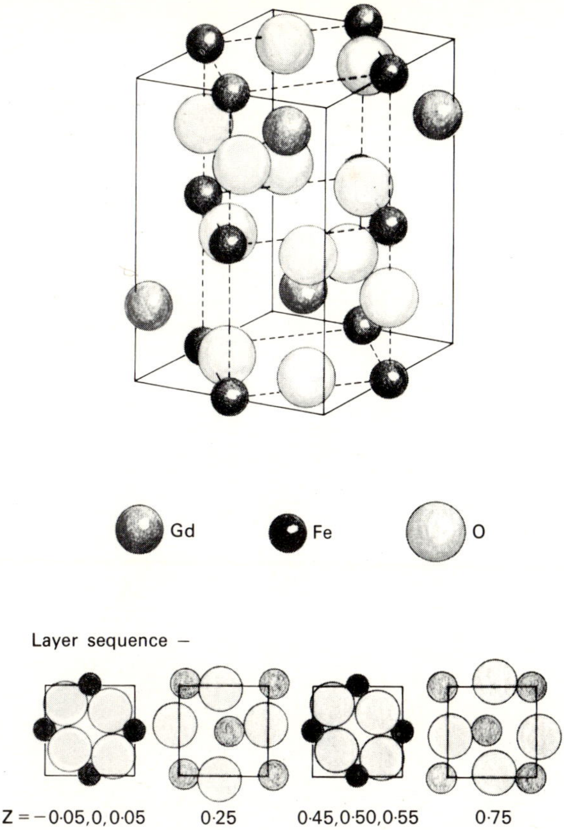

FIG. 7.4e The $GdFeO_3$ structure (two perovskite pseudo cells designated by dotted outline)

Ordered Perovskite Structure, Fm3m

The ordered perovskite structures which are formed by perovskite-type compounds having more than one B element can be described as alternating B′ and B″ atoms at the corners of eight perovskite cubic cells joined together to form a large cubic unit cell (see Fig. 7.4f). The larger the difference in the size of the B′ and B″ atoms or the size and charge of the B′ and B″ ions the greater the chance the compounds will have an ordered structure. The ideal formula for compounds with this structure is $A(B'_{0.5}B''_{0.5})X_3$, but $A(B'_{0.67}B''_{0.33})X_3$ type compounds also form this structure by only partially ordering the B′ and B″ ions.

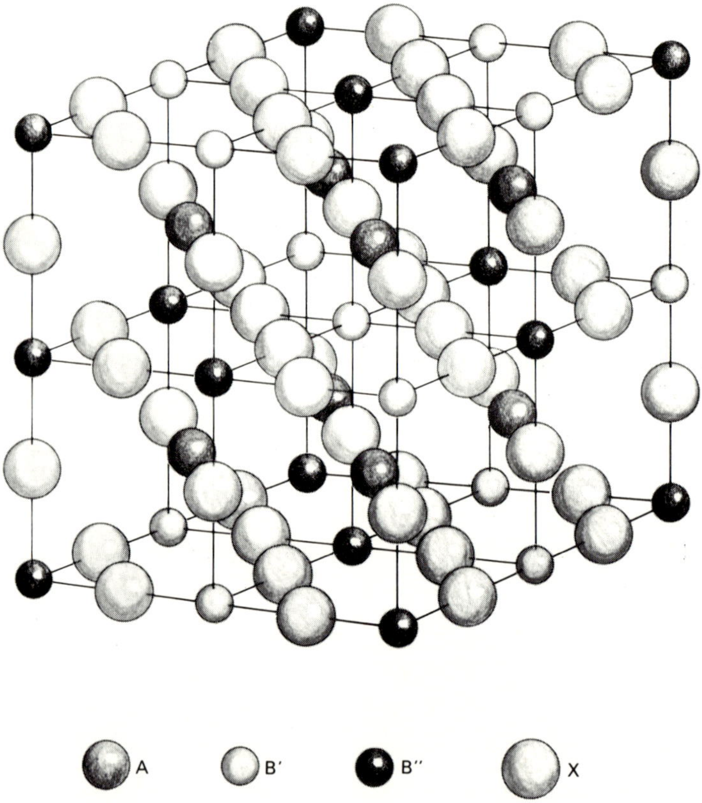

FIG. 7.4f The ordered Perovskite structure $A(B'_{0.5}\,B''_{0.5})X_3$

Properties

The B′—O—B″ bond is ideal for interaction of electrons on the B′ and B″ ions. Thus, many of them are ferromagnetic. Table 7.4b lists these as well as other magnetic compounds with the perovskite structure.

The $Ba(Sr_{0.33}Ta_{0.67})O_3$ *Type Ordered Perovskite Structure*, P3ml

Oxides which contain twice as much of the higher valence state B″ element as the lower valence state B′ element may form the $Ba(Sr_{0.33}Ta_{0.67})O_3$ type structure. This structure can be described by a hexagonal unit cell whose *c*-axis is equivalent to the ⟨111⟩ directions in the original perovskite cell. There are two layers of B″ cations and a layer of B′ cations in each unit cell

TABLE 7.4b. *Ferromagnetic Compounds*

	Curie point (°C)	Refs.
Ba_2FeMoO_6	64	1, 2
Sr_2FeMoO_6	146	1, 2
Ca_2FeMoO_6	104	1
Sr_2CrMoO_6	200	1
Ca_2CrMoO_6	−125	1
Sr_2CrWO_6	180	1
Ca_2CrWO_6	−130	1
Ba_2FeReO_6	43	3
Sr_2FeReO_6	128	3
Ca_2FeReO_6	265	3
Sr_2CrReO_6	magnetic R.T.	4
Ca_2CrReO_6	magnetic R.T.	4
$BiMnO_3$	−170	5
Fe_4N	488	6
Mn_4N	465	6
Fe_3NiN	487	6
Fe_3PtN	369	6

(see Fig. 7.4g). This figure shows the relation of the simple perovskite unit cell, outlined by dotted lines, to the hexagonal unit cell of $Ba(Sr_{0.33}Ta_{0.67})O_3$. One Ba atom is at 0, 0, 0, two more are at $\frac{1}{3}, \frac{2}{3}, z$; $\frac{2}{3}, \frac{1}{3}, \bar{z}$ with $z = \frac{2}{3}$. One

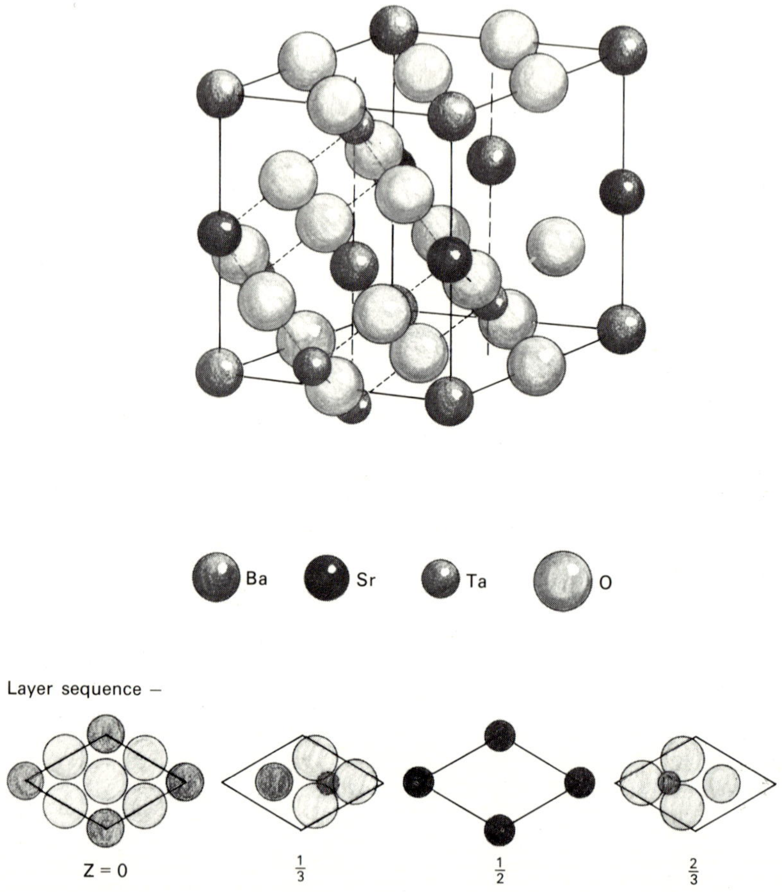

FIG. 7.4g The hexagonal $Ba(Sr_{0.33}Ta_{0.67})O_3$ structure

Sr atom is at 0, 0, $\frac{1}{2}$ and two Ta atoms are at $\frac{1}{3}, \frac{2}{3}, z$; $\frac{2}{3}, \frac{1}{3}, \bar{z}$ with $z = \frac{1}{6}$. Six O are at $x, \bar{x}, z$; $x, 2x, z$; $2\bar{x}, \bar{x}, z$; $\bar{x}, x, \bar{z}$; $\bar{x}, 2\bar{x}, \bar{z}$; $2x, x, \bar{z}$ with $x = \frac{1}{6}$, $z = \frac{1}{3}$ and 3 O at $\frac{1}{2}, 0, 0$; $0, \frac{1}{2}, 0$; $\frac{1}{2}, \frac{1}{2}, 0$.

A list of perovskite-type compounds and cell data are given in Table 7.4c.

TABLE 7.4c. *Unit Cell Parameters for Perovskite-type Phases*

Phases	Cell size (Å)		Remarks	Refs.
	a_0	c_0		
Carbides				
$AlFe_3C$	3.719			1
$AlMn_3C$	3.869			2
Fe_3SnC	3.85			3
$GaMn_3C$	8.376			1
Mn_3ZnC	3.92			4
Halides				
$CsCaF_3$	4.522			3
$CsCdBr_3$	10.70			5
$CsCdCl_3$	5.20			5
$CsFeF_3$	6.158	14.855	hexagonal	6
$CsGeCl_3$	5.47			5, 7
$CsHgBr_3$	5.77			5
$CsHgCl_3$	5.44	8.72	tetragonal	5
$CsMgF_3$	9.39	8.39	tetragonal	3
$CsPbBr_3$	5.874			
$CsPbCl_3$	5.590	5.630	tetragonal	
$CsZnF_3$	9.90	9.05	tetragonal	3
$KCaF_3$	8.742			3
$KCdF_3$	4.293			8
$KCoF_3$	4.071			9
$KCrF_3$	4.274	4.019	tetragonal	9
$KCuF_3$	4.140	3.926	tetragonal	9
$KFeF_3$	4.120			9
$KMgF_3$	3.973			3
$KMnF_3$	4.182			9
$KNiF_3$	4.012			9
$KZnF_3$	4.055			9
$LiBaF_3$	3.996			3
$NaZnF_3$	7.76	8.75	tetragonal	3
$RbCaF_3$	4.452			3
$RbCoF_3$	4.062			10
$RbFeF_3$	4.174			6
$RbMgF_3$	8.19		$\beta = 98°30'$ monocl.	3
$RbMnF_3$	4.250			11
$RbZnF_3$	8.71	8.01	tetragonal	3
$K(Cr_{0.5}Na_{0.5})F_3$	8.266			12
$K(Fe_{0.5}Na_{0.5})F_3$	8.323			12
$K(Ga_{0.5}Na_{0.5})F_3$	8.246			12
Hydrides				
$LiBaH_3$	4.023			13
$LiEuH_3$	3.796			14
$LiSrH_3$	3.833			13
Nitrides				
Fe_4N	3.795			15
Mn_4N	3.857			15
Fe_3NiN	3.790			15
Fe_3PtN	3.857			15

TABLE 7.4c *(cont.)*

$A^{1+}B^{5+}O_3$

Compound	a (Å)	b (Å)	c (Å)	Remarks	Refs.
Oxides					
		$A^{1+}B^{5+}O_3$			
$AgNbO_3$	7.888	15.660	7.888	$\beta = 90.57°$	16
$AgTaO_3$	3.931	3.914	3.931	$\beta = 90.35°$	16
$CsIO_3$	9.324 or 4.66			cubic	17, 18
KIO_3	8.92 or 4.410			cubic $\alpha = 80°24'$ rhombohedral	18, 19, 20, 21, 22
$KNbO_3$	3.9714	5.6946	5.7203	orthorhombic	23
$KTaO_3$	3.9885			cubic	18, 24, 25
$NaNbO_3$	5.512	5.577	3.885	orthorhombic	23
$NaTaO_3$	3.8851	5.4778	5.5239	orthorhombic	27
$RbIO_3$	4.52 or 9.04			cubic	18, 28
$TlIO_3$	4.510			$\alpha = 89.34°$ rhombohedral	19, 29, 30
		$A^{2+}B^{4+}O_3$			
$BaCeO_3$	4.397				20, 31, 23, 32
$BaFeO_3$	3.98		4.01	tetragonal	33, 34
$BaMoO_3$	4.0404				35
$BaPbO_3$	4.273				36
$BaPrO_3$	8.708 or 4.354				18, 37
$BaPuO_3$	4.39				38
$BaSnO_3$	4.117				39, 20, 31
$BaThO_3$	4.480 or 8.985			monoclinic	20, 23, 32, 37
$BaTiO_3$	3.989		4.029	tetragonal	31
$BaUO_3$	4.387			pseudocubic	40
$BaZrO_3$	4.192			cubic	31
$CaCeO_3$	7.70			cubic	18, 41
$CaHfO_3$				orthorhombic	42
$CaMnO_3$	10.683	7.449	10.476	orthorhombic	43, 44, 45
$CaMoO_3$	7.80	7.77	7.80	$\beta = 91°23'$ monoclinic	46
$CaSnO_3$	5.518	7.884	5.664	orthorhombic	20, 47
$CaThO_3$	8.74			monoclinic, pseudocubic	18
$CaTiO_3$	5.381	7.645	5.443	orthorhombic	47
$CaUO_3$	5.78	8.29	5.97	orthorhombic	40
$CaVO_3$	5.326	7.547	5.352	orthorhombic	45
$CaZrO_3$	5.587	8.008	5.758	orthorhombic	48
$CdCeO_3$	7.65			orthorhombic or cubic	18, 41
$CdSnO_3$	5.547	5.577	7.867	orthorhombic	31
$CdThO_3$	8.74			pseudocubic	18
$CdTiO_3$	5.301	7.606	5.419	orthorhombic	31
$CdZrO_3$				orthorhombic	18
$EuTiO_3$	3.897				49
$MgCeO_3$	8.54				18
$PbCeO_3$	7.62			orthorhombic	18, 41
$PbHfO_3$				pseudotetragonal	50
$PbSnO_3$	7.86		8.13	tetragonal	18
$PbTiO_3$	3.896		4.136	tetragonal	51
$PbZrO_3$	9.28			pseudocubic, orthorhombic	39
$SrCeO_3$	5.986	8.531	6.125	orthorhombic	20, 31, 18
$SrCoO_3$	7.725				43
$SrFeO_3$	3.869				43
$SrHfO_3$	4.069 or 8.138			orthorhombic	18
$SrMoO_3$	3.9751				35

TABLE 7.4c *(cont.)*

Compound	a(Å)	b(Å)	c(Å)	Remarks	Refs.
$SrPbO_3$	5.864	5.949	8.336	orthorhombic	36
$SrRuO_3$				cubic	52
$SrSnO_3$	4.0334 or 8.070			cubic	39, 15, 53
$SrThO_3$	8.84			pseudocubic	18
$SrTiO_3$	3.904			cubic	31
$SrUO_3$	6.01	8.60	6.17	orthorhombic	40
$SrZrO_3$	5.792 or 8.218	8.189	5.818	orthorhombic cubic	19, 31, 18
		$A^{3+}B^{3+}O_3$			
$BiAlO_3$	7.61		7.94	tetragonal	18
$BiCrO_3$	3.90	3.87	3.90	$\alpha = \gamma = 90° 35'$ triclinic $\beta = 89° 10'$	54
$BiMnO_3$	3.93	3.98	3.93	$\alpha = \gamma = 91° 25'$ triclinic $\beta = 90° 55'$	54
$CeAlO_3$	3.767		3.794	tetragonal	55, 56
$CeCrO_3$	3.866				56, 57, 58
$CeFeO_3$	3.900			pseudocubic, orthorhombic	56, 57
$CeGaO_3$	3.879			cubic, orthorhombic	57
$CeScO_3$				orthorhombic	57
$CeVO_3$	3.90				58, 59
$CrBiO_3$	7.77		8.08	tetragonal	18
$DyAlO_3$	5.21	5.31	7.40	orthorhombic	60
$DyFeO_3$	5.30	5.60	7.62	orthorhombic	60
$DyMnO_3$	3.70			cubic	61
$EuAlO_3$	5.271	5.292	7.458	$GdFeO_3$ strucutre	56, 62
$EuCrO_3$	3.803				56
$EuFeO_3$	5.371	5.611	7.686	$GdFeO_3$ structure	63, 56, 62
$FeBiO_3$	7.64 or 3·965			$\alpha = 89° 28'$ rhombohedral	64, 65, 66
$GdAlO_3$	5.247	5.304	7.447	$GdFeO_3$ structure	56, 60, 62
$GdCoO_3$	3.732	3.807	3.676	orthorhombic	67
$GdCrO_3$	5.312	5.514	7.611	$GdFeO_3$ structure	68
$GdFeO_3$	5.346	5.616	7.668	orthorhombic	63, 69
$GdMnO_3$	3.82				44
$GdScO_3$	5.487	5.756	7.925	$GdFeO_3$ structure	68
$GdVO_3$	5.345	5.623	7.638	$GdFeO_3$ structure	68
$LaAlO_3$	3.788			$\alpha = 90° 4'$ rhombohedral	31
$LaCoO_3$	3.824 or 7.651			$\alpha = 90° 42'$ rhombohedral	43, 56, 58, 70, 71, 72
$LaCrO_3$	5.477	5.514	7.755	$GdFeO_3$ structure	68
$LaFeO_3$	5.556	5.565	7.862	$GdFeO_3$ structure	31. 63
$LaGaO_3$	5.496	5.524	7.787	$GdFeO_3$ structure	68
$LaInO_3$	5.723	8.207	5.914	orthorhombic	32, 57
$LaNiO_3$	7.676			$\alpha = 90° 41'$ pseudocubic	72, 73
$LaRhO_3$	3.94				70, 74
$LaScO_3$	5.678	5.787	8.098	$GdFeO_3$ structure	26, 68
$LaTiO_3$	3.92				75, 76
$LaVO_3$	3.99 or 7.842			cubic	58, 59, 77
$LaYO_3$				orthorhombic	78
$NdAlO_3$	3.752			rhombohedral	56, 60, 62
$NdCoO_3$	3.777				58, 67
$NdCrO_3$	5.412	5.494	7.695	$GdFeO_3$ structure	68
$NdFeO_3$	5.441	5.573	7.753	$GdFeO_3$ structure	63, 57

TABLE 7.4c *(cont.)*

Compound	a (Å)	b (Å)	c (Å)	Remarks	Refs.
$NdGaO_3$	5.426	5.502	7.706	$GdFeO_3$ structure	68
$NdInO_3$	6.627	8.121	5.891	orthorhombic	31, 57
$NdMnO_3$	3.80				61
$NdScO_3$	5.574	5.771	7.998	$GdFeO_3$ structure	57, 68
$NdVO_3$	5.440	5.589	7.733	$GdFeO_3$ structure	68
$PrAlO_3$	3.757 or 5.31			$\alpha = 60° 20'$ rhombohedral	38, 56, 62
$PrCoO_3$	3.787			$\alpha = 90° 13'$ rhombohedral	58, 67
$PrCrO_3$	5.444	5.484	7.710	$GdFeO_3$ structure	68
$PrFeO_3$	5.495	5.578	7.810	$GdFeO_3$ structure	68
$PrGaO_3$	5.465	5.495	7.729	$GdFeO_3$ structure	68
$PrMnO_3$	3.82				61
$PrScO_3$	5.615	5.776	8.027	$GdFeO_3$ structure	68
$PrVO_3$	5.477	5.545	7.759	$GdFeO_3$ structure	68
$PuAlO_3$	5.33			$\alpha = 56° 4'$ rhombohedral	38
$PuCrO_3$	5.46	5.51	7.76	$GdFeO_3$ structure	38
$PuMnO_3$	3.86			pseudocubic	38
$PuVO_3$	5.48	5.61	7.78	$GdFeO_3$ structure	38
$SmAlO_3$	5.285	5.290	7.473	$GdFeO_3$ structure	57, 56, 62, 67
$SmCoO_3$	3.747	3.803	3.728	orthorhombic	58, 79
$SmCrO_3$	5.372	5.502	7.650	$GdFeO_3$ structure	68
$SmFeO_3$	5.394	5.592	7.711	$GdFeO_3$ structure	63, 26
$SmInO_3$	5.589	8.802	5.886	orthorhombic	31, 26
$SmVO_3$	3.89				58
$YAlO_3$	5.179	5.329	7.370	$GdFeO_3$ structure	63
$YCrO_3$	5.247	5.518	7.540	$GdFeO_3$ structure	63, 79, 57, 68
$YFeO_3$	5.302	5.589	7.622	$GdFeO_3$ structure	58
$YScO_3$	5.431	5.712	7.894	$GdFeO_3$ structure	68
		A_xBO_3 and ABO_{3-x}			
$Ce_{0.33}NbO_3$	3.89	3.91	7.86	orthorhombic	80
$Ce_{0.33}TaO_3$	3.90	3.91	7.86	orthorhombic	80
$Dy_{0.33}TaO_3$	3.83	3.83	7.75	$\gamma = 90.8°$ monoclinic	80
$Gd_{0.33}TaO_3$	3.87	3.89	7.73	orthorhombic	80
$La_{0.33}NbO_3$	3.91		7.90	tetragonal	80
$La_{0.33}TaO_3$	3.92		7.88	tetragonal	80
$Nd_{0.33}NbO_3$	3.90	3.91	7.76	orthorhombic	80
$Nd_{0.33}TaO_3$	3.91		7.77	tetragonal	80
$Pr_{0.33}NbO_3$	3.91	3.92	7.77	orthorhombic	80
$Pr_{0.33}TaO_3$	3.91	3.92	7.78	orthorhombic	80
$Sm_{0.33}TaO_3$	3.89		7.75	tetragonal	80
$Y_{0.33}TaO_3$	3.82	3.83	7.74	$\gamma = 90.9°$ monoclinic	80
$Yb_{0.33}TaO_3$	3.79	3.80	7.70	$\gamma = 91.6°$ monoclinic	80
$Ca_{0.5}TaO_3$	11.068	7.505	5.378	orthorhombic	81
Li_xWO_3	($x = 1$) 3.72			cubic $x = 0.35$–0.57	82
Na_xWO_3	($x = 1$) 3.8622			cubic $x = 0.7$–1.0	83, 41, 84, 85, 86, 87, 88, 89
$Sr_{0.5+x}Nb^{4+}_{2x}Nb^{5+}_{1-2x}O_3$	($x = 0.2$) 3.981			cubic	90
	($x = 0.45$) 4.016			$x = 0.7$–0.9	90
$CaMnO_{3-x}$					43, 44
$SrCoO_{3-x}$					43
$SrFeO_{3-x}$					91, 92, 43
$SrTiO_{3-x}$					77
$SrVO_{3-x}$					77

TABLE 7.4c *(cont.)*

Compound	*a* (Å)	*b* (Å)	*c* (Å)	Remarks	Refs.
		$A(B'_{0.67}B''_{0.33})O_3$			
$Ba(Al_{0.67}W_{0.33})O_3$					93
$Ba(Dy_{0.67}W_{0.33})O_3$	8.386			$(NH_4)_3FeF_6$ structure	93
$Ba(Er_{0.67}W_{0.33})O_3$	8.386			$(NH_4)_3FeF_6$ structure	93
$Ba(Eu_{0.67}W_{0.33})O_3$	8.605			$(NH_4)_3FeF_6$ structure	93
$Ba(Fe_{0.67}U_{0.33})O_3$	8.232			$(NH_4)_3FeF_6$ structure	94
$Ba(Gd_{0.67}W_{0.33})O_3$	8.411			$(NH_4)_3FeF_6$ structure	93, 95
$Ba(In_{0.67}U_{0.33})O_3$	8.512			$(NH_4)_3FeF_6$ structure	94
$Ba(In_{0.67}W_{0.33})O_3$	8.321			ordered structure	93
$Ba(La_{0.67}W_{0.33})O_3$	8.58			$(NH_4)_3FeF_6$ structure	95
$Ba(Lu_{0.67}W_{0.33})O_3$					93
$Ba(Nd_{0.67}W_{0.33})O_3$	8.513			$(NH_4)_3FeF_6$ structure	93
$Ba(Sc_{0.67}U_{0.33})O_3$	8.49			$(NH_4)_3FeF_6$ structure	94
$Ba(Sc_{0.67}W_{0.33})O_3$	8.24			$(NH_4)_3FeF_6$ structure	96, 93
$Ba(Y_{0.67}U_{0.33})O_3$	8.70			$(NH_4)_3FeF_6$ structure	94
$Ba(Y_{0.67}W_{0.33})O_3$	8.374			$(NH_4)_3FeF_6$ structure	93
$Ba(Yb_{0.67}W_{0.33})O_3$					93
$Pb(Fe_{0.67}W_{0.33})O_3$					97, 98
$Sr(Cr_{0.67}Re_{0.33})O_3$	8.01			$(NH_4)_3FeF_6$ structure	99
$Sr(Cr_{0.67}U_{0.33})O_3$	8.00			$(NH_4)_3FeF_6$ structure	94
$Sr(Fe_{0.67}Re_{0.33})O_3$	7.89			$(NH_4)_3FeF_6$ structure	99
$Sr(Fe_{0.67}W_{0.33})O_3$	3.945		3.951	tetragonal	37, 95
$Sr(In_{0.67}Re_{0.33})O_3$	8.297			$(NH_4)_3FeF_6$ structure	99
$La(Co_{0.67}Nb_{0.33})O_3$	5.58	5.58	7.89	orthorhombic	95
$La(Co_{0.67}Sb_{0.33})O_3$	5.57	5.57	7.87	orthorhombic	95
		$A(B'_{0.33}B''_{0.67})O_3$			
$Ba(Ca_{0.33}Nb_{0.67})O_3$	5.92		7.25	hexagonal ordered $Ba(Sr_{0.33}Ta_{0.67})O_3$ structure	100
$Ba(Ca_{0.33}Ta_{0.67})O_3$	5.895		7.284	hexagonal ordered $Ba(Sr_{0.33}Ta_{0.67})O_3$ structure	100, 102, 103
$Ba(Cd_{0.33}Nb_{0.67})O_3$	4.168				100
$Ba(Cd_{0.33}Ta_{0.67})O_3$	4.167				104
$Ba(Co_{0.33}Nb_{0.67})O_3$	4.09				105
$Ba(Co_{0.33}Ta_{0.67})O_3$	5.776		7.082	hexagonal ordered $Ba(Sr_{0.33}Ta_{0.67})O_3$ structure	106, 104
$Ba(Cu_{0.33}Nb_{0.67})O_3$	8.04		8.40	tetragonal	95
$Ba(Fe_{0.33}Nb_{0.67})O_3$	4.085				100
$Ba(Fe_{0.33}Ta_{0.67})O_3$	4.10				105
$Ba(Mg_{0.33}Nb_{0.67})O_3$	5.77		7.08	hexagonal ordered $Ba(Sr_{0.33}Ta_{0.67})O_3$ structure	105, 100, 95
$Ba(Mg_{0.33}Ta_{0.67})O_3$	5.782		7.067	hexagonal ordered $Ba(Sr_{0.33}Ta_{0.67})O_3$ structure	104, 96
$Ba(Mn_{0.33}Nb_{0.67})O_3$					93
$Ba(Mn_{0.33}Ta_{0.67})O_3$	5.819		7.127	hexagonal ordered $Ba(Sr_{0.33}Ta_{0.67})O_3$ structure	104
$Ba(Ni_{0.33}Nb_{0.67})O_3$	4.074				106, 100, 98
$Ba(Ni_{0.33}Ta_{0.67})O_3$	5.758		7.052	hexagonal ordered $Ba(Sr_{0.33}Ta_{0.67})O_3$ structure	106, 104, 103
$Ba(Pb_{0.33}Nb_{0.67})O_3$	4.26				100
$Ba(Pb_{0.33}Ta_{0.67})O_3$	4.25				104
$Ba(Sr_{0.33}Ta_{0.67})O_3$	5.95		7.47	hexagonal ordered $Ba(Sr_{0.33}Ta_{0.67})O_3$ structure	105, 101
$Ba(Zn_{0.33}Nb_{0.67})O_3$	4.094				105, 100, 98
$Ba(Zn_{0.33}Ta_{0.67})O_3$	5.782		7.097	hexagonal ordered $Ba(Sr_{0.33}Ta_{0.67})O_3$ structure	105, 104, 103
$Ca(Ni_{0.33}Nb_{0.67})O_3$	3.88				98
$Ca(Ni_{0.33}Ta_{0.67})O_3$	3.93				106

TABLE 7.4c *(cont.)*

Compound	*a* (Å)	*b* (Å)	*c* (Å)	Remarks	Refs.
$Pb(Co_{0.33}Nb_{0.67})O_3$	4.04				98, 107
$Pb(Co_{0.33}Ta_{0.67})O_3$	4.01				107
$Pb(Mg_{0.33}Nb_{0.67})O_3$	4.041				98, 108
$Pb(Mg_{0.33}Ta_{0.67})O_3$	4.02				98, 107
$Pb(Mn_{0.33}Nb_{0.67})O_3$					98
$Pb(Ni_{0.33}Nb_{0.67})O_3$	4.025				98, 108
$Pb(Ni_{0.33}Ta_{0.67})O_3$	4.01				98, 107
$Pb(Zn_{0.33}Nb_{0.67})O_3$	4.04				107
$Sr(Ca_{0.33}Nb_{0.67})O_3$	5.76		7.16	hexagonal ordered $Ba(Sr_{0.33}Ta_{0.67})O_3$ structure	100
$Sr(Ca_{0.33}Sb_{0.67})O_3$	8.17			$(NH_4)_3FeF_6$ structure	96
$Sr(Ca_{0.33}Ta_{0.67})O_3$	5.764		7.096	hexagonal ordered $Ba(Sr_{0.33}Ta_{0.67})O_3$ structure	104
$Sr(Cd_{0.33}Nb_{0.67})O_3$	4.089				100
$Sr(Co_{0.33}Nb_{0.67})O_3$	8.01			$(NH_4)_3FeF_6$ structure	95
$Sr(Co_{0.33}Sb_{0.67})O_3$	7.99			$(NH_4)_3FeF_6$ structure	95
$Sr(Co_{0.33}Ta_{0.67})O_3$	5.630		6.937	hexagonal ordered $Ba(Sr_{0.33}Ta_{0.67})O_3$ structure	105, 104
$Sr(Cu_{0.33}Sb_{0.67})O_3$	7.84		8.19	tetragonal	95
$Sr(Fe_{0.33}Nb_{0.67})O_3$	3.997		4.018	tetragonal	100
$Sr(Mg_{0.33}Nb_{0.67})O_3$	5.66		6.98	hexagonal ordered $Ba(Sr_{0.33}Ta_{0.67})O_3$ structure	100
$Sr(Mg_{0.33}Sb_{0.67})O_3$	7.96			$(NH_4)_3FeF_6$ structure	95
$Sr(Mg_{0.33}Ta_{0.67})O_3$	5.652		6.951	hexagonal ordered $Ba(Sr_{0.33}Ta_{0.67})O_3$ structure	106, 104
$Sr(Mn_{0.33}Nb_{0.67})O_3$					93
$Sr(Mn_{0.33}Ta_{0.67})O_3$					93
$Sr(Ni_{0.33}Nb_{0.67})O_3$	5.64		6.90	hexagonal ordered $Ba(Sr_{0.33}Ta_{0.67})O_3$ structure	100, 98
$Sr(Ni_{0.33}Ta_{0.67})O_3$	5.607		6.923	hexagonal ordered $Ba(Sr_{0.33}Ta_{0.67})O_3$ structure	105, 104
$Sr(Pb_{0.33}Nb_{0.67})O_3$					93
$Sr(Pb_{0.33}Ta_{0.67})O_3$					93
$Sr(Zn_{0.33}Nb_{0.67})O_3$	5.66		6.95	hexagonal ordered $Ba(Sr_{0.33}Ta_{0.67})O_3$ structure	105, 100
$Sr(Zn_{0.33}Ta_{0.67})O_3$	5.664		6.951	hexagonal ordered $Ba(Sr_{0.33}Ta_{0.67})O_3$ structure	105, 104
			$A^{2+}(B^{3+}_{0.5}B^{5+}_{0.5})O_3$		
$Ba(Bi_{0.5}Nb_{0.5})O_3$	8.630			$(NH_4)_3FeF_6$ structure	109
$Ba(Bi_{0.5}Ta_{0.5})O_3$	8.568			$(NH_4)_3FeF_6$ structure	109
$Ba(Ce_{0.5}Nb_{0.5})O_3$	4.293				110
$Ba(Ce_{0.5}Pa_{0.5})O_3$	8.800			$(NH_4)_3FeF_6$ structure	111
$Ba(Co_{0.5}Nb_{0.5})O_3$	4.06			$(NH_4)_3FeF_6$ structure	95
$Ba(Co_{0.5}Re_{0.5})O_3$	8.086			$(NH_4)_3FeF_6$ structure	99
$Ba(Cr_{0.5}U_{0.5})O_3$					112
$Ba(Cu_{0.5}W_{0.5})O_3$	7.88		8.61	tetragonal	95
$Ba(Dy_{0.5}Nb_{0.5})O_3$	8.437			$(NH_4)_3FeF_6$ structure	113, 110
$Ba(Dy_{0.5}Pa_{0.5})O_3$	8.740			$(NH_4)_3FeF_6$ structure	111
$Ba(Dy_{0.5}Ta_{0.5})O_3$	8.545			$(NH_4)_3FeF_6$ structure	114
$Ba(Er_{0.5}Nb_{0.5})O_3$	8.427			$(NH_4)_3FeF_6$ structure	113, 110
$Ba(Er_{0.5}Pa_{0.5})O_3$	8.716			$(NH_4)_3FeF_6$ structure	111
$Ba(Er_{0.5}Re_{0.5})O_3$	8.354			$(NH_4)_3FeF_6$ structure	99
$Ba(Er_{0.5}Ta_{0.5})O_3$	8.423			$(NH_4)_3FeF_6$ structure	114
$Ba(Er_{0.5}U_{0.5})O_3$	8.67			$(NH_4)_3FeF_6$ structure	112
$Ba(Eu_{0.5}Nb_{0.5})O_3$	8.507			$(NH_4)_3FeF_6$ structure	113, 110

TABLE 7.4c *(cont.)*

Compound	a (Å)	b (Å)	c (Å)	Remarks	Refs.
$Ba(Eu_{0.5}Pa_{0.5})O_3$	8.783			$(NH_4)_3FeF_6$ structure	111
$Ba(Eu_{0.5}Ta_{0.5})O_3$	8.506				114
$Ba(Fe_{0.5}Mo_{0.5})O_3$	8.08			$(NH_4)_3FeF_6$ structure	115
$Ba(Fe_{0.5}Nb_{0.5})O_3$	4.06				105, 113, 116
$Ba(Fe_{0.5}Re_{0.5})O_3$	8.05			$(NH_4)_3FeF_6$ structure	99
$Ba(Fe_{0.5}Ta_{0.5})O_3$	4.056				105, 111
$Ba(Gd_{0.5}Nb_{0.5})O_3$	8.496			$(NH_4)_3FeF_6$ structure	113, 110
$Ba(Gd_{0.5}Pa_{0.5})O_3$	8.774			$(NH_4)_3FeF_6$ structure	111
$Ba(Gd_{0.5}Re_{0.5})O_3$	8.431			$(NH_4)_3FeF_6$ structure	99
$Ba(Gd_{0.5}Sb_{0.5})O_3$	8.44			$(NH_4)_3FeF_6$ structure	95
$Ba(Gd_{0.5}Ta_{0.5})O_3$	8.487		8.513	tetragonal	114
$Ba(Ho_{0.5}Nb_{0.5})O_3$	8.434			$(NH_4)_3FeF_6$ structure	113, 110
$Ba(Ho_{0.5}Pa_{0.5})O_3$	8.730			$(NH_4)_3FeF_6$ structure	111
$Ba(Ho_{0.5}Ta_{0.5})O_3$	8.442			$(NH_4)_3FeF_6$ structure	114
$Ba(In_{0.5}Nb_{0.5})O_3$	8.279			$(NH_4)_3FeF_6$ structure	113, 117
$Ba(In_{0.5}Os_{0.5})O_3$	8.224			$(NH_4)_3FeF_6$ structure	99
$Ba(In_{0.5}Pa_{0.5})O_3$	8.596			$(NH_4)_3FeF_6$ structure	111
$Ba(In_{0.5}Re_{0.5})O_3$	8.258			$(NH_4)_3FeF_6$ structure	99
$Ba(In_{0.5}Sb_{0.5})O_3$	8.269			$(NH_4)_3FeF_6$ structure	95, 118
$Ba(In_{0.5}Ta_{0.5})O_3$	8.280			$(NH_4)_3FeF_6$ structure	93
$Ba(In_{0.5}U_{0.5})O_3$	8.52			$(NH_4)_3FeF_6$ structure	112
$Ba(La_{0.5}Nb_{0.5})O_3$	8.607		8.690	tetragonal	113, 110, 116
$Ba(La_{0.5}Pa_{0.5})O_3$	8.885			$(NH_4)_3FeF_6$ structure	111
$Ba(La_{0.5}Re_{0.5})O_3$	8.58			$(NH_4)_3FeF_6$ structure	99
$Ba(La_{0.5}Ta_{0.5})O_3$	8.611	8.639	8.764	$(NH_4)_3FeF_6$ structure	119, 93, 116
$Ba(Lu_{0.5}Nb_{0.5})O_3$	8.364			$(NH_4)_3FeF_6$ structure	113, 110
$Ba(Lu_{0.5}Pa_{0.5})O_3$	8.666			$(NH_4)_3FeF_6$ structure	111
$Ba(Lu_{0.5}Ta_{0.5})O_3$	8.372			$(NH_4)_3FeF_6$ structure	114
$Ba(Mn_{0.5}Nb_{0.5})O_3$	5.083				116
$Ba(Mn_{0.5}Re_{0.5})O_3$	8.18			$(NH_4)_3FeF_6$ structure	99
$Ba(Mn_{0.5}Ta_{0.5})O_3$	4.076				116
$Ba(Nd_{0.5}Nb_{0.5})O_3$	8.540			$(NH_4)_3FeF_6$ structure	113, 97, 116
$Ba(Nd_{0.5}Pa_{0.5})O_3$	8.840			$(NH_4)_3FeF_6$ structure	111
$Ba(Nd_{0.5}Re_{0.5})O_3$	8.51			$(NH_4)_3FeF_6$ structure	99
$Ba(Nd_{0.5}Ta_{0.5})O_3$	8.556				114, 116
$Ba(Ni_{0.5}Nb_{0.5})O_3$	4.1				95
$Ba(Pr_{0.5}Nb_{0.5})O_3$	4.27				110, 116
$Ba(Pr_{0.5}Pa_{0.5})O_3$	8.862			$(NH_4)_3FeF_6$ structure	111
$Ba(Pr_{0.5}Ta_{0.5})O_3$	4.27				118
$Ba(Rh_{0.5}Nb_{0.5})O_3$	8.17			$(NH_4)_3FeF_6$ structure	95
$Ba(Rh_{0.5}U_{0.5})O_3$				hexagonal $BaTiO_3$	112
$Ba(Sc_{0.5}Nb_{0.5})O_3$	4.121				98, 110
$Ba(Sc_{0.5}Os_{0.5})O_3$	8.152			$(NH_4)_3FeF_6$ structure	99
$Ba(Sc_{0.5}Pa_{0.5})O_3$	8.549			$(NH_4)_3FeF_6$ structure	111
$Ba(Sc_{0.5}Re_{0.5})O_3$	8.163			$(NH_4)_3FeF_6$ structure	99
$Ba(Sc_{0.5}Sb_{0.5})O_3$	8.197			$(NH_4)_3FeF_6$ structure	118
$Ba(Sc_{0.5}Ta_{0.5})O_3$	8.222			$(NH_4)_3FeF_6$ structure	93, 98
$Ba(Sc_{0.5}U_{0.5})O_3$	8.49			$(NH_4)_3FeF_6$ structure	112
$Ba(Sm_{0.5}Nb_{0.5})O_3$	8.518			$(NH_4)_3FeF_6$ structure	113, 89, 116
$Ba(Sm_{0.5}Pa_{0.5})O_3$	8.792			$(NH_4)_3FeF_6$ structure	111
$Ba(Sm_{0.5}Ta_{0.5})O_3$	8.519				114, 116
$Ba(Tb_{0.5}Nb_{0.5})O_3$	4.229				110
$Ba(Tb_{0.5}Pa_{0.5})O_3$	8.753			$(NH_4)_3FeF_6$ structure	111
$Ba(Tl_{0.5}Ta_{0.5})O_3$	8.42			$(NH_4)_3FeF_6$ structure	118
$Ba(Tm_{0.5}Nb_{0.5})O_3$	8.408			$(NH_4)_3FeF_6$ structure	113, 110
$Ba(Tm_{0.5}Pa_{0.5})O_3$	8.692			$(NH_4)_3FeF_6$ structure	111

TABLE 7.4c *(cont.)*

Compound	a(Å)	b(Å)	c(Å)	Remarks	Refs.
$Ba(Tm_{0.5}Ta_{0.5})O_3$	8.406			$(NH_4)_3FeF_6$ structure	114
$Ba(Y_{0.5}Nb_{0.5})O_3$	4.200				110, 116
$Ba(Y_{0.5}Pa_{0.5})O_3$	8.718			$(NH_4)_3FeF_6$ structure	111
$Ba(Y_{0.5}Re_{0.5})O_3$	8.372			$(NH_4)_3FeF_6$ structure	99
$Ba(Y_{0.5}Ta_{0.5})O_3$	8.433			$(NH_4)_3FeF_6$ structure	114, 116
$Ba(Y_{0.5}U_{0.5})O_3$	8.69			$(NH_4)_3FeF_6$ structure	112
$Ba(Yb_{0.5}Nb_{0.5})O_3$	8.374			$(NH_4)_3FeF_6$ structure	113, 98, 110
$Ba(Yb_{0.5}Pa_{0.5})O_3$	8.678			$(NH_4)_3FeF_6$ structure	111
$Ba(Yb_{0.5}Ta_{0.5})O_3$	8.390			$(NH_4)_3FeF_6$ structure	98, 114
$Ca(Al_{0.5}Nb_{0.5})O_3$	3.81	3.80	3.81	β = 90° 15′ monoclinic	120
$Ca(Al_{0.5}Ta_{0.5})O_3$	3.81	3.80	3.81	β = 90° 17′ monoclinic	120
$Ca(Co_{0.5}W_{0.5})O_3$	5.60	5.43	7.73	orthorhombic	95
$Ca(Cr_{0.5}Mo_{0.5})O_3$	5.49	7.70	5.36	orthorhombic	115
$Ca(Cr_{0.5}Nb_{0.5})O_3$	3.85	3.85	3.85	β = 90° 47′ monoclinic	120
$Ca(Cr_{0.5}Os_{0.5})O_3$	5.38	7.66	5.47	orthorhombic	99
$Ca(Cr_{0.5}Re_{0.5})O_3$	5.38	7.67	5.47	orthorhombic	99
$Ca(Cr_{0.5}Ta_{0.5})O_3$	3.85	3.85	3.85	β = 90° 45′ monoclinic	120
$Ca(Cr_{0.5}W_{0.5})O_3$	5.47	7.70	5.35	orthorhombic	115
$Ca(Dy_{0.5}Nb_{0.5})O_3$	4.03	4.03	4.03	β = 92° 25′ monoclinic	120
$Ca(Dy_{0.5}Ta_{0.5})O_3$	4.03	4.03	4.03	β = 92° 24′ monoclinic	120
$Ca(Er_{0.5}Nb_{0.5})O_3$	4.02	4.01	4.02	β = 92° 11′ monoclinic	120
$Ca(Er_{0.5}Ta_{0.5})O_3$	4.02	4.01	4.02	β = 92° 10′ monoclinic	120
$Ca(Fe_{0.5}Mo_{0.5})O_3$	5.53	7.73	5.42	orthorhombic	115
$Ca(Fe_{0.5}Nb_{0.5})O_3$	3.89	3.88	3.89	β = 91° 2′ monoclinic	120
$Ca(Fe_{0.5}Sb_{0.5})O_3$	5.54	5.47	7.74	orthorhombic	95
$Ca(Fe_{0.5}Ta_{0.5})O_3$	3.89	3.88	3.89	β = 91° 7′ monoclinic	120
$Ca(Gd_{0.5}Nb_{0.5})O_3$	4.03	4.04	4.03	β = 92° 42′ monoclinic	120
$Ca(Gd_{0.5}Ta_{0.5})O_3$	4.03	4.04	4.05	β = 92° 41′ monoclinic	120
$Ca(Ho_{0.5}Nb_{0.5})O_3$	4.02	4.02	4.02	β = 92° 41′ monoclinic	120
$Ca(Ho_{0.5}Ta_{0.5})O_3$	4.03	4.02	4.03	β = 92° 16′ monoclinic	120
$Ca(In_{0.5}Nb_{0.5})O_3$	3.97	3.95	3.97	β = 91° 53′ monoclinic	120
$Ca(In_{0.5}Ta_{0.5})O_3$	3.97	3.96	3.97	β = 91° 51′ monoclinic	120
$Ca(La_{0.5}Nb_{0.5})O_3$	4.07	4.07	4.07	β = 92° 8′ monoclinic	120
$Ca(La_{0.5}Ta_{0.5})O_3$	4.07	4.07	4.07	β = 92° 9′ monoclinic	120
$Ca(Mn_{0.5}Ta_{0.5})O_3$	3.90	3.87	3.90	β = 91° 9′ monoclinic	120
$Ca(Nd_{0.5}Nb_{0.5})O_3$	4.05	4.05	4.05	β = 92° 28′ monoclinic	120
$Ca(Nd_{0.5}Ta_{0.5})O_3$	4.05	4.05	4.05	β = 92° 25′ monoclinic	120
$Ca(Ni_{0.5}W_{0.5})O_3$	5.55	5.40	7.70	orthorhombic	95
$Ca(Pr_{0.5}Nb_{0.5})O_3$	4.06	4.05	4.06	β = 92° 25′ monoclinic	120
$Ca(Pr_{0.5}Ta_{0.5})O_3$	4.06	4.05	4.06	β = 92° 22′ monoclinic	120
$Ca(Sc_{0.5}Re_{0.5})O_3$	5.49	7.86	5.63	orthorhombic	99
$Ca(Sm_{0.5}Nb_{0.5})O_3$	4.04	4.04	4.04	β = 92° 42′ monoclinic	120
$Ca(Sm_{0.5}Ta_{0.5})O_3$	4.05	4.04	4.05	β = 92° 28′ monoclinic	120
$Ca(Tb_{0.5}Nb_{0.5})O_3$	4.03	4.03	4.03	β = 92° 35′ monoclinic	120
$Ca(Tb_{0.5}Ta_{0.5})O_3$	4.03	4.03	4.03	β = 92° 36′ monoclinic	120
$Ca(Y_{0.5}Nb_{0.5})O_3$	4.03	4.02	4.03	β = 92° 23′ monoclinic	120
$Ca(Y_{0.5}Ta_{0.5})O_3$	4.03	4.02	4.03	β = 92° 23′ monoclinic	120
$Ca(Yb_{0.5}Nb_{0.5})O_3$	4.01	4.00	4.01	β = 92° 0′ monoclinic	120
$Ca(Yb_{0.5}Ta_{0.5})O_3$	4.01	4.00	4.01	β = 92° 3′ monoclinic	120
$Pb(Fe_{0.5}Nb_{0.5})O_3$	4.017				121, 122
$Pb(Fe_{0.5}Ta_{0.5})O_3$	4.011				121
$Pb(In_{0.5}Nb_{0.5})O_3$	4.11				123
$Pb(Ho_{0.5}Nb_{0.5})O_3$	4.160		4.106	monoclinic	123
$Pb(Lu_{0.5}Nb_{0.5})O_3$	4.152		4.098	monoclinic	123
$Pb(Lu_{0.5}Ta_{0.5})O_3$	4.153		4.107	monoclinic	123
$Pb(Sc_{0.5}Nb_{0.5})O_3$	4.078		4.083	tetragonal	121, 124

TABLE 7.4c *(cont.)*

Compound	a(Å)	b(Å)	c(Å)	Remarks	Refs.
$Pb(Sc_{0.5}Ta_{0.5})O_3$	4.072				121, 124
$Pb(Yb_{0.5}Nb_{0.5})O_3$	4.15				98, 123, 124
$Pb(Yb_{0.5}Ta_{0.5})O_3$	4.13				98, 123
$Sr(Co_{0.5}Nb_{0.5})O_3$	3.93				95
$Sr(Co_{0.5}Sb_{0.5})O_3$	7.88			$(NH_4)_3FeF_6$ structure	95
$Sr(Cr_{0.5}Mo_{0.5})O_3$	7.82			$(NH_4)_3FeF_6$ structure	115
$Sr(Cr_{0.5}Nb_{0.5})O_3$	3.9421				95, 117
$Sr(Cr_{0.5}Os_{0.5})O_3$	7.84			$(NH_4)_3FeF_6$ structure	99
$Sr(Cr_{0.5}Re_{0.5})O_3$	7.82			$(NH_4)_3FeF_6$ structure	99
$Sr(Cr_{0.5}Sb_{0.5})O_3$	7.862			$(NH_4)_3FeF_6$ structure	95, 118
$Sr(Cr_{0.5}Ta_{0.5})O_3$	3.94				106
$Sr(Cr_{0.5}W_{0.5})O_3$	7.82			$(NH_4)_3FeF_6$ structure	115
$Sr(Dy_{0.5}Ta_{0.5})O_3$					93
$Sr(Er_{0.5}Ta_{0.5})O_3$					93
$Sr(Eu_{0.5}Ta_{0.5})O_3$					93
$Sr(Fe_{0.5}Mo_{0.5})O_3$	7.89			$(NH_4)_3FeF_6$ structure	115
$Sr(Fe_{0.5}Nb_{0.5})O_3$	3.97				105
$Sr(Fe_{0.5}Sb_{0.5})O_3$	7.916			$(NH_4)_3FeF_6$ structure	95, 118
$Sr(Fe_{0.5}Ta_{0.5})O_3$	3.96		3.981	tetragonal	125
$Sr(Ga_{0.5}Nb_{0.5})O_3$	3.946				106
$Sr(Ga_{0.5}Os_{0.5})O_3$	7.82			$(NH_4)_3FeF_6$ structure	99
$Sr(Ga_{0.5}Re_{0.5})O_3$	7.843			$(NH_4)_3FeF_6$ structure	99
$Sr(Ga_{0.5}Sb_{0.5})O_3$	7.84		7.91	tetragonal	118
$Sr(Gd_{0.5}Ta_{0.5})O_3$					93
$Sr(Ho_{0.5}Ta_{0.5})O_3$					93
$Sr(In_{0.5}Nb_{0.5})O_3$	4.0569				117
$Sr(In_{0.5}Os_{0.5})O_3$	8.06			$(NH_4)_3FeF_6$ structure	99
$Sr(In_{0.5}Re_{0.5})O_3$	8.071			$(NH_4)_3FeF_6$ structure	99
$Sr(In_{0.5}U_{0.5})O_3$	8.33			$(NH_4)_3FeF_6$ structure	112
$Sr(La_{0.5}Ta_{0.5})O_3$	8.27			$(NH_4)_3FeF_6$ structure	119
$Sr(Lu_{0.5}Ta_{0.5})O_3$					93
$Sr(Mn_{0.5}Mo_{0.5})O_3$	7.98			$(NH_4)_3FeF_6$ structure	115
$Sr(Mn_{0.5}Sb_{0.5})O_3$					126
$Sr(Nd_{0.5}Ta_{0.5})O_3$					93
$Sr(Ni_{0.5}Sb_{0.5})O_3$				tetragonal	95
$Sr(Rh_{0.5}Sb_{0.5})O_3$	5.77	5.55	7.99	orthorhombic	95
$Sr(Sc_{0.5}Os_{0.5})O_3$	8.02			$(NH_4)_3FeF_6$ structure	99
$Sr(Sc_{0.5}Re_{0.5})O_3$	8.02			$(NH_4)_3FeF_6$ structure	99
$Sr(Sm_{0.5}Ta_{0.5})O_3$					93
$Sr(Tm_{0.5}Ta_{0.5})O_3$					93
$Sr(Yb_{0.5}Ta_{0.5})O_3$					93
		$A^{2+}(B^{2+}_{0.5}B^{6+}_{0.5})O_3$			
$Ba(Ba_{0.5}Os_{0.5})O_3$	8.66		8.34	tetragonal	99
$Ba(Ba_{0.5}Re_{0.5})O_3$	8.65		8.33	tetragonal	99
$Ba(Ba_{0.5}U_{0.5})O_3$	8.89			$(NH_4)_3FeF_6$ structure	94
$Ba(Ba_{0.5}W_{0.5})O_3$	8.6			$(NH_4)_3FeF_6$ structure	127
$Ba(Ca_{0.5}Mo_{0.5})O_3$	8.355			$(NH_4)_3FeF_6$ structure	127
$Ba(Ca_{0.5}Os_{0.5})O_3$	8.362			$(NH_4)_3FeF_6$ structure	99
$Ba(Ca_{0.5}Re_{0.5})O_3$	8.356			$(NH_4)_3FeF_6$ structure	99, 128
$Ba(Ca_{0.5}Te_{0.5})O_3$	8.393			$(NH_4)_3FeF_6$ structure	118
$Ba(Ca_{0.5}U_{0.5})O_3$	8.67			$(NH_4)_3FeF_6$ structure	94
$Ba(Ca_{0.5}W_{0.5})O_3$	8.39			$(NH_4)_3FeF_6$ structure	96, 127
$Ba(Cd_{0.5}Os_{0.5})O_3$	8.325			$(NH_4)_3FeF_6$ structure	99
$Ba(Cd_{0.5}Re_{0.5})O_3$	8.322			$(NH_4)_3FeF_6$ structure	99, 128
$Ba(Cd_{0.5}U_{0.5})O_3$	6.13	8.64	6.07	orthorhombic	94
$Ba(Co_{0.5}Mo_{0.5})O_3$	4.0429				117
$Ba(Co_{0.5}Re_{0.5})O_3$	8.086			$(NH_4)_3FeF_6$ structure	99, 128

TABLE 7.4c *(cont.)*

Compound	a (Å)	b (Å)	c (Å)	Remarks	Refs.
$Ba(Co_{0.5}U_{0.5})O_3$	8.374			$(NH_4)_3FeF_6$ structure	94
$Ba(Co_{0.5}W_{0.5})O_3$	8.098			$(NH_4)_3FeF_6$ structure	96, 117
$Ba(Cr_{0.5}U_{0.5})O_3$	8.297			$(NH_4)_3FeF_6$ structure	94
$Ba(Cu_{0.5}U_{0.5})O_3$	8.18		8.84	tetragonal	94
$Ba(Fe_{0.5}Re_{0.5})O_3$	8.05			$(NH_4)_3FeF_6$ structure	99, 128
$Ba(Fe_{0.5}U_{0.5})O_3$	8.312			$(NH_4)_3FeF_6$ structure	94
$Ba(Fe_{0.5}W_{0.5})O_3$	8.133			$(NH_4)_3FeF_6$ structure	96
$Ba(Mg_{0.5}Os_{0.5})O_3$	8.08			$(NH_4)_3FeF_6$ structure	99
$Ba(Mg_{0.5}Re_{0.5})O_3$	8.082			$(NH_4)_3FeF_6$ structure	99, 128
$Ba(Mg_{0.5}Te_{0.5})O_3$	8.13			$(NH_4)_3FeF_6$ structure	118, 129
$Ba(Mg_{0.5}U_{0.5})O_3$	8.381			$(NH_4)_3FeF_6$ structure	94
$Ba(Mg_{0.5}W_{0.5})O_3$	8.099			$(NH_4)_3FeF_6$ structure	96, 127
$Ba(Mn_{0.5}Re_{0.5})O_3$	8.18			$(NH_4)_3FeF_6$ structure	99, 128
$Ba(Mn_{0.5}U_{0.5})O_3$	8.52			$(NH_4)_3FeF_6$ structure	94
$Ba(Ni_{0.5}Mo_{0.5})O_3$	4.0225				117
$Ba(Ni_{0.5}Re_{0.5})O_3$	8.04			$(NH_4)_3FeF_6$ structure	99, 128
$Ba(Ni_{0.5}U_{0.5})O_3$	8.336			$(NH_4)_3FeF_6$ structure	94
$Ba(Ni_{0.5}W_{0.5})O_3$	8.066			$(NH_4)_3FeF_6$ structure	96, 117
$Ba(Pb_{0.5}Mo_{0.5})O_3$					96, 117
$Ba(Sr_{0.5}Os_{0.5})O_3$	8.43		8.72	tetragonal	130
$Ba(Sr_{0.5}Re_{0.5})O_3$	8.60		8.29	tetragonal	99
$Ba(Sr_{0.5}U_{0.5})O_3$	8.84			$(NH_4)_3FeF_6$ structure	99, 128
$Ba(Sr_{0.5}W_{0.5})O_3$	8.5			$(NH_4)_3FeF_6$ structure	127
$Ba(Zn_{0.5}Os_{0.5})O_3$	8.095			$(NH_4)_3FeF_6$ structure	99
$Ba(Zn_{0.5}Re_{0.5})O_3$	8.106			$(NH_4)_3FeF_6$ structure	99, 128
$Ba(Zn_{0.5}U_{0.5})O_3$	8.397			$(NH_4)_3FeF_6$ structure	94
$Ba(Zn_{0.5}W_{0.5})O_3$	8.116			$(NH_4)_3FeF_6$ structure	96
$Ca(Ca_{0.5}Os_{0.5})O_3$	5.73	7.87	5.80	orthorhombic	99
$Ca(Ca_{0.5}Re_{0.5})O_3$	5.67	8.05	5.78	orthorhombic	99
$Ca(Ca_{0.5}W_{0.5})O_3$	8.0			$(NH_4)_3FeF_6$ structure	127
$Ca(Cd_{0.5}Re_{0.5})O_3$	5.64	7.99	5.77	orthorhombic	99
$Ca(Co_{0.5}Os_{0.5})O_3$	5.47	7.70	5.59	orthorhombic	99
$Ca(Co_{0.5}Re_{0.5})O_3$	5.46	7.71	5.58	orthorhombic	99
$Ca(Fe_{0.5}Re_{0.5})O_3$	5.41	7.69	5.53	orthorhombic	99, 128
$Ca(Mg_{0.5}Re_{0.5})O_3$	5.48	7.77	5.56	orthorhombic	99
$Ca(Mg_{0.5}W_{0.5})O_3$	7.7			$(NH_4)_3FeF_6$ structure	127
$Ca(Mn_{0.5}Re_{0.5})O_3$	5.52	7.82	5.55	orthorhombic	99
$Ca(Ni_{0.5}Re_{0.5})O_3$	5.45	7.67	5.55	orthorhombic	99
$Ca(Sr_{0.5}W_{0.5})O_3$	8.1			$(NH_4)_3FeF_6$ structure	127
$Pb(Ca_{0.5}W_{0.5})O_3$					98
$Pb(Cd_{0.5}W_{0.5})O_3$	4.150	4.101	4.150	$\beta = 90° 57'$	131
$Pb(Co_{0.5}W_{0.5})O_3$					132
$Pb(Mg_{0.5}Te_{0.5})O_3$	7.99			$(NH_4)_3FeF_6$ structure	129
$Pb(Mg_{0.5}W_{0.5})O_3$	4.0				97, 98
$Sr(Ca_{0.5}Mo_{0.5})O_3$					130
$Sr(Ca_{0.5}Os_{0.5})O_3$	8.21			$(NH_4)_3FeF_6$ structure	99
$Sr(Ca_{0.5}Re_{0.5})O_3$	5.76	8.21	5.85	orthorhombic	99
$Sr(Ca_{0.5}U_{0.5})O_3$	6.06	8.46	5.93	orthorhombic	94
$Sr(Ca_{0.5}W_{0.5})O_3$	8.2			$(NH_4)_3FeF_6$ structure	127
$Sr(Cd_{0.5}Re_{0.5})O_3$	5.73	8.16	5.81	orthorhombic	99
$Sr(Cd_{0.5}U_{0.5})O_3$	6.03	8.42	5.91	orthorhombic	94
$Sr(Co_{0.5}Mo_{0.5})O_3$	3.9367		3.9764		99, 125
$Sr(Co_{0.5}Os_{0.5})O_3$	7.86		7.92	tetragonal	99
$Sr(Co_{0.5}Re_{0.5})O_3$	7.88		7.98	tetragonal	99
$Sr(Co_{0.5}U_{0.5})O_3$	8.19			$(NH_4)_3FeF_6$ structure	94
$Sr(Co_{0.5}W_{0.5})O_3$	7.89		7.98	tetragonal	96, 125

TABLE 7.4c *(cont.)*

Compound	*a* (Å)	*b* (Å)	*c* (Å)	Remarks	Refs.
$Sr(Cr_{0.5}U_{0.5})O_3$	8.09			$(NH_4)_3FeF_6$ structure	94
$Sr(Cu_{0.5}W_{0.5})O_3$	7.66		8.40	tetragonal	95
$Sr(Fe_{0.5}Os_{0.5})O_3$	7.85			$(NH_4)_3FeF_6$ structure	99
$Sr(Fe_{0.5}Re_{0.5})O_3$	7.86		7.89	tetragonal	99, 128
$Sr(Fe_{0.5}U_{0.5})O_3$	8.11			$(NH_4)_3FeF_6$ structure	94
$Sr(Fe_{0.5}W_{0.5})O_3$	7.96			$(NH_4)_3FeF_6$ structure	95
$Sr(Mg_{0.5}Mo_{0.5})O_3$					130
$Sr(Mg_{0.5}Os_{0.5})O_3$	7.86		7.92	tetragonal	99
$Sr(Mg_{0.5}Re_{0.5})O_3$	7.88		7.94	tetragonal	99
$Sr(Mg_{0.5}Te_{0.5})O_3$	7.94			$(NH_4)_3FeF_6$ structure	129
$Sr(Mg_{0.5}U_{0.5})O_3$	8.19			$(NH_4)_3FeF_6$ structure	94
$Sr(Mg_{0.5}W_{0.5})O_3$	7.9			$(NH_4)_3FeF_6$ structure	127
$Sr(Mn_{0.5}Re_{0.5})O_3$	8.01			$(NH_4)_3FeF_6$ structure	99
$Sr(Mn_{0.5}U_{0.5})O_3$	8.28			$(NH_4)_3FeF_6$ structure	94
$Sr(Mn_{0.5}W_{0.5})O_3$	8.01			$(NH_4)_3FeF_6$ structure	95
$Sr(Ni_{0.5}Mo_{0.5})O_3$	3.9237		3.9474	tetragonal	117, 125
$Sr(Ni_{0.5}Re_{0.5})O_3$	7.85		7.92	tetragonal	99
$Sr(Ni_{0.5}U_{0.5})O_3$	8.15			$(NH_4)_3FeF_6$ structure	94
$Sr(Ni_{0.5}W_{0.5})O_3$	7.86		7.91	tetragonal	96, 117, 125
$Sr(Pb_{0.5}Mo_{0.5})O_3$					130
$Sr(Sr_{0.5}Os_{0.5})O_3$	8.32		8.12	tetragonal	99
$Sr(Sr_{0.5}Re_{0.5})O_3$	8.41		8.13	tetragonal	99
$Sr(Sr_{0.5}U_{0.5})O_3$	6.22	8.65	6.01	orthorhombic	94
$Sr(Sr_{0.5}W_{0.5})O_3$	8.2			$(NH_4)_3FeF_6$ structure	127
$Sr(Zn_{0.5}Re_{0.5})O_3$	7.89		8.01	tetragonal	99
$Sr(Zn_{0.5}W_{0.5})O_3$	7.92		8.01	tetragonal	96, 125
		$A^{2+}(B^{1+}_{0.5}B^{7+}_{0.5})O_3$			
$Ba(Ag_{0.5}I_{0.5})O_3$	8.46			$(NH_4)_3FeF_6$ structure	118
$Ba(Li_{0.5}Os_{0.5})O_3$	8.100			$(NH_4)_3FeF_6$ structure	99
$Ba(Li_{0.5}Re_{0.5})O_3$	8.118			$(NH_4)_3FeF_6$ structure	99
$Ba(Na_{0.5}I_{0.5})O_3$	8.33			$(NH_4)_3FeF_6$ structure	99, 118
$Ba(Na_{0.5}Os_{0.5})O_3$	8.281			$(NH_4)_3FeF_6$ structure	99
$Ba(Na_{0.5}Re_{0.5})O_3$	8.296			$(NH_4)_3FeF_6$ structure	99
$Ca(Li_{0.5}Os_{0.5})O_3$	7.83			$(NH_4)_3FeF_6$ structure	99
$Ca(Li_{0.5}Re_{0.5})O_3$	7.83			$(NH_4)_3FeF_6$ structure	99
$Sr(Li_{0.5}Os_{0.5})O_3$	7.86			$(NH_4)_3FeF_6$ structure	99
$Sr(Li_{0.5}Re_{0.5})O_3$	7.87			$(NH_4)_3FeF_6$ structure	99
$Sr(Na_{0.5}Os_{0.5})O_3$	8.13			$(NH_4)_3FeF_6$ structure	99
$Sr(Na_{0.5}Re_{0.5})O_3$	8.13			$(NH_4)_3FeF_6$ structure	99
		$A^{3+}(B^{2+}_{0.5}B^{4+}_{0.5})O_3$			
$La(Co_{0.5}Ir_{0.5})O_3$				orthorhombic	95
$La(Cu_{0.5}Ir_{0.5})O_3$				monoclinic	95
$La(Mg_{0.5}Ge_{0.5})O_3$	3.90				106
$La(Mg_{0.5}Ir_{0.5})O_3$	7.92				93
$La(Mg_{0.5}Nb_{0.5})O_3$					121, 95
$La(Mg_{0.5}Ru_{0.5})O_3$	7·91				121
$La(Mg_{0.5}Ti_{0.5})O_3$	3.932				106, 98
$La(Mn_{0.5}Ir_{0.5})O_3$	7.86			$(NH_4)_3FeF_6$ structure	121
$La(Mn_{0.5}Ru_{0.5})O_3$	7.84				121
$La(Ni_{0.5}Ir_{0.5})O_3$	7.90				121, 95
$La(Ni_{0.5}Ru_{0.5})O_3$	7.90				121
$La(Ni_{0.5}Ti_{0.5})O_3$	3.93				106
$La(Zn_{0.5}Ru_{0.5})O_3$	7.97				121
$Nd(Mg_{0.5}Ti_{0.5})O_3$	3.90				106

TABLE 7.4c *(cont.)*

Compound	a (Å)	b (Å)	c (Å)	Remarks	Refs.
		$A^{2+}(B^{1+}_{0.25}B^{5+}_{0.75})O_3$			
$Ba(Na_{0.25}Ta_{0.75})O_3$	4.137				132
$Sr(Na_{0.25}Ta_{0.75})O_3$	4.055				132
		$A(B^{3+}_{0.5}B^{4+}_{0.5})O_{2.75}$			
$Ba(In_{0.5}U_{0.5})O_{2.75}$	8.551			$(NH_4)_3FeF_6$ structure	112
		$A(B^{2+}_{0.5}B^{5+}_{0.5})O_{2.75}$			
$Ba(Ba_{0.5}Ta_{0.5})O_{2.75}$	8.69			$(NH_4)_3FeF_6$ structure	119, 105
$Ba(Fe_{0.5}Mo_{0.5})O_{2.75}$	8.08			$(NH_4)_3FeF_6$ structure	105
$Sr(Sr_{0.5}Ta_{0.5})O_{2.75}$	8.34			$(NH_4)_3FeF_6$ structure	119, 105

7.5. K_2NiF_4 Structure, *I4/mmm*, Tetragonal

The K_2NiF_4 structure can be derived from one unit cell of the perovskite structure with the B atom at the center combined with two perovskite unit cells with the A atom at the center and with a BX_3 layer removed (see Fig. 7.5). One of the cut cells is placed on the complete B-type unit cell of the perovskite structure and the other is placed below it. In this manner a tetragonal unit cell with $c \cong 3a$, where a is the length of the unit cell edge, is formed. The A atoms are located at 4(*e*) 0, 0, z; $\frac{1}{2}, \frac{1}{2}, \frac{1}{2} + z$; 0, 0, $\bar{z}$; $\frac{1}{2}, \frac{1}{2}, \frac{1}{2} - z$ with $z = 0.35$. The B atoms are at 0, 0, 0 and $\frac{1}{2}, \frac{1}{2}, \frac{1}{2}$, four X atoms are at 0, $\frac{1}{2}$, 0; $\frac{1}{2}$, 0, 0; $\frac{1}{2}$, 0, $\frac{1}{2}$; 0, $\frac{1}{2}, \frac{1}{2}$ and four more X atoms in (4e) with $z = 0.15$. The B atoms have the same environment as they have in the perovskite structure, but the A atoms have only nine X ions around them. A number of oxides, fluorides and oxyfluorides listed in Table 7.5 have this structure.

7.6. $Sr_3Ti_2O_7$ and $Sr_4Ti_3O_{10}$, Tetragonal

The structure of $Sr_3Ti_2O_7$ can be described by first taking two B-type unit cells of the perovskite structure joined together. The BO_3 layer is removed from two pairs of A-type unit cells of the perovskite structure and one pair is placed on top of the B-type pair and the other pair is placed below (see Fig. 7.6).

The details of the structure of $Sr_3Ti_2O_7$ are listed below.

Two Sr atoms are at 0, 0, $\frac{1}{2}$; $\frac{1}{2}, \frac{1}{2}$, 0;
two more are at $\pm(0, 0, z; \frac{1}{2}, \frac{1}{2}, \frac{1}{2} + z)$
with $z = 0.312$;
four Ti atoms are at $\pm$ $(0, 0, z; \frac{1}{2}, \frac{1}{2}, \frac{1}{2} + z)$,
$z = 0.094$;
two O are at 0, 0, 0; $\frac{1}{2}, \frac{1}{2}, \frac{1}{2}$;

eight O are at $\pm$ $(0, \frac{1}{2}, z; \frac{1}{2}, 0, z; \frac{1}{2}, 0, \frac{1}{2} + z; 0, \frac{1}{2}, \frac{1}{2} + z)$
with $z = 0.094$;
and four O are $\pm$ $(0, 0, z; \frac{1}{2}, \frac{1}{2}, \frac{1}{2} + z)$
with $z = 0.188$.

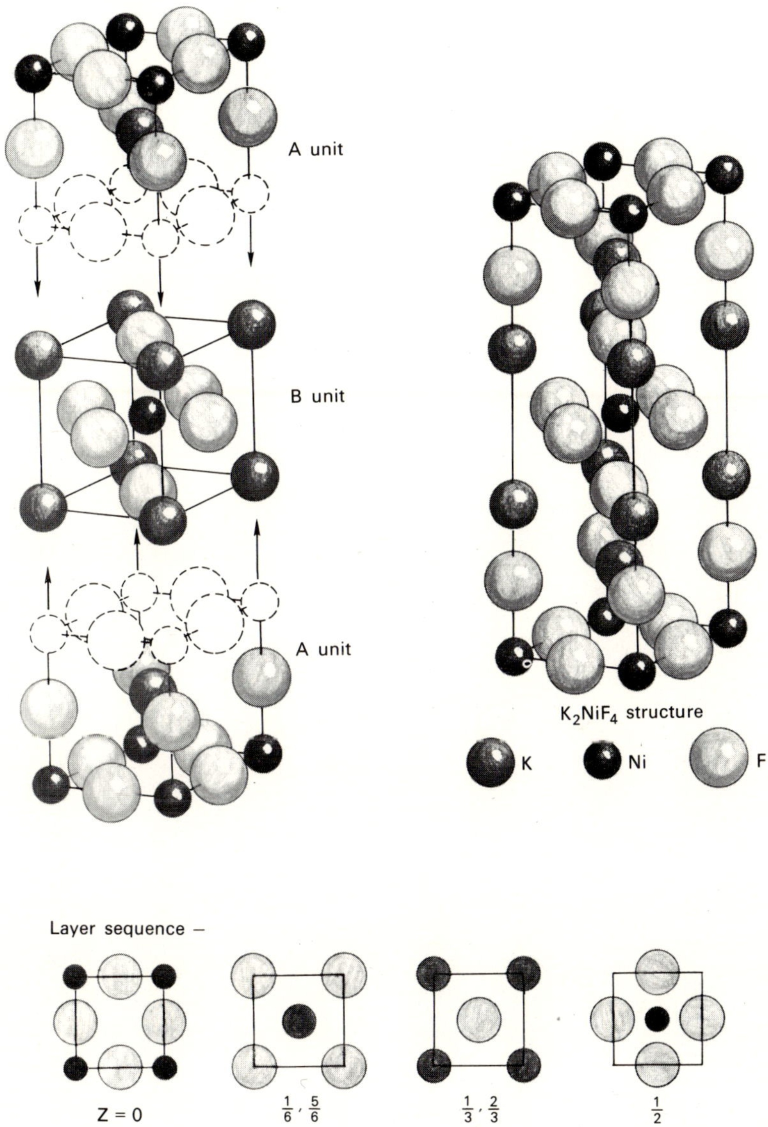

FIG. 7.5 K_2NiF_4 structure

TABLE 7.5. *Phases with the K_2NiF_4 Structure*

Phases	Cell size (Å)		Atomic parameters		Refs.
	a_0	c_0	z(A cation)	z(anion)	
Halides, Oxyhalides					
Cs_2CrCl_4	5.215	16.46			1
K_2CoF_4	4.074	13.08			2
K_2CuF_4	4.155	12.74	0.356	0.153	3
K_2MgF_4	3.977	13.16	0.35	0.15	4
K_2NiF_4	4.01	13.08	0.352	0.151	5
K_2NbO_3F	3.96	13.67			6
K_2ZnF_4	4.017	13.05			7
$(NH_4)_2NiF_4$	4.084	13.79			8
Rb_2NiF_4	4.087	13.71			8
Rb_2ZnF_4	4.104	13.28			7
Sr_2FeO_3F	3.84	12.98			9
Tl_2CoF_4	4.10	14.1			2
Tl_2NiF_4	4.051	14.22			8
Oxides					
Ba_2PbO_4	4.296	13.30	0.355	0.155	10
Ba_2SnO_4	4.130	13.27	0.355	0.155	10
Ca_2MnO_4	3.67	12.08			11
Cs_2UO_4	4.38	14.79			12
Gd_2CuO_4	3.89	11.85			13
K_2UO_4	4.34	13.10	0.36	0.145	14
La_2NiO_4	3.855	12.652	0.360	1.170	15
Nd_2CuO_4	3.94	12.15			13
Nd_2NiO_4	3.81	12.31			13
Rb_2UO_4	4.345	13.83			12
Sm_2CuO_4	3.91	11.93			13
Sr_2IrO_4	3.89	12.92	0.347	0.151	16
Sr_2MnO_4	3.79	12.43			5
Sr_2MoO_4	3.92	12.84			5
Sr_2RhO_4	3.85	12.90			17
Sr_2RuO_4	3.870	12.74			17
Sr_2SnO_4	4.037	12.53	0.353	0.153	10
Sr_2TiO_4	3.884	12.60	0.355	0.152	18
Complex Oxides					
$La_2(Li_{0.5}Co_{0.5})O_4$	3.77	12.58			19
$La_2(Li_{0.5}Ni_{0.5})O_4$	3.75	12.89			19
$SrLaAlO_4$	3.75	12.5			11
$SrLaCoO_4$	3.80	12.50			19
$(Sr_{1.5}La_{0.5})(Co_{0.5}Ti_{0.5})O_4$	3.85	12.62			19
$SrLaCrO_4$	3.84	12.52			19
$(Sr_{0.5}La_{1.5})(Mg_{0.5}Co_{0.5})O_4$	3.82	12.58			19
$SrLaFeO_4$	3.86	12.69			19
$SrLaGaO_4$	3.84	12.71			19
$SrLaMnO_4$	3.88	12.5			19
$SrLaNiO_4$	3.80	12.51			19
$SrLaRhO_4$	3.92	12.78			19

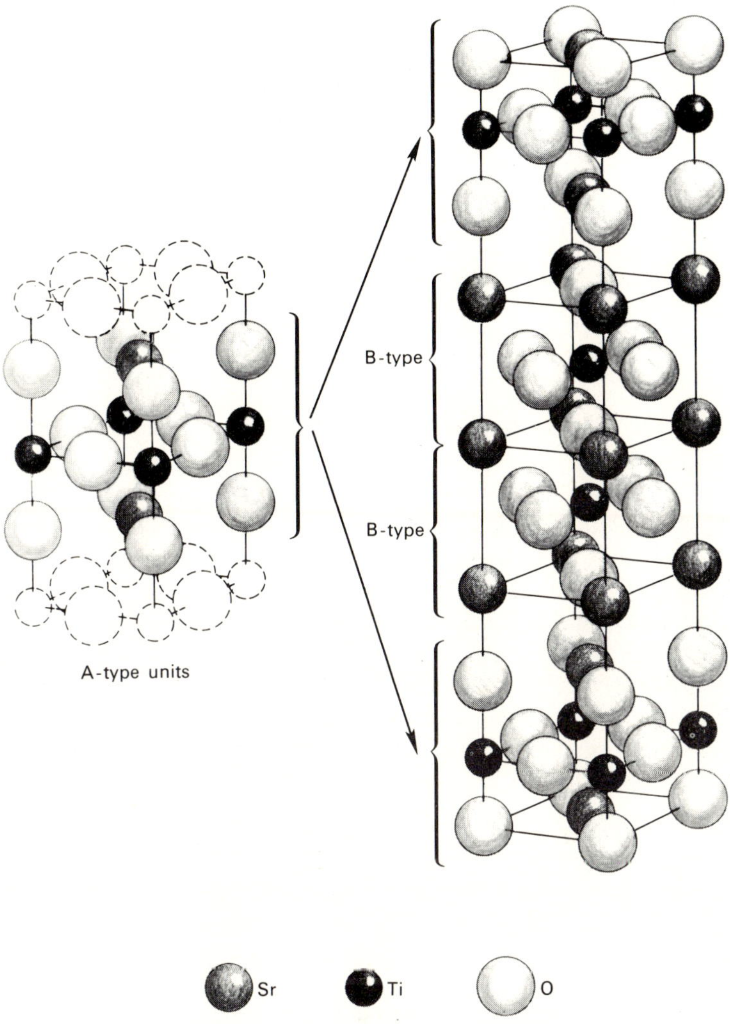

FIG. 7.6 The $Sr_3Ti_2O_7$ structure

The structure of $Sr_4Ti_3O_{10}$ is similar but requires a longer *c*-axis to describe it. The cell sizes are listed in Table 7.6.

TABLE 7.6. *Series of* Sr–Ti–O *Compounds*

Compound	Cell size (Å)		Refs.
	a_0	c_0	
Sr_2TiO_4	3.884	12.60	1
$Sr_3Ti_2O_7$	3.90	20.38	2
$Sr_4Ti_3O_{10}$	3.90	28.1	2
$K_3Zn_2F_7$	4.063	21.22	3

7.7. $Bi_4Ti_3O_{12}$ Structure, *Fmmm*, Orthorhombic

$Bi_4Ti_3O_{12}$ is one of a series of ferroelectric compounds which can be best described by unit cells of the perovskite structure stacked on one another and separated by bismuth oxygen layers. The structures of Bi_2NbO_5F, Bi_3NbTiO_9 and $BaBi_4TiO_{15}$ have been characterized, but in this book, only one, $Bi_4Ti_3O_{12}$, will be described. Half the unit cell consists of one and one-half A type perovskite unit cells with an oxygen layer on top and one and one-half B type perovskite unit cells on top of the oxygen layer. The top half of the cell is the mirror image of this one. The unit cell just described is shown in Fig. 7.7a inside the real cell. The actual *a* and *b* edges are taken as the face diagonals of the small unit cell of the perovskite structure. The layer sequence is shown in Fig. 7.7b. The atomic positions are given below:

		x	*y*	*z*
Bi(1)	(8i)	0	0	0.067
(2)	(8i)	0	0	0.211
Ti(1)	(4b)	0	0	0.50
(2)	(8i)	0	0	0.372
O(1)	(8e)	0.25	0.25	0
(2)	(8f)	0.25	0.25	0.25
(3)	(8i)	0	0	0.436
(4)	(8i)	0	0	0.308
(5)	(16j)	0.25	0.25	0.128

(4b) $0, 0, \frac{1}{2}; \frac{1}{2}, \frac{1}{2}, \frac{1}{2}; \frac{1}{2}, 0, 0; 0, \frac{1}{2}, 0$

(8e) $\frac{1}{4}, \frac{1}{4}, 0; \frac{1}{4}, \frac{1}{4}, \frac{1}{2}$
$\frac{3}{4}, \frac{3}{4}, 0; \frac{3}{4}, \frac{3}{4}, \frac{1}{2}$
$\frac{3}{4}, \frac{1}{4}, \frac{1}{2}; \frac{3}{4}, \frac{1}{4}, 0$
$\frac{1}{4}, \frac{3}{4}, \frac{1}{2}; \frac{1}{4}, \frac{3}{4}, 0$

(8f) $\frac{1}{4}, \frac{1}{4}, \frac{1}{4}; \frac{1}{4}, \frac{1}{4}, \frac{3}{4}$
$\frac{3}{4}, \frac{3}{4}, \frac{1}{4}; \frac{3}{4}, \frac{3}{4}, \frac{3}{4}$
$\frac{3}{4}, \frac{1}{4}, \frac{3}{4}; \frac{3}{4}, \frac{1}{4}, \frac{1}{4}$
$\frac{1}{4}, \frac{3}{4}, \frac{3}{4}; \frac{1}{4}, \frac{3}{4}, \frac{1}{4}$

(8i) $0, 0, z; \frac{1}{2}, \frac{1}{2}, z; \frac{1}{2}, 0, \frac{1}{2} + z; 0, \frac{1}{2}, \frac{1}{2} + z$

$0, 0, z; \frac{1}{2}, \frac{1}{2}, z; \frac{1}{2}, 0, \frac{1}{2} - z; 0, \frac{1}{2}, \frac{1}{2} - z$

(16j) $\frac{1}{4}, \frac{1}{4}, z; \frac{1}{4}, \frac{1}{4}, \bar{z}; \frac{1}{4}, \frac{3}{4}, z; \frac{1}{4}, \frac{3}{4}, \bar{z}$

$\frac{3}{4}, \frac{3}{4}, z; \frac{3}{4}, \frac{3}{4}, \bar{z}; \frac{3}{4}, \frac{1}{4}, z; \frac{3}{4}, \frac{1}{4}, \bar{z}$

$\frac{3}{4}, \frac{1}{4}, \frac{1}{2} + z; \frac{3}{3}, \frac{1}{4}, \frac{1}{2} - z; \frac{3}{4}, \frac{3}{4}, \frac{1}{2} + z; \frac{3}{4}, \frac{3}{4}, \frac{1}{2} - z$

$\frac{1}{4}, \frac{3}{4}, \frac{1}{2} + z; \frac{1}{4}, \frac{3}{4}, \frac{1}{2} - z; \frac{1}{4}, \frac{1}{4}, \frac{1}{2} + z; \frac{1}{4}, \frac{1}{4}, \frac{1}{2} - z$

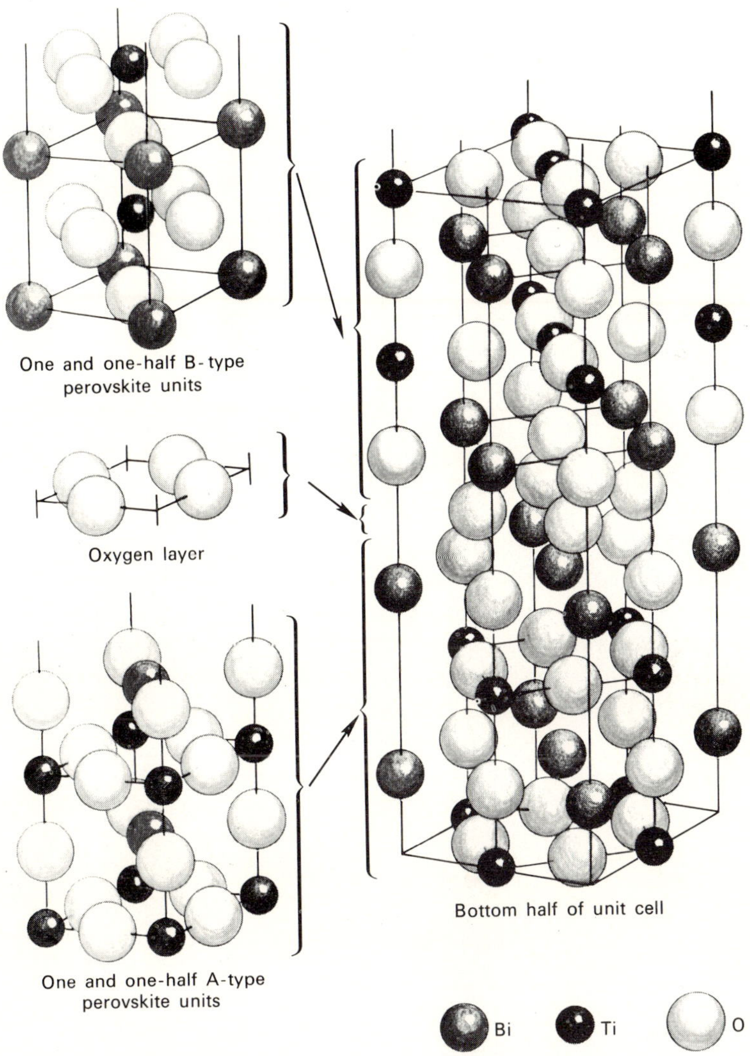

FIG. 7.7a The $Bi_4Ti_3O_{12}$ structure (one-half the unit cell)

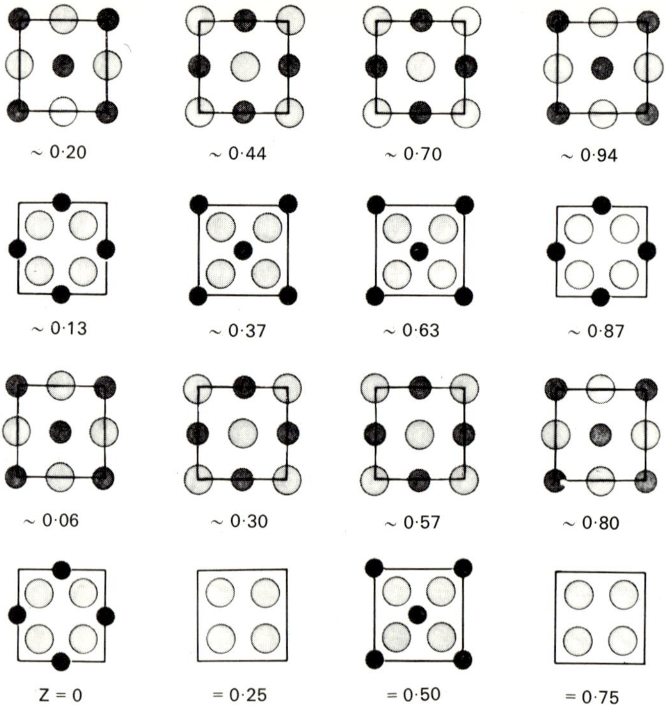

FIG. 7.7b The $Bi_4Ti_3O_{12}$ structure (layer sequence)

Properties

Other compounds with related structures are listed in Table 7.7. They contain $(Bi_2O_2)^{2+}$ layers and $(Me_{x-1}R_xO_{3x+1})^{2-}$ perovskite-type units where Me is a mono-, di-, or trivalent ion, or a mixture, and R represents Ti^{4+}, Nb^{5+}, Ta^{5+}, etc. They are ferroelectric and their Curie temperatures are listed in Table 7.7.

TABLE 7.7. $Bi_4Ti_3O_{12}$ *and Related Compounds*

Compound	Cell size (Å)		Ferroelectric Curie temperature (°C)	Ref.
	a_0	c_0		
$Bi_4Ti_3O_{12}$	5.510	32.84	675	1, 2
	$b_0 = 5.4487$			
$PbBi_2Nb_2O_9$	3.88	25.5	550	3
$PbBi_2Ta_2O_9$	3.88	25.3	430	3
$BaBi_3Ti_2NbO_{12}$	3.87	33.7	270	3
$PbBi_3Ti_2NbO_{12}$	3.86	33.5	290	3
$BaBi_4Ti_4O_{15}$	3.84	41.6	395	3
$PbBi_4Ti_4O_{15}$	3.84	41.3	570	3

7.8. Tungsten Bronzes, *P4/mbm*, Tetragonal, *P6/mcm*, Hexagonal

The sodium tungsten bronzes can be represented by the formula Na_xWO_3, where x varies between 0 and 1.0. They are golden yellow when $x = 1.0$ and red with $x = 0.6$. Phases with a sodium content of between 0.4 and 1.0 have the perovskite structure, where the A ions are surrounded with rings of four corner-sharing WO_6 octahedra. When potassium is the A ion, with x varying between 0.475 and 0.57, a tetragonal bronze structure is formed. In the structure of K_xWO_3 the potassium ion is surrounded by four- and five-membered rings of corner-sharing WO_6 octahedra (see Fig. 7.8a). K, Rb and Cs ions with $x = 0.3$ are in A_xWO_3 phases with the hexagonal bronze structure having six-membered rings of octahedra around them (see Fig. 7.8b). The unit cell of the tetragonal bronze structure is one octahedron high, while the unit cell of the hexagonal tungsten bronze structure

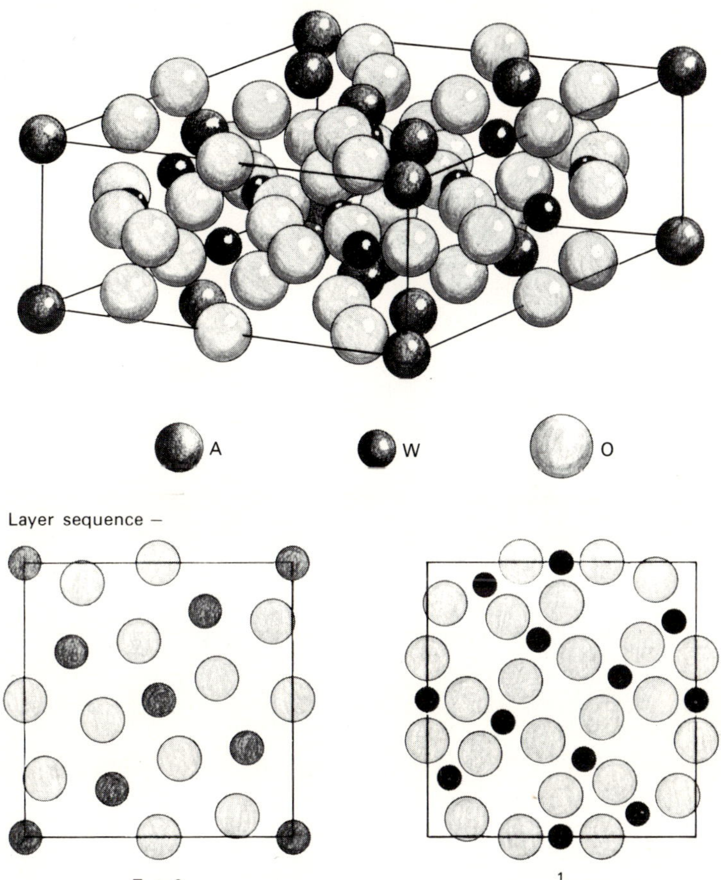

FIG. 7.8a The tetragonal tungsten bronze structure

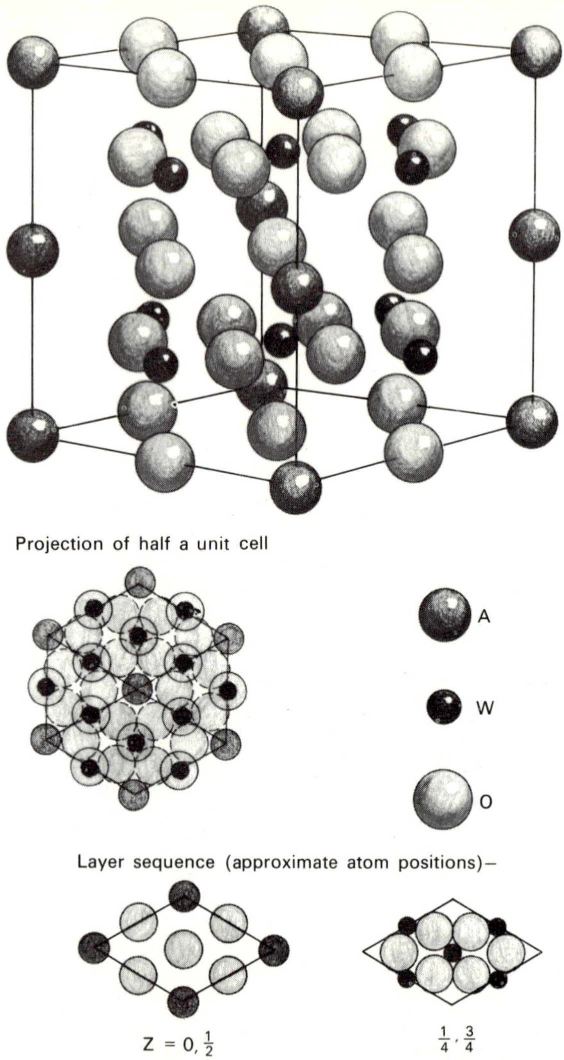

FIG. 7.8b The hexagonal tungsten bronze structure

is two octahedra high. The positions of the atoms in these structures are listed below.

In the tetragonal tungsten bronze structure, the following atomic positions are used in space group *P4/mbm*:

2 W (e) $0, \frac{1}{2}, \frac{1}{2}; \frac{1}{2}, 0, \frac{1}{2}$
8 W (f) $0.078, 0.209, \frac{1}{2}; 0.578, 0.291, \frac{1}{2}; 0.791, 0.078, \frac{1}{2}; 0.709, 0.578, \frac{1}{2};$
$0.922, 0.791, \frac{1}{2}; 0.422, 0.709, \frac{1}{2}; 0.209, 0.922, \frac{1}{2}; 0.291, 0.422, \frac{1}{2}$

2 K (a) 0, 0, 0; $\frac{1}{2}$, $\frac{1}{2}$, 0
4 K (g) 0.175, 0.675, 0; 0.675, 0.825, 0;
0.825, 0.325, 0; 0.325, 0.175, 0
2 O (d) 0, $\frac{1}{2}$, 0; $\frac{1}{2}$, 0, 0
8 O (i) 0.075, 0.215, 0; 0.575, 0.285, 0; 0.785, 0.075, 0; 0.715, 0.575, 0;
0.925, 0.785, 0; 0.425, 0.715, 0; 0.215, 0.925, 0; 0.285, 0.425, 0
4 O (h) 0.30, 0.80, $\frac{1}{2}$; 0.80, 0.70, $\frac{1}{2}$;
0.70, 0.20, $\frac{1}{2}$; 0.20, 0.30, $\frac{1}{2}$
8 O (j) 0, 0.345, $\frac{1}{2}$; $\frac{1}{2}$, 0.155, $\frac{1}{2}$; 0.655, 0, $\frac{1}{2}$; 0.845, $\frac{1}{2}$, $\frac{1}{2}$;
0, 0.655, $\frac{1}{2}$; $\frac{1}{2}$, 0.845, $\frac{1}{2}$; 0.345, 0, $\frac{1}{2}$; 0.155, $\frac{1}{2}$, $\frac{1}{2}$
8 O (j) 0.15, 0.07, $\frac{1}{2}$; 0.65, 0.43, $\frac{1}{2}$; 0.93, 0.15, $\frac{1}{2}$; 0.57, 0.65, $\frac{1}{2}$;
0.85, 0.93, $\frac{1}{2}$; 0.35, 0.57, $\frac{1}{2}$; 0.07, 0.85, $\frac{1}{2}$; 0.43, 0.35, $\frac{1}{2}$

In the hexagonal tungsten bronze structure, the atomic positions in space group *P6/mcm* which are occupied are:

W in (6g) 0.48, 0, $\frac{1}{4}$; 0, 0.48, $\frac{1}{4}$; 0.52, 0.52, $\frac{1}{4}$;
0.52, 0, $\frac{3}{4}$; 0, 0.52, $\frac{3}{4}$; 0.48, 0.48, $\frac{3}{4}$
O in (6f) $\frac{1}{2}$, 0, 0; 0, $\frac{1}{2}$, 0; $\frac{1}{2}$, $\frac{1}{2}$, 0; $\frac{1}{2}$, 0, $\frac{1}{2}$; 0, $\frac{1}{2}$, $\frac{1}{2}$; $\frac{1}{2}$, $\frac{1}{2}$, $\frac{1}{2}$
O in (12j) 0.42, 0.22, $\frac{1}{4}$; 0.78, 0.20, $\frac{1}{4}$; 0.80, 0.58, $\frac{1}{4}$;
0.22, 0.42, $\frac{1}{4}$; 0.58, 0.80, $\frac{1}{4}$; 0.20, 0.78, $\frac{1}{4}$
0.58, 0.78, $\frac{3}{4}$; 0.22, 0.80, $\frac{3}{4}$; 0.20, 0.42, $\frac{3}{4}$;
0.78, 0.58, $\frac{3}{4}$; 0.42, 0.20, $\frac{3}{4}$; 0.80, 0.22, $\frac{3}{4}$
Rb in (2b); 0, 0, 0; 0, 0, $\frac{1}{2}$

Properties

$PbNb_2O_6$ and $PbTa_2O_6$ are ferroelectrics with Curie temperatures of 570° and 260°C, respectively. Their unit cells are related to that of the tetragonal tungsten bronzes. These unit cells are orthorhombic with $a_0 = 17.51$ Å, $b_0 = 17.81$ Å and $c_0 = 7.62$ Å for the former compound and $a_0 = 17.68$ Å, $b_0 = 17.72$ Å and $c_0 = 7.95$ Å for the latter, which are approximately $\sqrt{2}$. a' and $2c'$, where a' and c' are the unit cell edges for the tetragonal tungsten bronzes.

All of these phases are highly conducting, which has been attributed to the presence of W^{5+}. Phases have been prepared in which W^{5+} has been replaced by Ta^{5+}, resulting in materials which are not colored and have high resistivities. See the compounds in Table 7.8.

7.9. Layer Structures Related to the Perovskite Structure, AO_3 Layers

These structures consist of AO_3 close-packed layers with metal ions situated in the octahedrally coordinated holes between the layers. In each layer, the A ions and oxygen ions alternate in every other row as shown in Fig. 7.9a. The unit cell edge a is then selected so that the axes connect A

TABLE 7.8. *Compounds with the Tetragonal and Hexagonal Tungsten Bronze Structures*

Tetragonal bronze	Cell size (Å) a_0	Cell size (Å) c_0	Ferroelectric Curie temperature (°C)	Refs.
$Ba_{0.5}NbO_3$†				1
$Ba_{0.5}TaO_3$	12.60	3.95		2
$K_4Nb_6O_7$	12.525	3.763		3
$K_{0.5}WO_3$	12.3	3.8		3
$K_{0.5}(W_{0.5}Ta_{0.5})O_3$	12.36	3.90		2
$Nb_2W_3O_{14}$	12.190	3.934		5
$Pb_{0.5}NbO_3$† (570°C)	12.46	3.907	570	1
$Pb_{0.5}TaO_3$†			260	1
$Sr_{0.5}TaO_3$	12.41	3.90		2

† These compounds have structures which are closely related to the tetragonal tungsten bronze structure.

Hexagonal bronze	Cell size (Å) a_0	Cell size (Å) c_0	Refs.
$K_{0.3}WO_3$	7.385	7.513	6
$K_{0.3}(Ta_{0.3}W_{0.7})O_3$	7.333	7.685	2
$Rb_{0.27}WO_3$	7.394	7.516	6
$Rb_{0.3}(Ta_{0.3}W_{0.7})O_3$	7.342	7.715	2
$Cs_{0.3}WO_3$	7.406	7.608	6
$Cs_{0.3}(Ta_{0.3}W_{0.7})O_3$	7.450	7.821	2

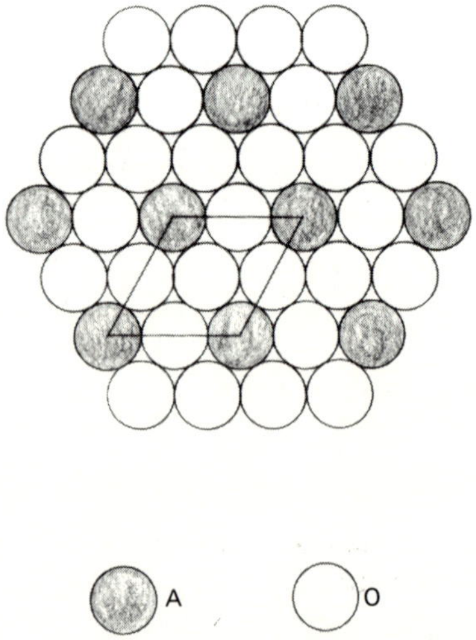

FIG. 7.9a AO_3 layer

ions, and therefore lie parallel to the close-packed layers. Figure 7.9b shows the method used in this book to illustrate these layer structures. The atomic arrangement of three perovskite unit cells taken perpendicular to the⟨111⟩ directions are shown. The layer sequence is ABCA for the AO_3 layers and the B ions are situated one-third of the distance up the $\frac{1}{3}, \frac{2}{3}$ line, half-way up the 0, 0 line and two-thirds of the way up the $\frac{2}{3}, \frac{1}{2}$ line. If two of these B ions are replaced by one B″ ion and one B′ ion, then the structure is that of $Ba(Sr_{0.33}Ta_{0.67})O_3$ (see Fig. 7.9c). In both structures each BO_6 octahedron is joined to six others by corners.

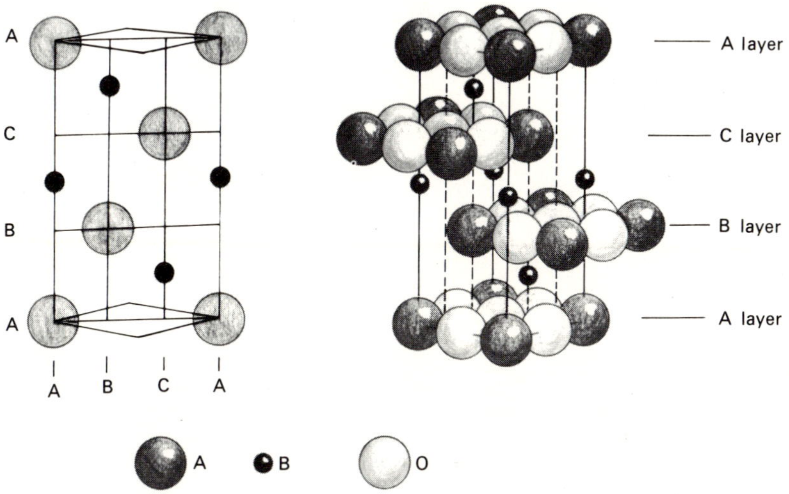

FIG. 7.9b Layer structures. Simple Perovskite structure Cubic (– – –ABC– – –) stacking of AO_3 layers

The $BaNiO_3$ structure has a hexagonal close-packed arrangement of BaO_3 layers (see Fig. 7.9d). One A ion is one-quarter the way up the $\frac{1}{3}, \frac{2}{3}$ line and three-quarters the way up the $\frac{2}{3}, \frac{1}{3}$ line. The A ion is at $\frac{1}{3}, \frac{2}{3}, \frac{1}{4}; \frac{2}{3}, \frac{1}{3}, \frac{3}{4}$, two B ions are at 0, 0, 0; 0, 0, $\frac{1}{2}$, and six oxygen ions are at $x, \bar{x}, z$; $x, 2x, z$; $2\bar{x}, \bar{x}, z$; $\bar{x}, x, z + \frac{1}{2}$; $\bar{x}, 2\bar{x}, z + \frac{1}{2}$; $2x, x, z + \frac{1}{2}$ with $x = \frac{1}{6}$, $z = \frac{3}{4}$. In this two-layer unit cell the BO_6 octahedra share opposite faces. The closeness of the B ions in this structure suggests the possibility of metal-metal bonding.

The $Ba_5Ta_4O_{15}$ structure shown in Fig. 7.9e has a five-layer sequence, ABCBC, two tantalum ions ~0.3 and 0.7 of the c-axis distance up the cell edge, one tantalum ion, 0.1c up the $\frac{1}{3}, \frac{2}{3}$ line and 0.9c up the $\frac{2}{3}, \frac{1}{3}$ line. Face sharing of the TaO_6 octahedra is prevented by the absence of a tantalum ion between one pair of AO_3 layers. There is one Ba ion at 0, 0, 0; two Ba ions at $\frac{1}{3}, \frac{2}{3}, z$; $\frac{2}{3}, \frac{1}{3}, \bar{z}$ with $z = 0.793$ and two more Ba ions at $\frac{1}{3}, \frac{2}{3}, z$; $\frac{2}{3}, \frac{1}{3}, \bar{z}$ with

$z = 0.425$. Two Ta ions are at $0, 0, z$ and $0, 0, \bar{z}$ with $z = 0.313$, and two more Ta ions at $\frac{1}{3}, \frac{2}{3}, z$; $\frac{2}{3}, \frac{1}{3}, z$ with $z = 0.104$. Three oxygen ions are at $\frac{1}{2}, 0, 0$; $0, \frac{1}{2}, 0$; $\frac{1}{2}, \frac{1}{2}, 0$; six oxygen ions at $x, \bar{x}, z$; $x, 2x, z$; $2\bar{x}, \bar{x}, \bar{z}$; $\bar{x}, x, \bar{z}$; $\bar{x}, 2\bar{x}, \bar{z}$; and $2x, x, \bar{z}$ with $x = 0.17$, $z = 0.20$, and six more oxygen ions in these same positions but with $x = 0.17$ and $z = 0.60$.

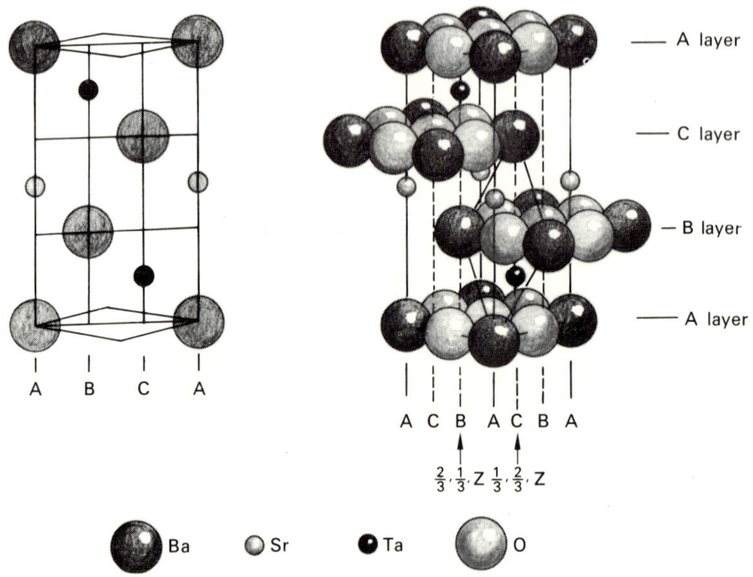

FIG. 7.9c Layer structures. $Ba(Sr_{0.33}Ta_{0.67})O_3$ structure. Cubic (- - -ABC- - -) stacking of AO_3 layers

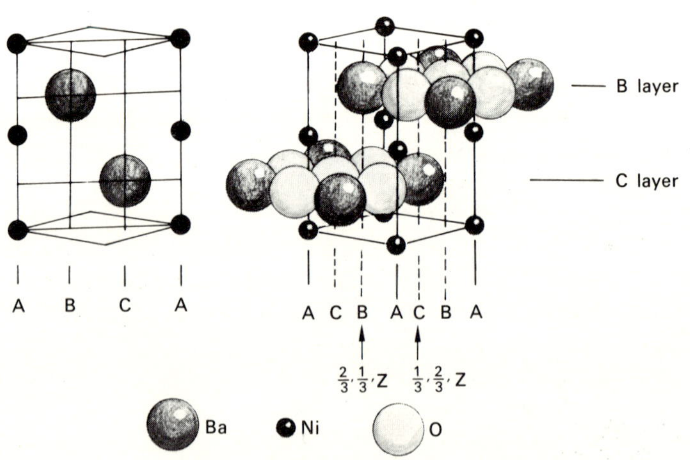

FIG. 7.9d Layer structures. $BaNiO_3$ structure. Hexagonal (- - -BC- - -) stacking of AO_3 layers

The hexagonal $BaTiO_3$ structure is a six-layered structure with a layer sequence of ABCACB as shown in Fig. 7.9f. In the unit cell, the Ti ions are situated $\frac{2}{7}$ and $\frac{4}{7}c$ up the vertical edges of the unit cell, $\frac{1}{7}$ and $\frac{6}{7}c$ up the $\frac{1}{3}, \frac{2}{3}, z$ line and $\frac{4}{7}$ and $\frac{5}{7}c$ up the $\frac{2}{3}, \frac{1}{3}, z$ line. Some of the TiO_6 octahedra share faces in this structure. Two barium ions are at $0, 0, \frac{1}{4}$ and $0, 0, \frac{3}{4}$, four others are

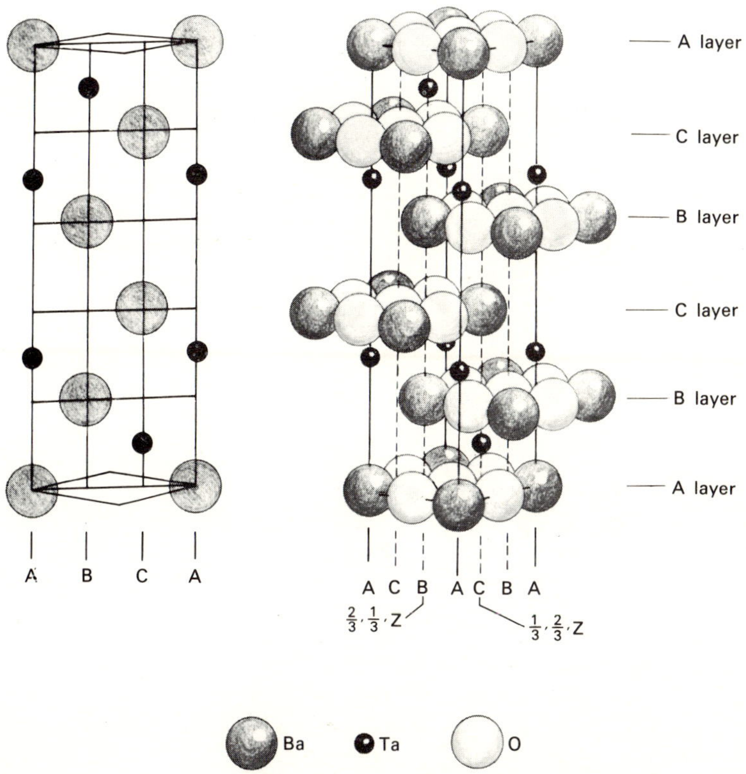

FIG. 7.9e Layer structures. $Ba_5Ta_4O_{15}$ structure.
--- ABCBC --- stacking of AO_3 layers

at $\pm(\frac{1}{3}, \frac{2}{3}, z; \frac{2}{3}, \frac{1}{3}, z + \frac{1}{2})$ with $z = 0.090$, two titanium ions are at $0, 0, 0$ and $0, 0, \frac{1}{2}$ and four more at $\pm(\frac{1}{3}, \frac{2}{3}, z; \frac{2}{3}, \frac{1}{3}, z + \frac{1}{2})$ with $z = 0.845$, six oxygen ions are at $\pm(x, 2x, \frac{1}{4}; x, \bar{x}, \frac{1}{4})$ with $x = 0.522$ and twelve more oxygen ions at $\pm(x, 2x, z; 2\bar{x}, \bar{x}, z; x, x, z; x, 2x, \frac{1}{2} - z; 2\bar{x}, \bar{x}, \frac{1}{2} - z; x, \bar{x}, \frac{1}{2} - z)$ with $x = 0.386$ and $z = 0.076$.

Compounds with the perovskite and $Ba(Sr_{0.33}Ta_{0.67})O_3$ structure are listed in Table 7.4c. Compounds with other layered structures described above are listed in Table 7.9.

TABLE 7.9. *Compounds with Layer-type Structures*

	Cell size (Å)		Refs.
	a_0	c_0	
$BaNiO_3$ type			
$BaCoO_{3-x}$	5.59	4.82	1
$BaMnO_3$ (L.T.)	5.7	4.8	1
$BaNiO_3$	5.58	4.832	1
$CsNiF_3$	6.23	5.22	2
$Ba_5Ta_4O_{15}$ type			
$Ba_5Nb_4O_{15}$	5.79	11.75	3
$Ba_5Ta_4O_{15}$	5.79	11.75	3
$Sr_5Ta_4O_{15}$	5.67	11.42	3
Hex.$BaTiO_3$ type			
Ba_2CoOsO_6	5.74	14.1	4
$Ba_3CoTi_2O_9$	5.74	14.1	5
$Ba_3Cr_2MoO_9$	5.72	14.0	6
Ba_2CrOsO_6			7
$Ba_3Cr_2ReO_9$	5.70	13.8	4
$Ba(Cr, Ti)O_3$	5.74	14.1	5
Ba_2CrTaO_6			7
$Ba_3Cr_2UO_9$	5.83	14.4	8
$Ba_3Cr_2WO_9$	5.75	14.3	6
Ba_2ErIrO_6	5.91	14.6	7
$Ba(Fe, Ir)O_3$			9
Ba_2FeOsO_6	5.76	14.1	4
$Ba_3Fe_2ReO_9$	5.81	14.1	4
Ba_2FeSbO_6	5.76	14.5	7
$Ba_3FeTi_2O_9$	5.74	14.1	5
Ba_2InIrO_6	5.83	14.7	7
$Ba_3IrTi_2O_9$	5.74	14.2	5
Ba_2MnOsO_6	5.82	14.2	4
$Ba_3MnTi_2O_9$	5.74	14.1	5
Ba_2NiOsO_6	5.75	14.1	4
$Ba_3OsTi_2O_9$	5.76	14.3	7
$BaPt_{0.1}Ti_{0.9}O_3$	5.74	14.1	5
$Ba(Rh, Ti)O_3$	5.74	14.1	5
$Ba_3RuTi_2O_9$	5.74	14.1	5
Ba_2RhUO_6	5.84	14.9	8
Ba_2ScIrO_6	5.79	14.6	7

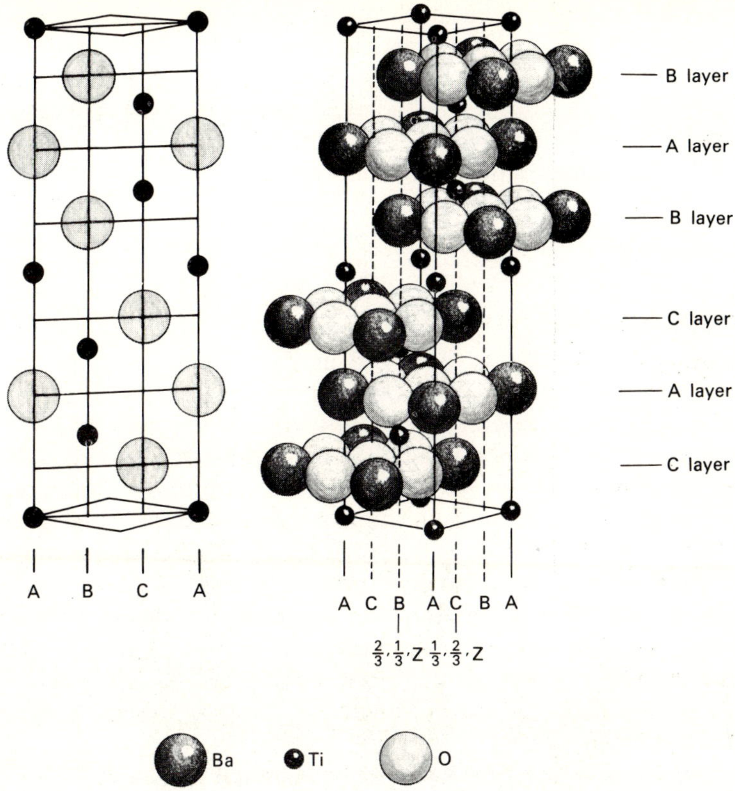

FIG. 7.9f Layer structure. Hexagonal $BaTiO_3$ structure.
– – – ABCACB – – – stacking of AO_3 layers

7.10. Discussion

Structures (Relations)

The perovskite-type structures are derived from either the cubic close-packed or the simple cubic structure. By ordering the atomic positions in a cubic close-packed structure, the ordered Cu_3Au structure is obtained. Ordering of the cubic close-packed structure in another manner and moving atoms in two sets of faces of the unit cell in opposite direction produces a unit cell of the PbO structure. The B-type unit cell of the perovskite structure is formed by placing a small atom in the center of the unit cell of the ordered Cu_3Au structure.

The A-type unit cell of the perovskite structure is formed by placing a large A atom in the center of the unit cell of the ReO_3 structure. The A- and B-type unit cells, of course, represent the same structure except that the origin is taken at a B atom in the former structure and at an A atom in the latter. The $Bi_4Ti_3O_{12}$, K_2NiF_4 and $Sr_3Ti_2O_7$ unit cells contain several unit cells of the perovskite structure, the tungsten bronze structures contain BO_6 octahedra with various sized holes between them for A atoms and the layer structures are built up of AO_3 layers with B atoms in between as is found in the $\langle 111 \rangle$ directions of the perovskite structure.

Oxides

The group of compounds described in this chapter are probably best known for their ferroelectric and related properties. Barium titanate has a dielectric constant of several thousand, and is ferroelectric with a Curie temperature of 120°C. Solid solution compositions of lead zirconate and titanate are some of the best piezoelectric phases for transducer applications. Another solid solution of potassium tantalate and potassium niobate exhibits a large electro-optical effect and has, therefore, been used for laser modulation. The compound $Bi_4Ti_3O_{12}$ and some related phases, and some tetragonal tungsten bronze-type phases also have ferroelectric properties. All of the properties described above depend on slight distortions of the lattice, which produce dipoles in the structure. These dipoles can be reoriented by an applied electric field, just as the orientation of ferromagnetic domains is changed in a magnetic field.

The sodium tungsten bronzes represent unusual phases with very high conductivities. These materials change color from blue to red to yellow as the amount of sodium in the A position is increased from $x = 0.5$ to ~ 1 in the phases Na_xWO_3. The sodium tungsten bronzes show metallic conduction which depends on the sodium content.

There are a large number of complex perovskite-type compounds but their properties have only been touched upon here. There are several which are ferroelectric and they also represent the majority of ferromagnetic perovskite-type compounds. They provided the combinations of ions which allowed an understanding of ordering in oxides in terms of the size and charge differences of ions.

The related layer structures show how structures can be built up of a close-packed layer arrangement of anions and A ions with small cations in the holes. The two-layer structure is represented by $BaNiO_3$, the three-layer by the perovskite structure, the five-layer structure by $Ba_5Ta_4O_{15}$ and the six-layer structure by hexagonal $BaTiO_3$. The latter structure is extremely interesting in that the face-sharing of octahedra allows for metal—metal bonding. There are also a large number of other layer structures but these are beyond the scope of this book.

Table 7.1 *References*

1. G. Hagg and A. L. Kindstrom, *Z. Phys. Chem.* **B22**, 453 (1933).
2. F. Gronvold, H. Haroldsen and J. Vihovde, *Acta Chem. Scand.* **8**, 1927 (1954).
3. W. P. Binnie, *Acta Cryst.* **9**, 686 (1956).
4. H. Dachs, *Z. Krist.* **112**, 60 (1959).
5. W. J. Moore and L. Pauling, *J. Am. Chem. Soc.* **63**, 1392 (1941).
6. V. A. Izvozchekov, *Zh. Prikl. Spectroskopii, Akad. Nauk Beloussk, SSR* **4**, 282 (1966).
7. J. Leciejewicz, *Acta Cryst.* **14**, 1304 (1961).
Stds. Circ. 539, **IV**, 25 (1955).
8. H. E. Swanson, R. K. Fuyat and G. M. Ugrinic, NBS Rept. No. 2691 (1953), Natl. Bur.

Table 7.2 *References*

1. A. Schneider and U. Esch, *Z. Electrochem.* **49**, 72 (1943).
2. A. J. Bradley and G. C. Seager, *J. Inst. Met.* **64**, 81 (1939).
3. J. L. Moriarty, R. O. Gordon and J. E. Humphreys, *Acta Cryst.* **19**, 285 (1965).
4. A. Taylor and R. W. Floyd, *J. Inst. Met.* **81**, 25 (1952).
5. O. J. C. Runnalls, *J. Metals* **5**, 1460 (1953).
6. M. I. Ivanov, V. A. Tumbakov and N. S. Podolskaya, *Soviet J. At. Energy* **5**, 1007 (1959).
7. J. L. Moriarty, J. E. Humphreys, R. O. Gordon and W. C. Baenziger, *Acta Cryst.* **21**, 840 (1966).
8. J. H. Keeler and J. H. Mallery, *J. Metals* **7**, 374 (1955).
9. W. Betteridge, *J. Inst. Met.* **75**, 559 (1949).
10. G. Grube, A. Schneider and U. Esch, *Festschrift aus Anlass des* 100-*jahrigen Jubilaums der Firma W. C. Heraeus GmbH* 20 (1951).
11. F. Ruggiero and G. L. Olcese, *Accad. Naze. Lincei Rend., Classe Sc. Fis.* **37**, 169 (1964).
12. H. Holleck, *Acta Cryst.* **21**, 451 (1966).
13. H. Nowotny, E. Bauer, A. Stempfl and H. Bittner, *Monatsh. Chem.* **83**, 221 (1952).
14. R. Vogel and H. Klose, *Z. Metallk.* **45**, 633 (1954).
15. A. H. Geisler and D. L. Martin, *J. Appl. Phys.* **23**, 375 (1952).
16. E. Raub and W. Mahler, *Z. Metallk.* **46**, 210 (1955).
17. P. Greenfield and P. A. Beck, *J. Metals* **8**, 265 (1956).
18. M. Hirabayashi and S. Ogawa, *J. Phys. Soc. Japan* **12**, 259 (1957).
19. A. Schneider and U. Esch, *Z. Electrochem.* **50**, 290 (1944).
20. R. J. Wakelin and E. L. Yates, *Proc. Phys. Soc.* B **66**, 221 (1953).
21. R. Hultgren and C. A. Zapffe, *Z. Krist.* **A99**, 509 (1938).
22. W. B. Pearson, *Nature* **173**, 364 (1954).
23. B. R. T. Frost and J. T. Maskrey, *J. Inst. Met.* **82**, 171 (1953).
24. H. Pfisterer and K. Schubert, *Z. Metallk.* **41**, 358 (1950).
25. A. S. Coffinberry and F. H. Ellinger, U.S.A. Rep. A/conf. 81P/826 (1955).
26. E. S. Makarov and V. N. Bykov, *Soviet Phys. Cryst.* **4**, 164 (1959).
27. P. Pietrokowsky, *J. Metals* **6**, 219 (1954).
28. A. Iandelli, *Congr. Intern. Chim. Pure Appl.*, 16, Paris, 1957, Mem. Sect. Chim. Minerale, 35 (1958).
29. A. A. Bochvar, S. T. Konobeevsky, V. I. Kutaitsev, T. S. Menshikova and N. T. Chebotarev, *Proc. U.N. Intern. Conf. Peaceful Uses At. Energy*, 2d, Geneva, **6**, 184 (1958).
30. E. Parthe, D. Hohnke, W. Jeitschko and O. Schob, *Naturwiss.* **52**, 155 (1965).
31. E. Raub and W. Mahler, *Z. Metalk.* **46**, 282 (1955).
32. A. Rossi, *Gazz. Chim. Ital.*. **64**, 832 (1934).
33. T. H. Geballe, B. T. Matthias, V. B. Compton, E. Corenzwit, G. W. Hull, Jr. and L. D. Longinotti, *Phys. Rev.* **137**, A119 (1965).
34. E. Zintl and A. Harder, *J. Inst. Met.* **50**, 431 (1932).
35. F. Galasso and J. Pyle, *Acta Cryst.* **16**, 228 (1963).
36. A. Osawa and M. Okamoto, *Nippon Kink. Gakk.* **2**, 378 (1938).
37. H. Nowotny, K. Schubert and U. Dettinger, *Z. Metallk.* **37**, 137 (1946).

38. K. Schubert and K. Pfisterer, *Z. Metallk.* **40**, 405 (1949).
39. J. H. Wallbaum, *Naturwissenschaften* **31**, 91 (1943).
40. T. J. Heal and G. I. Williams, *Acta Cryst.* **8**, 494 (1955).
41. A. Iandelli and R. Ferro, *Ann. Chim. (Rome)* **42**, 598 (1952).

Table 7.3 References

1. R. dePape, *Compt. Rend.* **260**, 4527 (1965).
2. D. E. LaValle, R. M. Steele, M. K. Wilkinson and H. L. Yakel, Jr., *J. Am. Chem. Soc.* 82, 2433 (1960).
3. E. L. Muetterties and J. E. Castle, *J. Inorg. Nucl. Chem.* **18**, 148 (1961).
4. L. K. Frevel and H. W. Rinn, *Acta Cryst.* **9**, 626 (1956).
5. K. Vorres and J. Donohue, *Acta Cryst.* **8**, 25 (1955).
6. R. Juza, *Z. Anorg. Chem.* **248**, 118 (1941).
7. A. Magneli, *Acta Chem. Scand.* **11**, 28 (1957).
8. E. Wait, *J. Inorg. Nucl. Chem.* **1**, 309 (1955).
9. I. Morgenstein-Bordarau, Y. Billist, P. Poix and A. Michel, *Compt. Rend.* **260**, 3668 (1965).

Table 7.4a References

1. A. von Hippel, R. G. Breckenridge, F. G. Chesley and L. Tisza, *Ind. Eng. Chem.* **38**, 1097 (1946).
2. G. Shirane, S. Hoshino and K. Suzuki, *Phys. Rev.* **80**, 1105 (1950).
3. B. T. Matthias, *Phys. Rev.* **75**, 1771 (1949).
4. L. E. Cross and B. J. Nicholson, *Phil. Mag.* Ser. 7, **46**, 453 (1955).
5. G. A. Smolenskii, *Dokl. Akad. Nauk, SSSR*, **70**, 405 (1950).
6. V. A. Isupov and L. T. Elem'yanova, *Kristallografiya* **11**, 776 (1966).
7. G. A. Smolenskii, V. A. Isupov and A. I. Agranovskaya, *Sov. Phys., Solid State* **1**, 150 (1959).
8. A. I. Agranovskaya, *Bull. Acad. Sciences USSR, Phys. Series* **24**, 1271 (1960).
9. G. A. Smolenskii, A. I. Agranovskaya, S. N. Popov and V. A. Isupov, *Sov. Phys., Tech. Phys.* **3**, 1981 (1960).
10. G. A. Smolenskii, A. I. Agranovskaya and V. A. Isupov, *Sov. Phys., Solid State* **1**, 907 (1959).
11. V. A. Bokov and E. I. Myl'Nikova, *Sov. Phys., Solid State* **2**, 2428 (1961).

Table 7.4b References

1. F. K. Patterson, C. W. Moeller and R. Ward, *Inorg. Chem.* **2**, 196 (1962).
2. F. Galasso, F. C. Douglas and R. Kasper, *J. Chem. Phys.* **44**, 1672 (1966).
3. J. Longo and R. Ward, *J. Am. Chem. Soc.* **83**, 2816 (1961).
4. A. W. Sleight, J. Longo and R. Ward, *Inorg. Chem.* **1**, 245 (1962).
5. F. Sugawara and S. Iida, *J. Phys. Soc. Japan* **20**, 1529 (1965).
6. G. W. Wiener and J. A. Berger, *J. Met.* **7**, 360 (1955).

Table 7.4c References

1. H. P. Myers, Univ. of British Columbia (1956).
2. R. G. Butters and H. P. Myers, *Phil. Mag.* **46**, 895 (1955).

3. W. L. W. LUDEKENS and A. J. E. WELCH, *Acta Cryst.* **5**, 841 (1952).
4. R. G. BUTTERS and H. P. MYERS, *Phil. Mag.* **46**, 132 (1955).
5. I. NARÁY-SZABÓ, *Mulgyet Zoylemen* **1**, 30 (1947).
6. M. KESTIGIAN, F. D. LEIPZIGER, W. J. CROFT and R. GUIDOBONI, *Inorg. Chem.* **5**, 1462 (1966).
7. A. N. CHRISTENSEN, *Act. Chem. Scand.* **19**, 42 (1965).
8. C. BRISI, *Ann. Chem.* **42**, 356 (1952).
9. K. KNOX, *Acta Cryst.* **14**, 583 (1961).
10. W. RUDORFF, J. KANDLER, G. LINCKE and D. BABEL, *Angew. Chem.* **71**, 672 (1959).
11. A. OKAZAKI and Y. SUEMUNE, *J. Phys. Soc. Japan* **17**, 204 (1962).
12. K. KNOX and D. W. MITCHELL, *J. Nucl. Chem.* **21**, 253 (1961).
13. C. E. MESSER, J. C. EASTMAN, R. G. MERS and A. J. MAELAND, *Inorg. Chem.* **3**, 776 (1964).
14. C. E. MESSER and K. HARDCASTLE, *Inorg. Chem.* **3**, 1327 (1964).
15. G. W. WIENER and J. A. BERGER, *J. Metals* **7**, 360 (1955).
16. M. H. FRANCOMBE and B. LEWIS, *Acta Cryst.* **11**, 175 (1958).
17. W. H. ZACHARIASEN, *Skrifter Norske Videnskads— Akad. Oslo I. Mat.— Naturv. Kl.* No. 4 (1928).
18. I. NARÁY-SZABÓ, *Muegvetemi Kozlemenyek* **1**, 30 (1947).
19. R. P. OZEROV, N. V. RANNEV, V. I. PARHOMOV, I. S. REZ and G. S. ZHDANDOV, *Kristallografiya* **7**, 620 (1962).
20. A. J. SMITH and A. J. E. WELCH, *Acta Cryst.* **13**, 653 (1960).
21. I. NARAY-SZABO and A. KALMAN, *Acta Cryst.* **14**, 791 (1961).
22. J. H. SMITH, *Nature* **115**, 334 (1925).
23. E. A. WOOD, *Acta Cryst.* **4**, 353 (1951).
24. L. L. QUILL, *Z. Anorg. Allgem. Chem.* **208**, 257 (1932).
25. P. VOUSDEN, *Acta Cryst.* **4**, 373 (1951).
27. H. F. KAY and J. L. MILES, *Acta Cryst.* **10**, 213 (1957).
28. V. M. GOLDSCHMIDT, *Skrifter Norske Videnskads— Akad. Oslo I. Mat.— Naturv. Kl.* No. 8 (1926).
29. D. SANTANA, *Anales Real. Soc. Espan. Fis. Quim. (Madrid)* Ser. A, **44**, 557 (1948).
30. L. RIVOIR and M. ABBAD, *Anales. Fis. Quim. (Madrid)* **43**, 1051 (1947).
31. R. S. ROTH, *J. Research NBS*, RP2736, **58** (1957).
32. A. F. WELLS, *Structural Inorganic Chemistry*, Oxford University Press, London (1950).
33. W. W. MALINOFSKY and H. KEDESDY, *J. Am. Chem. Soc.* **76**, 3090 (1954).
34. S. W. DERBYSHIRE, A. C. FRAKER and H. H. STADELMAIER, *Acta Cryst.* **14**, 1293 (1961).
35. L. H. BRIXNER, *J. Inorg. Nucl. Chem.* **14**, 225 (1960).
36. R. WEISS, *Compt. Rend.* **246**, 3073 (1958).
37. A. HOFFMAN, *Z. Physik. Chem.* **28**, 65 (1935).
38. L. E. RUSSELL, J. D. L. HARRISON and N. H. BRETT, *J. Nucl. Mater.* **2**, 310 (1960).
39. H. D. MEGAW, *Proc. Phys. Soc.* **58**, 133, 326 (1946).
40. S. M. LANG, F. P. KUNDSEN, C. L. FILLMORE, R. S. ROTH, NBS Circ. 568 (1956).
41. I. NARÁY-SZABÓ, *Naturwiss.* **31**, 466 (1943).
42. C. E. CURTIS, L. M. DONEY and J. R. JOHNSON, Oak Ridge Natl. Lab. ORNL-1681 (1954).
43. H. L. YAKEL, *Acta Cryst.* **8**, 394 (1955).
44. G. H. JONKER and J. H. VAN SANTEN, *Physica* **16**, 337 (1950).
45. R. WARD, B. GUSHEE, W. MCCARROLL and D. H. RIDGELY, Univ. of Conn. 2d Tech. Rep., NR-052-268, Contract ONR-367(00) (1953).
46. W. H. MCCARROLL, R. WARD and L. KATZ, *J. Am. Chem. Soc.* **78**, 2909 (1956).
47. L. W. COUGHANOUR, R. S. ROTH, S. MARZULLO and F. E. SENNETT, *J. Research NBS*, RP 2576, **54**, 149 (1955).
48. L. W. COUGHANOUR, R. S. ROTH, S. MARZULLO and F. E. SENNETT, *J. Research NBS*, RP 2580, **54**, 191 (1955).
49. J. BROUS, I. FANKUCHEN and E. BANKS, *Acta Cryst.* **6**, 67 (1953).
50. G. SHIRANE and R. PEPINSKY, *Phys. Rev.* **91**, 812 (1953).
51. B. JAFFEE, R. S. ROTH and S. MARZULLO, *J. Research NBS*, RP 2626, **55**, 239 (1955).
52. J. J. RANDALL and R. WARD, *J. Am. Chem. Soc.* **81**, 2629 (1959).
53. W. W. COFFEEN, *J. Am. Ceram. Soc.* **36**, 207 (1953).

54. F. SUGAWARA and SHUICHI IIDA, *J. Phys. Soc. Japan* **20**, 1529 (1965).
55. W. H. ZACHARIASEN, *Acta Cryst.* **2**, 388 (1949).
56. A. RUGGIERO and R. FERRO, *Gazz. Chim. Ital.* **85**, 892 (1955).
57. M. L. KEITH and R. ROY, *Am. Mineralogist* **39**, 1 (1954).
58. A. WOLD and R. WARD, *J. Am. Chem. Soc.* **76**, 1029 (1954).
59. W. RUDORFF and H. BECKER, *Z. Naturforsch* **96**, 613 (1954).
60. J. A. W. DALZIEL and A. J. E. WELCH, *Acta Cryst.* **13**, 956 (1960).
61. R. C. VICKERY and A. KLANN, *J. Chem. Phys.* **27**, 1161 (1957).
62. S. GELLER and V. B. BALA, *Acta Cryst.* **9**, 1019 (1956).
63. S. GELLER and A. E. WOOD, *Acta Cryst.* **9**, 563 (1956).
64. V. S. FILIDEV, N. P. SMOLYANINOV, E. G. FESENKO and I. N. BELYAEV, *Kristallografiya* **5**, 958 (1960).
65. A. I. ZASLAVSKII and A. G. TUTOV, *Dokl. Akad. Nauk SSSR* **135**, 815 (1960).
66. S. A. FEDULOV, Y. N. VENEVTSEV, G. S. ZHDANOV and E. G. SMAZHEVSKAYA, *Kristallografiya* **6**, 795 (1961).
67. A. RUGGIERO and R. FERRO, *Atti. Accad. Nazl. Lincei. Rend., Classe Sci. Fis. Mat. Nat.* **17**, 254 (1954).
68. S. GELLER, *Acta Cryst.* **10**, 243 (1957).
69. S. GELLER, *J. Chem. Phys.* **24**, 6 (1956).
70. A. WOLD, B. POST and E. BANKS, *J. Am. Chem. Soc.* **79**, 6365 (1957).
71. F. ASKHAM, I. FANKUCHEN and R. WARD, *J. Am. Chem. Soc.* **72**, 3799 (1950).
72. W. C. KOEHLER and E. O. WOLLAN, *Phys. Chem. Solids* **2**, 100 (1957).
73. M. FOEX, A. MANCHERON and M. LINE, *Compt. Rend.* **250**, 3027 (1960).
74. D. D. KHANOLKAR, *Current Sci. (India)* **30**, 52 (1961).
75. M. KESTIGIAN and R. WARD, *J. Am. Chem. Soc.* **76**, 6027 (1954).
76. W. D. JOHNSON and D. SESTRICH, *J. Inorg. Nucl. Chem.* **20**, 32 (1961).
77. M. KESTIGIAN, J. G. DICKINSON and R. WARD, *J. Am. Chem. Soc.* **79**, 5598 (1957).
78. N. N. PADUROW and C. SCHUSTERIUS, *Ber. Deut. Keram. Ges.* **32**, 292 (1955).
79. J. T. LOOBY and L. KATZ, *J. Am. Chem. Soc.* **76**, 6029 (1954).
80. H. P. ROOKSBY, E. A. D. WHITE and S. A. LANGSTON, *J. Am. Ceram. Soc.* **48** (1965).
81. L. JAHNBERG, S. ANDERSSON and A. MAGNELI, *Acta Chem. Scand.* **13**, 1248 (1959).
82. A. MAGNELI and R. NILSSON, *Acta Chem. Scand.* **4**, 398 (1950).
83. W. F. DE JOHN, *Z. Krist.* **81**, 314 (1932).
84. G. HAGG, *Z. Physik. Chem.* **29B**, 192 (1935).
85. D. VAN DUYN, *Rec. Trav. Chim.* **61**, 669 (1942).
86. A. MAGNELI, *Arkiv. Kemi.* **1**, 269 (1949).
87. E. O. BRIMM, J. C. BRANTLEY, J. H. LORENZ and M. H. JELLINEK, *J. Am. Chem. Soc.* **73**, 5427 (1951).
88. B. W. BROWN and E. BANKS, *J. Am. Chem. Soc.* **76**, 963 (1954).
89. M. ATOJI and R. E. RUNDLE, *J. Chem. Phys.* **32**, 627 (1960).
90. D. RIDGELY and R. WARD, *J. Am. Chem. Soc.* 77, 6132 (1955).
91. C. BRISI, *Ricerca Sci.* **24**, 1858 (1954).
92. G. H. JONKER, *Physica* **20**, 1118 (1954).
93. F. GALASSO, W. DARBY and J. PINTO, prepared at the United Aircraft Corporation Research Laboratories—not reported.
94. A. W. SLEIGHT and R. WARD, *Inorg Chem.* **1**, 790 (1962).
95. G. BLASSE, *J. Inorg. Nucl. Chem.* **27**, 993 (1965).
96. E. J. FRESIA, L. KATZ and R. WARD, *J. Am. Chem. Soc.* **81**, 4783 (1959).
97. G. A. SMOLENSKII, A. I. AGRANOVSKAYA and V. A. ISUPOV, *Sov. Phys., Solid State* **1**, 907 (1959).
98. A. I. AGRANOVSKAYA, *Bulletin of Acad. Sciences of USSR Physics Series,* **24**, 1271 (1960).
99. A. W. SLEIGHT, J. LONGO and R. WARD, *Inorg. Chem.* **1**, 245 (1962).
100. F. GALASSO and J. PYLE, *J. Phys. Chem.* **67**, 1561 (1963).
101. F. GALASSO, J. R. BARRANTE and L. KATZ, *J. Am. Chem. Soc.* **83**, 2830 (1961).
102. F. GALASSO and J. PYLE, *J. Phys. Chem.* **67**, 533 (1963).
103. F. GALASSO and J. PINTO, *Nature* **207**, 70 (1965).

104. F. Galasso and J. Pyle, *Inorg. Chem.* **2**, 482 (1963).
105. F. Galasso, L. Katz and R. Ward, *J. Am. Chem. Soc.* **81**, 820 (1959).
106. R. Roy, *J. Am. Ceram. Soc.* **37**, 581 (1954).
107. V. A. Bokov and I. E. Mylnikova, *Sov. Phys., Solid State* **2**, 2428 (1961).
108. I. G. Ismailzade, *Sov. Phys. Cryst.* **5**, 292 (1960).
109. A. S. Viskov, Yu. N. Venevtsev and G. S. Zhdanov, *Sov. Phys., Dokl.* **10**, 391 (1965).
110. L. Brixner, *J. Inorg. Nucl. Chem.* **15**, 352 (1960).
111. C. Keller, *J. Inorg. Nucl. Chem.* **27**, 321 (1965).
112. A. W. Sleight and R. Ward, *Inorg. Chem.* **1**, 790 (1962).
113. F. Galasso and W. Darby, *J. Phys. Chem.* **66**, 131 (1962).
114. F. Galasso, G. Layden and D. Flinchbaugh, *J. Chem. Phys.* **44**, 2703 (1966).
115. F. K. Patterson, C. W. Moeller and R. Ward, *Inorg. Chem.* **2**, 196 (1963).
116. V. S. Filipev and E. G. Fesenko, *Sov. Phys., Cryst.* **6**, 616 (1962).
117. L. Brixner, *J. Phys. Chem.* **64**, 165 (1960).
118. A. W. Sleight and R. Ward, *Inorg. Chem.* **3**, 292 (1964).
119. L. Brixner, *J. Am. Chem. Soc.* **80**, 3215 (1958).
120. V. S. Filipev and E. G. Fesenko, *Sov. Phys., Cryst.* **10**, 243 (1965).
121. F. Galasso and W. Darby, *Inorg. Chem.* **4**, 71 (1965).
122. G. A. Smolenskii, A. I. Agranovskaya, S. N. Popov and V. A. Isupov, *Sov. Phys., Tech. Phys.* **3**, 1981 (1958).
123. M. F. Kupriyanov and E. G. Fesenko, *Sov. Phys., Cryst.* **10**, 189 (1965).
124. I. G. Ismailzade, *Sov. Phys., Cryst.* **4**, 389 (1960).
125. M. F. Kupriyanov and E. G. Fesenko, *Sov. Phys., Cryst.* **7**, 358 (1962).
126. G. Blasse, *Philips Research Reports* **20**, 327 (1965).
127. E. G. Steward and H. P. Rooksby, *Acta Cryst.* **4**, 503 (1951).
128. J. Longo and R. Ward, *J. Am. Chem. Soc.* **83**, 2816 (1961).
129. G. Bayer, *J. Am. Ceram. Soc.* **46**, 604 (1963).
130. I. N. Belyaev, L. I. Medvedeva, E. G. Fesenko and M. F. Kupriyanov, *Izv. Akad. Nauk SSSR, Neorg. Materialy* **1**, (6) (1965) (Russian).
131. Yu. E. Roginskaya and Yu. N. Venevtsev, *Sov. Phys., Cryst.* **10**, 275 (1965).
132. F. Galasso and J. (Pyle) Pinto, *Inorg. Chem.* **4**, 255 (1965).

Table 7.5 References

1. H. J. Seifert and K. Klatyk, *Naturwiss.* **49**, 539 (1962).
2. W. Ruedorff, J. Kändler, G. Linike and D. Babel, *Angew. Chem.* **71**, 672 (1959).
3. K. Knox, *J. Chem. Phys.* **30**, 991 (1959).
4. B. Brehler and H. G. F. Winkler, *Heidelberger Beitr. zur Mineral. O. Petrog.* **4**, 6 (1954).
5. D. Balz and K. Plieth, *Z. Elektrochem.* **59**, 545 (1955).
6. F. Galasso and W. Darby, *J. Phys. Chem.* **66**, 1318 (1962).
7. O. Schmitz Du Mont and H. Bornefeld, *Z. Anorg. Allgem. Chem.* **287**, 120 (1956).
8. W. Ruedorff, J. Kaendler and D. Babel, *Z. Anorg. Allgem. Chem.* **317**, 261 (1962).
9. F. Galasso and W. Darby, *J. Phys. Chem.* **67**, 1451 (1963).
10. R. Weiss and R. Faivre, *Compt. Rend.* **248**, 106 (1959).
11. S. N. Ruddlesden and P. Popper, *Acta Cryst.* **10**, 538 (1957).
12. L. M. Kovba, E. A. Ippolitova, Yu P. Simanov and V. A. Spitsyn, *Dokl. Akad. Nauk, SSSR* **120**, 1042 (1958).
13. M. Foex, *Bull. Soc. Chim. France* **1961**, 109 (1961).
14. L. M. Kovba, E. A. Ippolitova, Y. P. Simanov and V. A. Spitsyn, *Zh. Fiz. Khim.* **35**, 563 (1961).
15. A. Rabenau and P. Eckerlin, *Acta Cryst.* **11**, 304 (1958).
16. J. J. Randall, L. Katz and R. Ward, *J. Am. Chem. Soc.* **79**, 266 (1957).
17. J. J. Randall and R. Ward, *J. Am. Chem. Soc.* **81**, 2629 (1959).
18. K. Lukaszewica, *Roczniki Chemi* **33**, 239 (1959).
19. G. Blasse, *J. Inorg. Nucl. Chem.* **27**, 2683 (1965).

Table 7.6 References

1. S. N. Ruddlesden and P. Popper, *Acta Cryst.* **10**, 538 (1957).
2. S. N. Ruddlesden and P. Popper, *Acta Cryst.* **11**, 54 (1958).
3. C. Brisi and P. Rolando, *Ric. Sci.* **36**, 48 (1966).

Table 7.7 References

1. B. Aurivillius, *Arkiv. Kemi.* **1**, 499 (1949).
2. D. A. Tambovtsev, V. M. Skorikov and I. S. Zheludev, *Kristallografiya* **6**, 889 (1963).
3. E. C. Subbarro, *J. Chem. Phys.* **34**, 695 (1961).

Table 7.8 References

1. F. H. Francombe, *Acta Cryst.* **13**, 131 (1960).
2. F. Galasso, L. Katz and R. Ward, *J. Am. Chem. Soc.* **81**, 5898 (1959).
3. C. D. Whiston and A. J. Smith, *Acta Cryst.* **19**, 169 (1965).
4. A. Magneli, *Acta Chem. Scand.* **7**, 315 (1953).
5. L. J. Fregel, G. P. Mohanty and J. H. Healy, *J. Chem. Eng. Data* **9**, 365 (1964).
6. E. Werner, P. Kierkegaard and A. Magneli, *Acta Chem. Scand.* **15**, 427 (1961).

Table 7.9 References

1. B. E. Gushee, L. Katz and R. Ward, *J. Am. Chem. Soc.* **79**, 5601 (1957).
2. D. Babel, *Z. Naturforsch.* **20a**, 165 (1965).
3. F. Galasso and L. Katz, *Acta Cryst.* **14**, 647 (1961).
4. A. W. Sleight, J. Longo and R. Ward, *Inorg. Chem.* **1**, 245 (1963).
5. J. G. Dickinson, L. Katz and R. Ward, *J. Am. Chem. Soc.* **83**, 3026 (1961).
6. F. K. Patterson, C. W. Moeller and R. Ward, *Inorg. Chem.* **2**, 196 (1963).
7. A. W. Sleight, Ph.D. Thesis, Univ. of Conn. (1963).
8. R. Ward, *Inorg. Chem.* **1**, 723 (1962).
9. A. Paton, M.S. Thesis, Univ. of Conn. (1962).

CHAPTER 8

SPINEL AND RELATED STRUCTURES

IN THE spinel structure, oxygen ions are close-packed in the ⟨111⟩ directions of the unit cell with one-eighth of the tetrahedral holes filled with A ions and one-half the octahedral holes filled with B ions. The Cu_2Mg structure contains atoms in the same arrangement as the cations in the spinel structure, and the magnetoplumbite structure is made up of close-packed layers of the spinel structure with AO_3 layers in between. A flow sheet of these structures is presented in Table 8.0.

TABLE 8.0. *Spinel and Related Structures*

8.1. Cu_2Mg structure
|
8.2. Spinel structure
|
8.3. $BaFe_{12}O_{19}$ structure

8.1. Cu_2Mg Type, C15, *Fd3m*, Cubic

The Cu_2Mg structure is similar to the spinel structure without the oxygen ions. Magnesium atoms are at the corners, at the centers of the faces and in two diagonally opposite lower and two diagonally opposite upper octants of a cubic unit cell. These magnesium atoms are not in the octants directly above one another. The four other octants each contain four copper atoms arranged in tetrahedra which have their atoms in the same relative positions in the octants as the four copper atoms had inside the unit cell. Thus, the atomic positions are at $(0, 0, 0; 0, \frac{1}{2}, \frac{1}{2}; \frac{1}{2}, 0, \frac{1}{2}; \frac{1}{2}, \frac{1}{2}, 0) + 0, 0, 0$ and $\frac{1}{4}, \frac{1}{4}, \frac{1}{4}$ for the magnesium atoms and these same values of x, y, z in parentheses + $(\frac{5}{8}, \frac{5}{8}, \frac{5}{8}; \frac{5}{8}, \frac{7}{8}, \frac{7}{8}; \frac{7}{8}, \frac{5}{8}, \frac{7}{8}$ and $\frac{7}{8}, \frac{7}{8}, \frac{5}{8})$ for the copper atoms (Fig. 8.1a).

Phases with this structure are called Laves phases. Laves phases contain atoms with similar radius ratios. A list is given in Table 8.1a.

Properties

As is shown in Fig. 8.1b, a number of these phases are superconductors

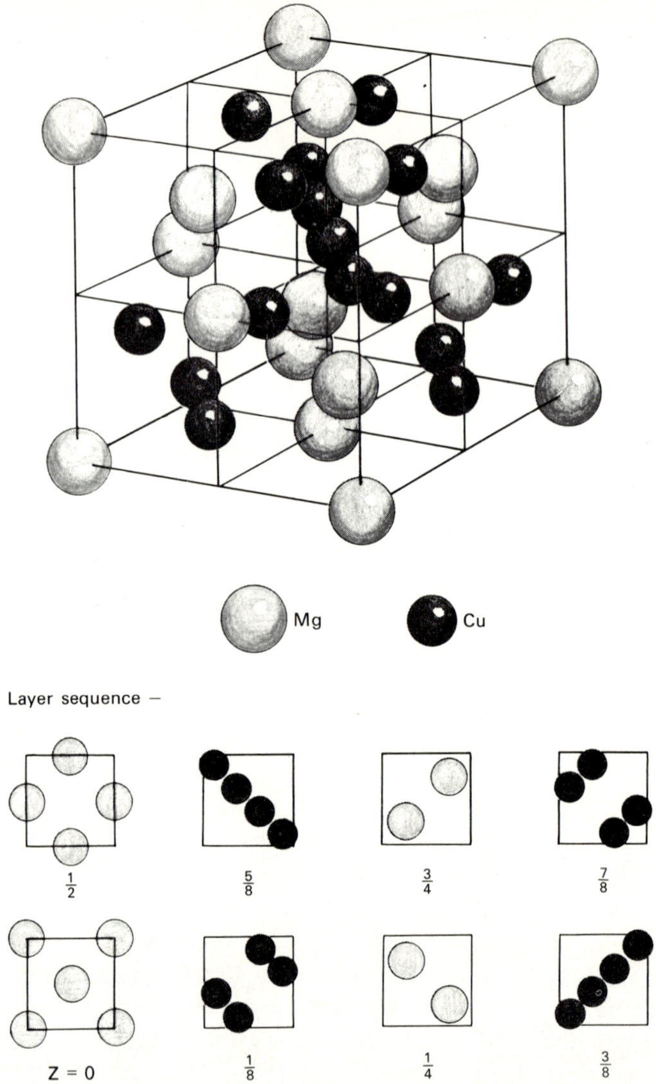

FIG. 8.1a The Cu_2Mg structure

with ZrV_2 having the highest transition temperature (8.5°K). Note that the electron per atom ratio for ZrV_2 is 5, a value which is common for high transition temperature superconductors.

A large number of rare earth phases listed in Table 8.1a are ferromagnetic. These phases, as well as those with other structures, are listed with their Curie points in Table 8.1b.

TABLE 8.1a. *Phases with the* Cu_2Mg *Structure*

Phase	Cell size, a_0 (Å)	Refs.
δ-$AgBe_2$	6.2997, 6.34	1, 2
Al_2Ca	8.038	3
Al_2Ce	8.064, 8.059	4, 5
Al_2La	8.145	5
Al_2Np	7.785	6
Al_2Pu	7.836(Al rich)–7.840(Pu rich)	7
Al_2U	7.744	8
Al_2Y	7.86	9
Au_2Bi	7.958	10
Au_2Na	7.803–7.817	11, 12
Au_2Pb	7.946	13
$BaPd_2$	7.953	14
$BaPt_2$	7.920	14
$BaRh_2$	7.852	14
Be_2Cu	5.952(Cu rich)–5.899(Be rich)	15
ε-Be_2Fe	5.884	16
Be_2K	—	17
Be_2Ti	6.448, 6.440	18
Bi_2Cs	9.760	19
Bi_2Rb	9.601	19
$CaPd_2$	7.665	14
$CaPt_2$	7.629	14
$CeCo_2$	7.161, 7.15	5, 20
$CeFe_2$	7.303	5
$CeIr_2$	7.571	21
$CeMg_2$	8.733	22
$CeNi_2$	7.202	5
$CeOs_2$	7.593	21
$CePt_2$	7.723	23
$CeRh_2$	7.538	24
$CeRu_2$	7.535	24
Co_2Dy	7.190, 7.187	25
Co_2Er	7.144	5
Co_2Gd	7.255, 7.23	20
Co_2Hf	6.918	24
Co_2Ho	7.168	5
Co_2Lu	7.123	26
Co_2Nb	6.758	27
Co_2Nd	7.300	28
Co_2Pr	7.312	5
Co_2Pu	7.075, 7.095–7.023	29, 30
Co_2Sc	6.921	23
Co_2Sm	7.260	5
β-Co_2Ta (H.T.)	6.720	31
Co_2Tb	7.206	5
Co_2Ti	6.704	32
Co_2Tm	7.121	28
Co_2U	6.9924	33
Co_2Y	7.216	34
Co_2Yb	7.060	35
Co_2Zr	6.90	23
Cr_2Hf (1200°)	7.15	36

TABLE 8.1a *(cont.)*

Phase	Cell size, a_0 (Å)	Refs.
Cr_2Nb	6.990	37
Cr_2Ta (L.T.)	6.961	37
Cr_2Ti (L.T.)	6.943	38
Cr_2Zr (H.T.)	7.207	39
Cu_2Mg	7.027	40
Cu_4MnSn	—	41
CuTaV	7.115	42
$DyFe_2$	7.325, 7.321	5, 25
$DyIr_2$	—	43
$DyMn_2$	7.564, 7.5731	25
$DyNi_2$	7.155, 7.142	5, 24, 25
$DyPt_2$	7.5966	25
$DyRh_2$	7.483	23
$ErFe_2$	7.273, 7.274	5, 23
$ErIr_2$	7.473	23
$ErMn_2$	7.07	44
$ErNi_2$	7.11	5
$ErRh_2$	7.444	23
$EuIr_2$	—	43
Fe_2Gd	7.36–7.53	5, 19, 25, 45, 46, 47
Fe_2Ho	7.30, 7.28	5, 20
Fe_2Lu	7.222	48
FeNiTa	6.717	42
FeNiU	7.054	33
Fe_2Pu	7.150–7.191	7, 29, 50, 50, 51
Fe_2Sm	7.415, 7.45	5, 20
Fe_2Tb	7.369	35
Fe_2Tm	7.247	28
Fe_2U	7.04–7.05	52, 53
Fe_2Y	7.355–7.357	23, 33, 54
Fe_2Zr	7.05	55
$GdIr_2$	7.550	23
$GdMn_2$	7.724–7.74	5, 25, 20, 56
$GdNi_2$	7.20–7.27	5, 25, 45, 57
$GdPt_2$	7.577	23, 24, 25, 58
$GdRh_2$	7.514	24
$HfMo_2$	7.555, 7.560	36, 59
HfV_2	7.400	36
HfW_2	7.591–7.599	36, 60, 61
$HoIr_2$	7.490	23
$HoMn_2$	7.507, 7.50	5, 20
$HoNi_2$	7.136–7.13	5, 20
$HoRh_2$	7.426	23
Ir_2La	7.686, 7.688	24, 52
Ir_2Nd	7.605	24
Ir_2Pr	7.621	24
Ir_2Sc	7.348, 7.346	58
Ir_2Sm	—	43
Ir_2Sr	7.849, 7.700	62, 63
Ir_2Tb	—	42
Ir_2Th	7.6615, 7.664	64, 65
Ir_2Tm	—	43

TABLE 8.1a *(cont.)*

Phase	Cell size, a_0 (Å)	Refs.
Ir_2V	7.4939	66
Ir_2Y	7.524, 7.550–7.520	23, 24
Ir_2Yb	—	43
Ir_2Zr	7.359	23, 65
$LaMg_2$	8.787	22
$LaOs_2$	7.737, 7.736	24, 58
$LaPt_2$	—	24, 58
$LaRh_2$	7.64	24, 58
$LaRu_2$	7.702	24, 58
$LuNi_2$	7.085	48
$LuRh_2$	7.412	48
MgNiZn	7.022	67
Mg_2Pr	8.689	22
Mg_2Th (H.T.)	8.570	68
Mn_2Pu	7.290, 7.292	28, 69
Mn_2Tb	7.620	5
Mn_2U	7.147	33
Mn_2Y	7.62–7.680	20, 23, 34
Mo_2Zr	7.60, 7.596	70, 71, 72
$NaPt_2$	7.48	73
$NdNi_2$	7.270	5
$NdPt_2$	7.694	24
$NdRh_2$	7.564	24
$NdRu_2$	7.614	24
Ni_2Lu	7.083	35
Ni_2Pr	7.285	50
Ni_2Pu	7.141–7.115 (600°C)	7, 29, 74
Ni_2Sc	6.926	23
Ni_2Sm	7.226	5
Ni_2Tb	7.160	5
Ni_2Tm	6.985	35
Ni_2Y	7.12–7.184	18, 20, 30, 40, 23, 34, 54
Ni_2Yb	7.060	28
Os_2Pr	7.663	24
Os_2Th	7.7050	64
Os_2U	7.512	66
Pd_2Sr	7.800, 7.826	62, 63
$PrPt_2$	7.709	24
$PrRh_2$	7.575	24
$PrRu_2$	7.624	24
Pt_2Sr	7.742, 7.777	62, 63
Pt_2Y	7.590, 7.607	32, 57, 22
$PuRu_2$	7.476, 7.474	50, 75
$PuZn_2$	7.760–7.747	76
Rh_2Sr	7.706, 7.695	57, 63
Rh_2Y	7.459, 7.488	24, 58
Ru_2Sm	7.580	23
Ru_2Th	7.649, 7.651	58, 65
V_2Zr	7.439	65
W_2Zr	7.615, 7.63	72, 77
Zn_2Zr	7.3958	78

TABLE 8.1b. *Rare Earth Ferromagnetic Materials*

Phase	Structure	T_c(°K)	Refs.
Ce_2Co_{17}	rhomb. Th_2Zn_{17}	1063	1
Pr_2Co_{17}		1153	1
Nd_2Co_{17}		1158	1
Sm_2Co_{17}		1198	1
Gd_2Co_{17}		1213	1
Tb_2Co_{17}		1198	1
Dy_2Co_{17}		1193	1
Y_2Co_{17}	hex. Th_2Ni_{17}	1198	1
Ho_2Co_{17}		1188	1
Er_2Co_{17}		1198	1
Tm_2Co_{17}		1188	1
Lu_2Co_{17}		1198	1
$CeFe_7$		100	2
$PrFe_7$		280	2
$NdFe_7$		326	2
Gd_2Fe_{17}		>420	2
Lu_2Fe_{17}		245	2
YCo_2	$MgCu_2$ structure	320	3
$PrCo_2$		50	3
$NdCo_2$		116	3
$SmCo_2$		259	3
$GdCo_2$			3
$TbCo_2$		256	3
$DyCo_2$		159	3
$HoCo_2$		95	3
$ErCo_2$		36	3
$TmCo_2$		18	3
$PrSi_2$		35	4, 5
$PrNi_2$		8	4, 5
$GdNi_2$		90	4, 5
$DyNi_2$		32	4, 5
$PrRu_2$		40	4, 5
$NdRu_2$		10.5	4, 5
$GdRu_2$		>77	4, 5
$PrRh_2$		8.6	4, 5
$NdRh_2$		8.1	4, 5
$PrOs_2$		35	4, 5
$PrIr_2$		18.5	4, 5
$NdIr_2$		11.8	4, 5
$GdIr_2$		>77	4, 5
$PrPt_2$		7.9	4, 5
$NdPt_2$		6.7	4, 5
$GdPt_2$		>77	4, 5
$ErRu_2$	hex. $MgZn_2$	13	4, 5
$GdRh_2$		>77	4, 5
$SmOs_2$		34	4, 5
$GdOs_2$		>77	4, 5
YCo_5	Cu_5Ca structure	975	6
$CeCo_5$		687	6
$SmCo_5$		1015	6
$GdCo_5$		1030	6
$DyCo_5$		1125	6
$HoCo_5$		1025	6

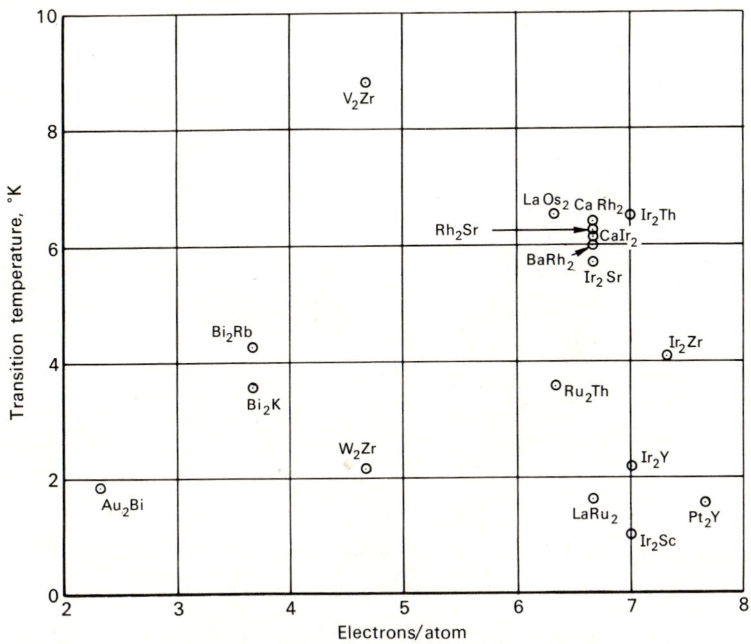

FIG. 8.1b T_c vs.ē/A for superconductors with the Cu_2Mg structure

8.2. Spinel Structure, $MgAl_2O_4$ Type, $H1_1$, *Fd3m*

The spinel structure can be described in terms of a close-packed cubic arrangement of anions with one-half the octahedral holes (B sites) and one-eighth of the tetrahedral holes (A sites) filled with cations. The general formula of compounds with the spinel structure is AB_2X_4, and there are eight of these units in a unit cell. The atoms are located in the following positions:

Equivalent positions $(0, 0, 0; 0, \frac{1}{2}, \frac{1}{2}; \frac{1}{2}, 0, \frac{1}{2}; \frac{1}{2}, \frac{1}{2}, 0)$ + 8 Mg at $0, 0, 0; \frac{1}{4}, \frac{1}{4}, \frac{1}{4}$

16 Al at $\frac{5}{8}, \frac{5}{8}, \frac{5}{8}; \frac{5}{8}, \frac{7}{8}, \frac{7}{8}; \frac{7}{8}, \frac{5}{8}, \frac{7}{8}; \frac{7}{8}, \frac{7}{8}, \frac{5}{8}$

32 O at $x, x, x; \frac{1}{4} - x, \frac{1}{4} - x, \frac{1}{4} - x$

$x, \bar{x}, \bar{x}; \frac{1}{4} - x, \frac{1}{4} + x, \frac{1}{4} + x$

$\bar{x}, x, \bar{x}; \frac{1}{4} + x, \frac{1}{4} - x, \frac{1}{4} + x$

$\bar{x}, \bar{x}, x; \frac{1}{4} + x, \frac{1}{4} + x, \frac{1}{4} - x$

In the nonideal structure the anions are moved from their ideal positions in ⟨111⟩ directions away from the nearest tetrahedron. The deviation can be seen from the anion parameter x, which is 0·375 in the ideal structure.

The spinel structure is quite complex and therefore can best be described

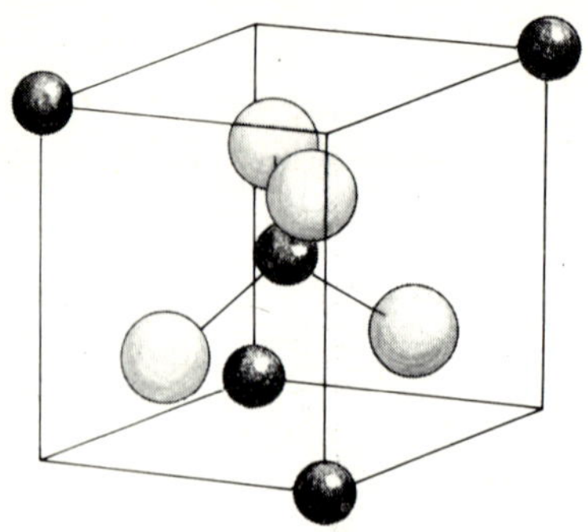

Type I octant containing A cations in tetrahedral co-ordination

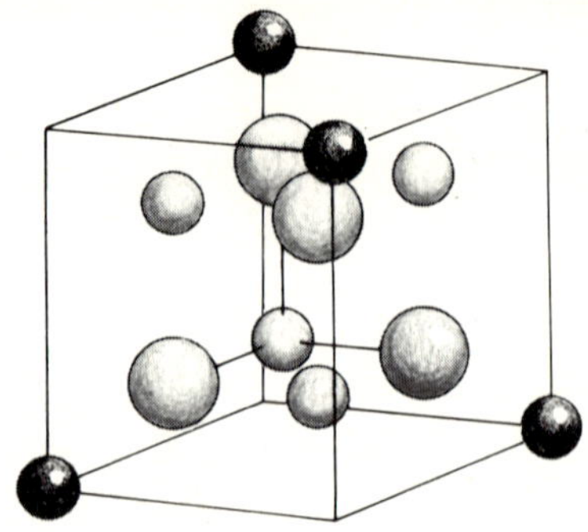

Type II octant containing B cations in octahedral co-ordination (octahedra extend beyond single octant)

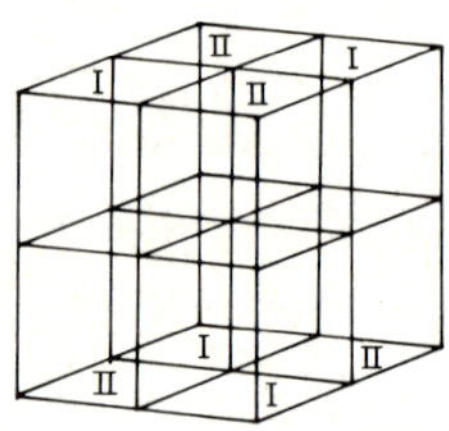

FIG. 8.2a The spinel structure

in steps. The A cations in the unit cell form a diamond cubic structure, with A ions at the corners, faces, and in half the corners inside the unit cell (see Fig. 8.1a). The large unit cell can be divided into eight small cells (octants), half of which contain a tetrahedral arrangement of anions around the A ions as is shown in Fig. 8.2a with the anions located in half the corners at the top and bottom of the small unit cell, type I.

In the other small unit cell, type II, the anions are located at sites similar to those occupied by the anions in type I cells, and the B ions are in other corners. When cells I and II are brought together the connection between the anion tetrahedra and octahedra can be seen, Fig. 8.2b. Note that the arrangement of metal ions is the same as the atoms in Cu_2Mg.

In the spinel structure there are one A and three B cations around each anion. The angle BXB is about 90° and angle AXB about 125°. The AX distance is $a(x - \frac{1}{4})\sqrt{3}$ and the BX distance is $a(\frac{5}{8} - x)$.

These angles and distances are important in determining the magnetic properties of the spinels. The exchange interactions in ferrites with this structure are of the indirect (superexchange) type. The strongest interaction is between the A ions in tetrahedral sites and the B ions in the octahedral sites, because the *AXB* distance is relatively small and the angle *AXB* is large. Thus the cation distribution in these sites also is important for magnetic materials.

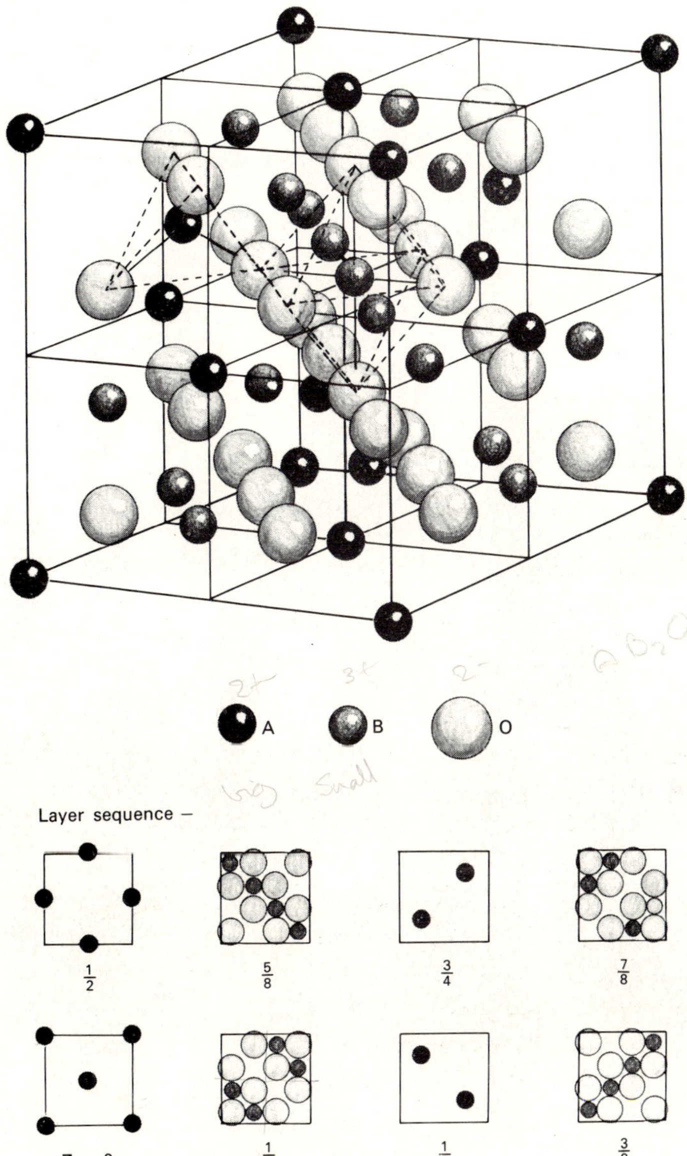

FIG. 8.2b The spinel structure

In the compound $MgAl_2O_4$ the Mg^{2+} ions are in tetrahedral sites and the Al^{3+} ions are in octahedral sites, an arrangement which is typical of a normal spinel. The distribution in $NiFe_2O_4$ is inverse, and can be written $Fe(NiFe)O_4$, indicating that the Ni^{2+} are in the octahedral sites with half the Fe^{3+} and the other half of the Fe^{3+} in tetrahedral sites. The distribution of cations also can be intermediate, such as is found in $Mn_{0.8}Fe_{0.2}(Mn_{0.2}Fe_{1.8})O_4$ and $Mg_{0.1}Fe_{0.9}(Mg_{0.9}Fe_{1.1})O_4$. A list of compounds with the spinel structure, cell sizes and distributions of ions is given in Table 8.2a(i).

TABLE 8.2a(i). *Compounds with the Spinel Structure*

Phase	Cell size a_0(Å)	Cation distribution	Oxygen parameter	Refs.
Oxides		1–3 type		
$Cu_{0.5}Fe_{2.5}O_4$	8.413			1
$Li_{0.5}Al_{2.5}O_4$	7.94	$Al[Li_{0.5}Al_{1.5}]$	0.385	2
$Li_{0.5}Fe_{2.5}O_4$	8.33	$Fe[Li_{0.5}Fe_{1.5}]$	0.382	3
$Li_{0.5}Ga_{2.5}O_4$	8.21	$Ga[Li_{0.5}Ga_{1.5}]$		4
		1–4 type		
$Li_{1.33}Mn_{1.67}O_4$	8.19	$Li[Li_{0.33}Mn_{1.67}]$		5
$Li_{1.33}Ti_{1.67}O_4$	8.36	$Li[_{0.33}Ti_{1.67}]$		6
		1–6 type		
Ag_2MoO_4	9.26	$Mo[Ag_2]$	0.364	7
Li_2MoO_4 (H.T.)				8
Li_2WO_4 (H.T.)				8
Na_2MoO_4	8.99	$Mo[Na_2]$		9
Na_2WO_4	8.99	$W[Na_2]$		9
$BaIn_2O_4$				10
$CaIn_2O_4$				
$CdCr_2O_4$	8.567	$Cd[Cr_2]$	0.385	2
$CdFe_2O_4$	8.69	$Cd[Fe_2]$		2
$CdGa_2O_4$	8.39	$Cd[Ga_2]$		11
$CdIn_2O_4$	9.167			5
$CdMn_2O_4$	8.22	$Cd[Mn_2]$		12
		2–3 type		
$CdRh_2O_4$	8.781	$Cd[Rh_2]$		13
CdV_2O_4	8.68			14
Co_3O_4	8.083	$Co^{2+}[Co_2{}^{3+}]$		15
$CoAl_2O_4$	8.105	$Co[Al_2]$		10
$CoCr_2O_4$	8.332	$Co[Cr_2]$		2
$CoFe_2O_4$	8.38	$Fe[CoFe]$		16
$CoGa_2O_4$	8.307	$Co_{0.1}Ga_{0.9}[Co_{0.9}Ga_{1.1}]$		15
$CoMn_2O_4$	8.1	$Co[Mn_2]$		18
$CoRh_2O_4$	8.495	$Co[Rh_2]$		13
CoV_2O_4	8.407	$Co[V_2]$		19
$CuAl_2O_4$	8.086	$Cu_{0.6}Al_{0.4}[Cu_{0.4}Al_{1.6}]$		20
$CuCo_2O_4$	8.054			21
$CuCr_2O_4$	a = 8.532 b = 7.788	$Cu[Cr_2]$		22

TABLE 8.2a(i) (*cont.*)

Phase	Cell size a_0(Å)	Cation distribution	Oxygen parameter	Refs.
$CuFe_2O_4$	$a = 8.22$ $c = 8.71$	Fe[CuFe]	0.380	23
$CuGa_2O_4$	8.39	Ga[CuGa]		24, 25
$CuMn_2O_4$	8.33	$Cu[Mn_2]$		12
$CuRh_2O_4$	$a = 8.702$ $c = 7.914$	$Cu[Rh_2]$		13
Fe_3O_4	8.394	$Fe[Fe^{2+}Fe^{3+}]$	0.379	26
$FeAl_2O_4$	8.10			27
$FeCo_2O_4$	8.254	$Co^{2+}[Fe^{3+}Co^{3+}]$		28
$FeCr_2O_4$	8.377	$Fe[Cr_2]$		2, 22
$FeGa_2O_4$	8.360	Ga[FeGa]		29
$FeMn_2O_4$	$a = 8.31$ $c = 8.85$	$Fe_{0.33}Mn_{0.67}[Fe_{0.67}Mn_{1.33}]$		12
FeV_2O_4	8.454			19
$MgAl_2O_4$	8.086	$Mg[Al_2]$	0.387	30
$MgCr_2O_4$	8.333	$Mg[Cr_2]$	0.385	21, 22
$MgFe_2O_4$	8.36	$Mg_{0.1}Fe_{0.9}[Mg_{0.9}Fe_{1.1}]$	0.382	31, 32, 33
$MgGa_2O_4$	8.286	$Mg_{0.33}Ga_{0.67}[Hg_{0.67}Ga_{1.33}]$	0.379	34, 35
$MgIn_2O_4$	8.81	In[MgIn]	0.372	36
$MgMn_2O_4$	$a = 8.07$ $c = 9.28$	$Mg[Mn_2]$	0.385	12, 37
$MgRh_2O_4$	8.530	$Mg[Rh_2]$		13
$MgTi_2O_4$	8.474	$Mg[Ti_2]$		38
MgV_2O_4	8.413	$Mg[V_2]$	0.385	39
Mn_3O_4	$a = 8.13$ $c = 9.43$	$Mn^{2+}[Mn^{23+}]$		10, 18
$MnAl_2O_4$	8.241	$Mn_{0.7}Al_{0.3}[Mn_{0.3}Al_{1.7}]$		40
$MnCo_2O_4$	8.153			28
$MnCr_2O_4$	8.437	$Mn[Cr_2]$		22
$MnFe_2O_4$	8.507	$Mn_{0.8}Fe_{0.2}[Mn_{0.2}Fe_{1.8}]$	0.385	41, 42
$MnGa_2O_4$	8.435	$Mn_{0.8}Ga_{0.2}[Mn_{0.2}Ga_{1.8}]$		17
$MnRh_2O_4$	8.613	$Mn[Rh_2]$		13
$MnTi_2O_4$	8.600	$Mn[Ti_2]$		38
MnV_2O_4	8.675	$Mn[V_2]$	0.388	43
$NiAl_2O_4$	8.046	$Ni_{0.25}Al_{0.75}[Ni_{0.75}Al_{1.25}]$	0.381	10
$NiCo_2O_4$	8.121	$Co^{2+}[Co^{3+}Ni^{3+}]$		28
$NiCo_2O_4$	8.252			45
$NiCr_2O_4$	$a - 8.248$ $c = 8.454$	$Ni[Cr_2]$		2, 22
$NiFe_2O_4$	8.325	Fe[NiFe]	0.381	44
$NiGa_2O_4$	8.258	Ga[NiGa]	0.387	40
$NiMn_2O_4$	8.39	$(Mn^{2+},Mn^{3+})[Ni(Mn^{3+},Mn^{4+})]$	0.383	46, 47
$NiRh_2O_4$	$a = 8.36$ $c = 8.67$	$Ni[Rh_2]$		13
$SnAl_2O_4$	8.12			48
$SrIn_2O_4$				
$ZnAl_2O_4$	8.086	$Zn[Al_2]$		16
$ZnCo_2O_4$	8.047	$Zn[Co_2]$		28
$ZnCr_2O_4$	8.327	$Zn[Cr_2]$		2, 28
$ZnFe_2O_4$	8.416	$Zn[Fe_2]$	0.380	2
$ZnGa_2O_4$	8.37	$Zn[Ga_2]$		49, 50

TABLE 8.2a(i) *(cont.)*

Phase	Cell size a_0(Å)	Cation distribution	Oxygen parameter	Refs.
$ZnMn_2O_4$	$a = 8.087$ $c = 9.254$	Zn[Mn_2]		10, 18
$ZnRh_2O_4$	8.54	Zn[Rh_2]		13
ZnV_2O_4	8.410	Zn[V_2]		11, 19
		2–4 type		
Cd_2SiO_4 (H.P.)	9.75			51
Co_2GeO_4	8.317	Ge[Co_2]	0.375	10
Co_2MnO_4	8.153			45
Co_2SiO_4 (H.P.)	8.140			52
Co_2SnO_4	8.644	Co[CoSn]		53
Co_2TiO_4	8.465	Co[CoTi]		10
Co_2VO_4	8.379	Co[CoV]		19
Fe_2GeO_4	8.411	Ge[Fe_2]	0.375	54
Fe_2SiO_4 (H.P.)				55
Fe_2TiO_4	8.521	Fe[FeTi]	0.386	56
Fe_2VO_4	8.421			
Mg_2GeO_4 (H.P.)	8.255	Ge[Mg_2]		57
Mg_2SnO_4	8.60	Mg[MgSn]		58
Mg_2TiO_4	8.445	Mg[MgTi]		10
Mg_2VO_4	8.39	Mg[MgV]	0.386	11
Mn_2SnO_4	8.865	Mn[MnSn]		29
Mn_2TiO_4	8.675	Mn[MnTi]		28, 43
Mn_2VO_4	8.575	$Mn_{0.8}V_{0.2}$[$Mn_{1.2}V_{0.8}$]	0.392	59
Ni_2GeO_4	8.221	Ge[Ni_2]	0.375	10
Ni_2SiO_4 (H.P.)	8·02			60
Zn_2SnO_4	8.665	Zn[ZnSn]		61
Zn_2TiO_4	8.445	Zn[ZnTi]	0.380	10
Zn_2VO_4	8.38	Zn[ZnV]		11
		2–5 type		
$Co_{2.33}Sb_{0.67}O_4$	8.523	Co[$Co_{1.33}Sb_{0.67}$]	0.380	62
$Zn_{2.33}Nb_{0.67}O_4$	8.589			63
$Zn_{2.33}Sb_{0.67}O_4$	8.594	Zn[$Zn_{1.33}Sb_{0.67}$]	0.390	62
Selenides				
$CdCr_2Se_4$	10.721	Cd[Cr_2]	0.383	64
$CuCr_2Se_4$	10.356	Cu[Cr_2]	0.380	15
$ZnCr_2Se_4$	10.443	Zn[Cr_2]	0.378	64
Sulfides		1–3 type		
$Ag_{0.5}Al_{2.5}S_4$	10.052	$Ag_{0.5}Al_{0.5}$[Al_2]	0.389	65
$Ag_{0.5}In_{2.5}S_4$	10.821	$Ag_{0.5}In_{0.5}$[In_2]	0.389	65
$Cu_{0.5}Al_{2.5}S_4$	9.856	$Cu_{0.5}Al_{0.5}$[Al_2]	0.387	65
$Cu_{0.5}In_{2.5}S_4$	10.698	$Cu_{0.5}In_{0.5}$[In_2]	0.390	65
$Li_{0.5}Al_{2.5}S_4$(L.T.)	9.996	$Li_{0.5}Al_{0.5}$[Al_2]	0.386	65
		2–3 type		
$CaIn_2S_4$	10.774	Ca[In_2]	<0.393	66
$CdCr_2S_4$		Cd[Cr_2]		67
$CdIn_2S_4$	10.797	Cd[In_2]	0.386	66
Co_3S_4	9.416			28

TABLE 8.2a(i) *(cont.)*

Phase	Cell size a_0(Å)	Cation distribution	Oxygen parameter	Refs.
$CoCr_2S_4$	9.934	Co[Cr_2]		67, 28
$CoIn_2S_4$	10.559	In[CoIn]		66
$CoRh_2S_4$	9.71	Co[Rh_2]		5
$CrAl_2S_4$	9.914	Al[CrAl]	0.384	28
$CrGa_2S_4$	9.95		0.387	68
$CrIn_2S_4$	10.59	In[CrIn]	0.386	68
$CuCo_2S_4$	9.482			28
$CuCr_2S_4$	9.629	Cu[Cr_2]	0.381	69
$CuRh_2S_4$	9.72	Cu[Rh_2]		5
$CuTi_2S_4$	9.880	Cu[Ti_2]	0.382	70
CuV_2S_4	9.824	Cu[V_2]	0.384	69
$FeCr_2S_4$	9.998	Fe[Cr_2]		67, 28
$FeIn_2S_4$	10.598	In[FeIn]	0.387	66
$HgCr_2S_4$	10.206	Hg[Cr_2]	0.392	67
$HgIn_2S_4$	10.812	Hg[In_2]	<0.403	66
$MgIn_2S_4$	10.687	In[MgIn]	<0.387	66
$MnCr_2S_4$	10.129	Mn[Cr_2]		67
$MnIn_2S_4$	10.694	$Mn_{0.33}In_{0.67}[Mn_{0.67}In_{0.33}]$	0.390	66
Ni_3S_4	9.65		0.375	71
$NiCo_2S_4$	9.392			28
$NiIn_2S_4$	10.464	In[NiIn]		66
$ZnAl_2S_4$(L.T.)	9.988	Zn[Al_2]	0.384	72
$ZrCr_2S_4$	9.983	Zn[Cr_2]		67
Tellurides				
$CuCr_2Te_4$	11.049	Cu[Cr_2]	0.379	69

TABLE 8.2a(ii). *Complex Spinels*

Phase	Cell size a_0(Å)	Cation distribution	Refs.
		$LiB^{2+}B^{5+}O_4$	
$LiCoSbO_4$	8.56	$Li_{1-x}Co_x[Li_xCo_{1-x}Sb]$	73
$LiCoVO_4$	8.28	V[LiCo]	5
$LiMgVO_4$ (H.P.)	8.27	V[LiMg]	5
$LiNiVO_4$	8.21	V[LiNi]	5
$LiZnNbO_4$	$a = 8.60$ $c = 8.40$	Zn[LiNb]	5
$LiZnSbO_4$	8.55	$Li_{1-x}Zn_x[Li_xZn_{1-x}Sb]$	73
$LiZnVO_4$ (H.P.)	8.31	V[LiZn]	5
		$LiB^{3+}B^{4+}O_4$	
$LiCrGeO_4$	8.20	Li[CrGe]	
$LiRhGeO_4$	8.39	Li[RhGe]	
$LiMn_2O_4$	8.25	Li[$Mn^{3+}Mn^{4+}$]	5
$LiAlMnO_4$	8.12	Li[AlMn]	5
$LiCoMnO_4$	8.07	Li[CoMn]	5
$LiCrMnO_4$	8.19	Li[CrMn]	5
$LiFeMnO_4$	8.26	$Li_{0.6}Fe_{0.4}[Li_{0.4}Fe_{0.6}Mn]$	5
$LiGaMnO_4$	8.23	$Li_{0.5}Ga_{0.5}[Li_{0.5}Ga_{0.5}Mn]$	5

TABLE 8.2a(ii) *(cont.)*

Phase	Cell size a_0(Å)	Cation distribution	Refs.
$LiRhMnO_4$	8.30	Li[RhMn]	5
$LiAlTiO_4$	8.30	Li[AlTi]	5
$LiCrTiO_4$	8.32	Li[CrTi]	5
$LiFeTiO_4$	8.35	$Li_{0.5}Fe_{0.5}[Li_{0.5}Fe_{0.5}Ti]$	5
$LiGaTiO_4$	8.29	$Li_{0.3}Ga_{0.7}[Li_{0.7}Ga_{0.3}Ti]$	5
$LiMnTiO_4$	8.30	Li[MnTi]	5
$LiRhTiO_4$	8.39	Li[RhTi]	5
$LiVTiO_4$	8.35	Li[VTi]	5
LiV_2O_4	8.22	$Li[V^{3+}V^{4+}]$	5
		$LiB^{2+}_{0.5}B^{4+}_{1.5}O$	
$LiCo_{0.5}Ge_{1.5}O_4$	8.20	$Li_{0.5}Co_{0.5}[Li_{0.5}Ge_{1.5}]$	5
$LiNi_{0.5}Ge_{1.5}O_4$	8.17	$Li[Ni_{0.5}Ge_{1.5}]$	5
$LiZn_{0.5}Ge_{1.5}O_4$	8.213	$Li_{0.5}Zn_{0.5}[Li_{0.5}Ge_{1.5}]$	74
$LiCo_{0.5}Mn_{1.5}O_4$	8.16	$Li[Co_{0.5}Mn_{1.5}]$	5
$LiCu_{0.5}Mn_{1.5}O_4$	8.23	$Li[Cu_{0.5}Mn_{1.5}]$	5
$LiMg_{0.5}Mn_{1.5}O_4$	8.18	$Li[Mg_{0.5}Mn_{1.5}]$	5
$LiNi_{0.5}Mn_{1.5}O_4$	8.17	$Li[Ni_{0.5}Mn_{1.5}]$	
$LiZn_{0.5}Mn_{1.5}O_4$	8.19	$Li_{0.5}Zn_{0.5}[Li_{0.5}Mn_{1.5}]$	5
$LiCd_{0.5}Ti_{1.5}O_4$	8.52	$Li_{0.5}Cd_{0.5}[Li_{0.5}Ti_{1.5}]$	75
$LiCo_{0.5}Ti_{1.5}O_4$	8.36	$Li_{0.5}Co_{0.5}[Li_{0.5}Ti_{1.5}]$	75
$LiCu_{0.5}Ti_{1.5}O_4$	8.39	$Li_{0.7}Cu_{0.5}[Li_{0.3}Ti_{1.5}]$	5
$LiMg_{0.5}Ti_{1.5}O_4$	8.37	$Li_{0.5}Mg_{0.5}[Li_{0.5}Ti_{1.5}]$	5
$LiMn_{0.5}Ti_{1.5}O_4$	8.44	$Li_{0.5}Mn_{0.5}[Li_{0.5}Ti_{1.5}]$	5
$LiZn_{0.5}Ti_{1.5}O_4$	8.190	$Li_{0.5}Zn_{0.5}[Li_{0.5}Ti_{1.5}]$	74
$LiCr_{1.5}Sb_{0.5}O_4$	8.37	$Li[Cr_{1.5}Sb_{0.5}]$	5
$LiRh_{1.5}Sb_{0.5}O_4$	8.54	$Li[Rh_{1.5}Sb_{0.5}]$	5
$Li_{1.5}TiSb_{0.5}O_4$	8.35	$Li[Li_{0.5}TiSb_{0.5}]$	5
$CuGaAlO_4$			25
$CuGaCrO_4$			25
$CuGaMnO_4$		Cu[GaMn]	25

Properties

Ferrimagnetic spinels and their Curie points are listed in Table 8.2b.

The melting point of $MgAl_2O_4$ is 2130°C, the thermal conductivity is 0.035 cal/cm-sec-°C and the coefficient of thermal expansion is 7.79×10^{-6}/°C. The melting points and coefficients of thermal expansion of other spinels also are given in Table 8.2c.

It should be noted that some of the most interesting magnetic materials are mixed ferrites. The MnZn ferrites have high values of initial permeability, the NiZn ferrites have high coercive values, high resistivities and high Curie temperatures and the MgMn ferrites exhibit square B—H loops. The

TABLE 8.2b. *Ferromagnetic Curie Points of Spinels*

Phase	Curie point (°C)	Refs.
$CdFe_2O_4$		1
$CoFe_2O_4$	520	1
$CuFe_2O_4$	455	1
Fe_3O_4	588	1
$FeCo_2O_4$	177	2
$Fe_{0.50}Li_{0.50}Cr_2O_4$	80	1
$Fe[Li_{0.5}Fe_{1.50}]O_4$	680	1
$MgFe_2O_4$	440	1
$MnCr_2O_4$	103	2
$MnCo_2O_4$	233	1
$MnFe_2O_4$	330	1
$Mn_{1.5}FeTi_{0.5}O_4$	89	1
$NiCo_2O_4$	77	2
$NiFe_2O_4$	580	1
$Ni_{1.5}FeTi_{0.5}O_4$	293	1
$NiZn_{0.5}FeTi_{0.5}O_4$	225	1
$ZnFe_2O_4$	60	1

TABLE 8.2c. *Properties of Phases with the Spinel Structure*

Phase	Melting point (°C)	Thermal expansion ($\times 10^{-6}$/°C)
$CoAl_2O_4$	1955[1]	8.52[4]
Fe_3O_4	1594[2]	
$FeAl_2O_4$		8.62[4]
$MgAl_2O_4$	2130[1]	7.79[4]
$MnAl_2O_4$		7.18[4]
$NiAl_2O_4$	2020[1]	8.41[4]
$ZnAl_2O_4$	1950[1]	8.72[4]
$MgCr_2O_4$	2000[1]	7.90[4]
Mn_3O_4	1564[2]	
$MnCr_2O_4$		9.34[4]
$ZnCr_2O_4$		8.72[4]
$MgFe_2O_4$		12.11[4]
$ZnFe_2O_4$		9.85[4]
$LiZn_{0.5}Ge_{1.5}O_4$	d650	
$LiZn_{0.5}Ti_{1.5}O_4$	995[2]	

constitution, as well as the distribution of ions in these phases, have to be carefully controlled to obtain reproducible properties.

Complex Spinel Compounds

A large number of phases of the $LiB^u_{2-x}B^v_xO_4$ (where u and v are the charges on different B ions) type were prepared by Blasse (see Table 8.2a(ii)). He assumed that the oxygen parameter x for these compounds was 0.375.

In his studies, Blasse found three types of ordering in the lithium spinels, a 1:1 ordering of Li^+ and Nb^{5+} on B sites with tetragonal distortion of the unit cell, another 1:1 ordering of B" and Sb^{5+} ions or Li^+ and B^{4+} on B sites for nearly all spinel compositions $LiB^{2+}_{0.5}B^{4+}_{1.5}O_4$.

8.3. $BaFe_{12}O_{19}$ and $KFe_{11}O_{17}$ Structures, Magnetoplumbite, $P6_3/mmc$

Both the $BaFe_{12}O_{19}$ and $KFe_{11}O_{17}$ structures contain four cubic-close-packed layers from the spinel structure. In the $BaFe_{12}O_{19}$ structure (see Fig. 8.3a), close-packed barium–oxygen layers are on the bottom and top

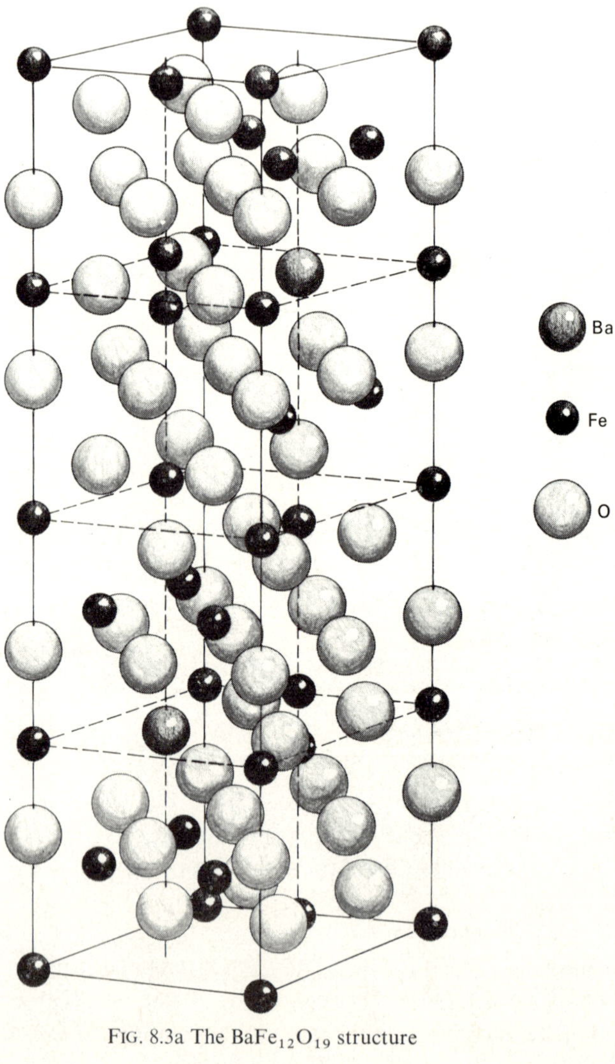

FIG. 8.3a The $BaFe_{12}O_{19}$ structure

of these layers, while in the $KFe_{11}O_{17}$ structure the potassium–oxygen layers on the top and bottom of the four close-packed layers are not close-packed. Then two more close-packed layers are mirrored across the barium–oxygen layers and potassium–oxygen layers. The layer structure for $BaFe_{12}O_{19}$ is shown in Fig. 8.3b; the layer sequence in Fig. 8.3c.

The atomic positions from space group $P6_3/mmc$ used in defining this structure are:

(2a) $0, 0, 0; 0, 0, \frac{1}{2}$
(2b) $0, 0, \frac{1}{4}; 0, 0, \frac{3}{4}$
(2d) $\frac{1}{3}, \frac{2}{3}, \frac{3}{4}; \frac{2}{3}, \frac{1}{3}, \frac{1}{4}$
(4e) $\pm(0, 0, z; 0, 0, \frac{1}{2} + z)$

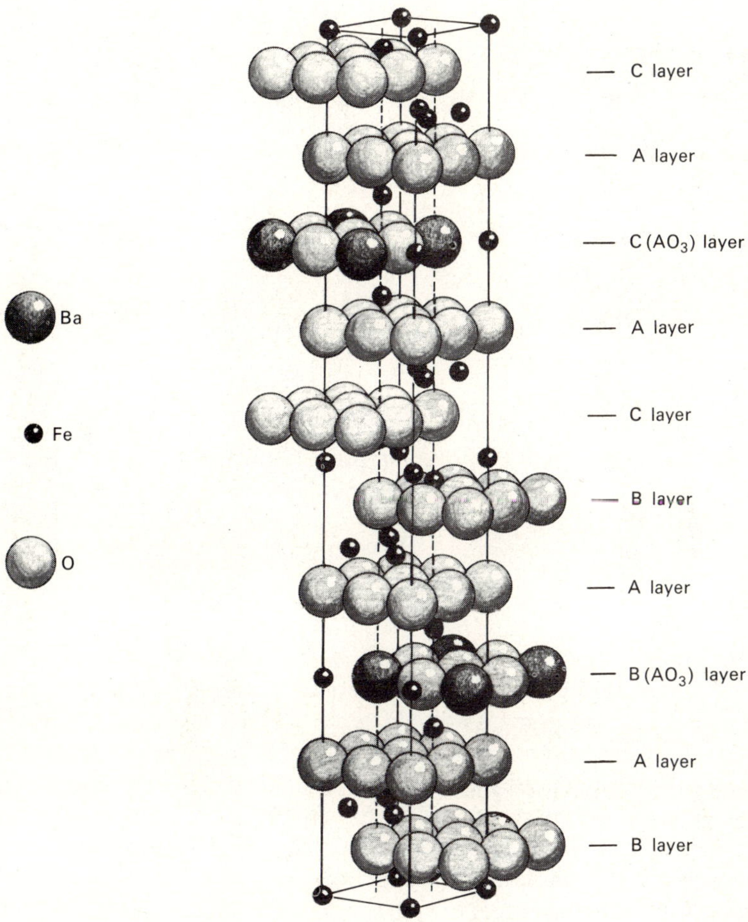

FIG. 8.3b The $BaFe_{12}O_{19}$ structure

(4f) $\pm(\frac{1}{3}, \frac{2}{3}, z; \frac{2}{3}, \frac{1}{3}, \frac{1}{2} + z)$
(6h) $\pm(x, 2x, \frac{1}{4}; 2x, x, \frac{3}{4}; x, \bar{x}, \frac{1}{4})$
(12k) $\pm(x, 2x, z; 2x, x, \bar{z}; x, \bar{x}, z; x, 2x, \frac{1}{2} - z; 2x, x, \frac{1}{2} + z; \bar{x}, x, \frac{1}{2} + z)$

For $BaFe_{12}O_{19}$ the atoms adopt the following positions: 2 Ba in (2d); 24 Fe in (2a), (sb), (4f) with $z = 0.028$, (4f) with $z = 0.189$, and (12k) with $x = 0.167, z = -0.108$; and 38 O in (4e) with $z = 0.150$, (4f) with $z = -0.050$, (6h) with $x = 0.186$, and (12k) with $x = 0.167$, $z = 0.050$, and again with

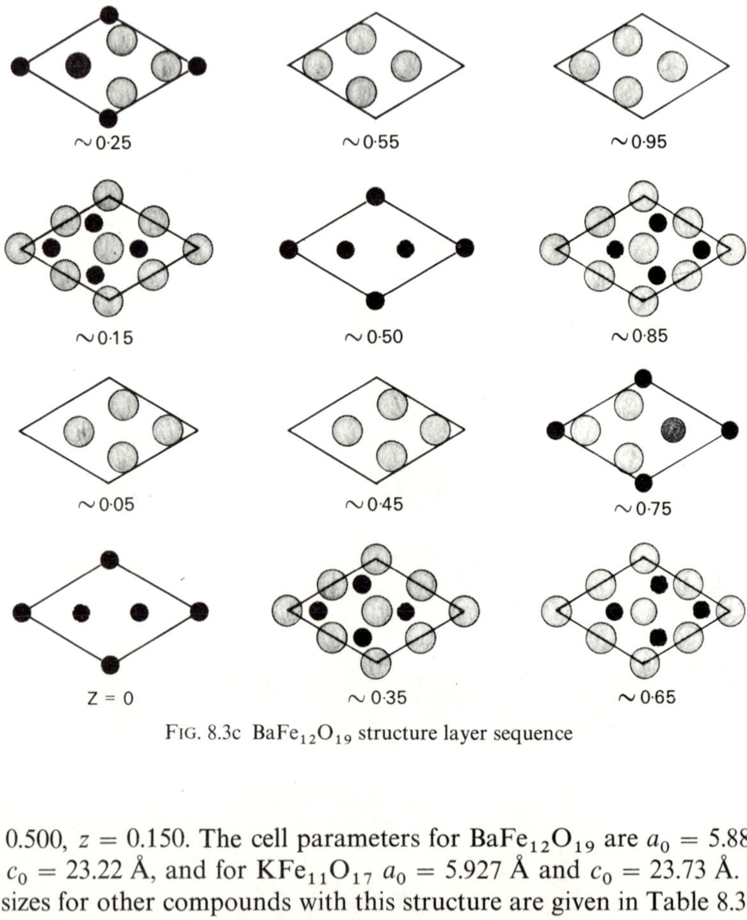

FIG. 8.3c $BaFe_{12}O_{19}$ structure layer sequence

$x = 0.500$, $z = 0.150$. The cell parameters for $BaFe_{12}O_{19}$ are $a_0 = 5.888$ Å and $c_0 = 23.22$ Å, and for $KFe_{11}O_{17}$ $a_0 = 5.927$ Å and $c_0 = 23.73$ Å. The cell sizes for other compounds with this structure are given in Table 8.3a.

These arrangements of paramagnetic metal ions allow a positive interaction across the barium–oxygen layers and a negative interaction across the potassium–oxygen layers. Thus, $BaFe_{12}O_{19}$ is a ferrimagnetic material while $KFe_{11}O_{17}$ is antiferromagnetic.

Other variations of $BaFe^{2+}_{12}O_{19}$ structure are formed by $BaFe^{2+}_{2}Fe^{3+}_{16}O_{27}$, $Ba_2Fe^{2+}_{2}Fe^{3+}_{28}O_{46}$, $Ba_2Zn^{2+}_{2}Fe^{3+}_{12}O_{22}$ and $Ba_3Co^{2+}_{2}Fe^{3+}_{24}O_{41}$. Each of these

TABLE 8.3a. *Compounds with the Magnetoplumbite Structure*

Compound	Cell size (Å)		Refs.
	a_0	c_0	
$BaAl_{12}O_{19}$	5.588	22.71	1
$BaCr_6Fe_6O_{19}$	5.844	22.81	1
$BaFe_{12}O_{19}$	5.888	23.22	2
$BaGa_{12}O_{19}$	5.818	23.00	1
$CaAl_{12}O_{19}$	5.547	21.87	3
$PbAl_{12}O_{19}$	5.563	22.033	4
$PbFe_{12}O_{19}$	5.889	23.07	1
$PbGa_{12}O_{19}$	5.820	23.056	4
$SrAl_{12}O_{19}$	5.568	21.99	2
$SrFe_{12}O_{19}$	5.876	23.08	1

compounds has a structure related to that of spinel. The *a*-axis is ~5.9 Å for all of these compounds but the *c*-axes are 32.8, 84.1, 43.6 and 52.3 Å, respectively. $NaAl_{11}O_{17}$, $KAl_{11}O_{17}$ and $RbFe_{11}O_{17}$ also have this structure with cell parameters of $a_0 = 5.59$ Å, $c_0 = 22.49$ Å; $a_0 = 5.598$ Å, $c_0 = 22.71$ Å;

TABLE 8.3b. *Ferromagnetic Curie Points of Magnetoplumbite and Related Phases*

Phase	Curie point (°C)	Refs.
$AFe_{12}O_{19}$		
$BaFe_{12}O_{19}$	452	1
$La(Fe^{2+}Fe^{3+}_{11})O_{19}$	422	2
$PbFe_{12}O_{19}$	452	3
$BaB^{2+}_2Fe_{16}O_{27}$		
$Ba(CoFe)Fe_{16}O_{27}$	430	4
$BaFe_2Fe_{16}O_{27}$	455	4
$BaMn_2Fe_{16}O_{27}$	415	4
$Ba(NiFe)Fe_{16}O_{27}$	523	4
$Ba(Ni_{0.5}Zn_{0.5}Fe)Fe_{16}O_{27}$	450	4
$Ba(ZnFe)Fe_{16}O_{27}$	430	4
$Ba_2B^{2+}_2Fe_{12}O_{22}$		
$Ba_2Co_2Fe_{12}O_{22}$	343	5
$Ba_2Mg_2Fe_{12}O_{22}$	373	5
$Ba_2Mn_2Fe_{12}O_{22}$	287	5
$Ba_2Ni_2Fe_{12}O_{22}$	383	5
$Ba_2Zn_2Fe_{12}O_{22}$	127	5
$Ba_3B^{2+}_2Fe_{24}O_{41}$		
$Ba_3Co_2Fe_{24}O_{41}$	403	5
$Ba_3Cu_2Fe_{24}O_{41}$	437	5
$Ba_3Zn_2Fe_{24}O_{41}$	357	5

and $a_0 = 5.927$ Å, $c_0 = 23.88$ Å, respectively. The atomic positions for $NaAl_{11}O_{17}$ are as follows: 2 Na in (2d); 22 Al in (2a), (4f) with $z = 0.022$ and with $z = 0.178$, (12k) with $x = 0.167$, $z = -0.106$; and 34 O in (4e) with $z = 0.144$, (4f) with $z = -0.050$, in half the (4f) positions with $z = 0.25$, and in (12k) with $x_1 = 0.167$, $z_1 = 0.050$ and with $x_2 = 0.500$, $z_2 = 0.144$.

Properties

The magnetic curie points for a number of magnetoplumbite and related structures are listed in Table 8.3b.

8.4. Discussion

Structure (Relations)

The Cu_2Mg structure has magnesium atoms in a diamond cubic arrangement and a tetrahedral arrangement of copper atoms in the unit cell octants not occupied by magnesium atoms. If oxygen atoms are placed in the unit cell of Cu_2Mg so that four oxygen atoms surround the magnesium atoms and six oxygen atoms are around each copper atom, then the spinel structure is formed. By taking the spinel structure in the $\langle 111 \rangle$ direction and dividing these units with AO_3 layers the magnetoplumbite structure is formed.

Intermetallics

Compounds with the Cu_2Mg structure like those with the $MgZn_2$ structure are called Laves phases. As was mentioned in Chapter 6, the formation of these phases depends only on the relative sizes of atoms. A number of these compounds are superconductors. The compound ZrV_2 has a transition temperature of $\sim 8°K$.

Oxides

The oxides with the spinel structure are important because of their ferromagnetic properties. By substituting various ions in the tetrahedral and octahedral holes the properties of these phases can be modified. The ferromagnetic magnetoplumbite type phases have been used as permanent magnet materials. The compound $BaFe_{12}O_{19}$ is extremely useful, since the conductivity is low and losses are minimized. By orienting the grains in this material, coercive forces as great as 2500 oersteds have been obtained. Excluding a platinum cobalt alloy and MnBi, which have coercive forces of 4300 and 3600 oersteds, respectively, and some rare earth cobalt phases, this coercive force is one of the largest for permanent magnet materials.

Table 8.1a References

1. L. Misch, *Metallwirt* **15**, 163 (1936).
2. O. Hajicek, *Hutnické Listy* **3**, 265 (1948).
3. H. Nowotny and A. Mohrnheim, *Z. Krist.* **A100**, 540 (1939)
4. J. H. N. van Vucht, *Philips Res. Rept.* **16**, 1 (1961).
5. J. H. Wernick and S. Geller, *Trans. AIME* **218**, 866 (1960).
6. D. C. Hamilton, C. S. Raub, B. T. Matthias, E. Corenzwit and G. W. Hull, Jr., *J. Phys. Chem. Solids* **26**, 665 (1965).
7. O. J. C. Runnalls, *Trans. AIME* **197**, 1460 (1953).
8. A. A. Bochvar *et al.*, *Proc. U.N. Conf. Peaceful Uses At. Energy*, 2d, Geneva, **6**, 184 (1958).
9. M. I. Ivanov, V. A. Tumbakov and N. S. Podolskaya, *Soviet J. At. Energy* **5**, 1007 (1959).
10. F. Jurriaane, *Z. Krist.* **90**, 322 (1935).
11. H. Perlitz and E. Aruja, *Z. Krist.* **100**, 157 (1938).
12. W. Haucke, *Z. Electrochem.* **43**, 712 (1937).
13. H. J. Wallbaum, *Z. Metallk.* **35**, 218 (1943).
14. E. A. Wood and V. B. Compton, *Acta Cryst.* **11**, 429 (1958).
15. L. Losana and G. Venturello, *Aluminio* **11**, 8 (1942).
16. R. J. Teitel and M. Cohen, *Trans. AIME* **188**, 1028 (1950).
17. E. Janecke, Karl Winter Univ., Heidelberg, 1949; quoted by A. E. Vol, *Constitution and Prop. of Binary Metallic Systems*, Vol. I, 588, Gosudarst. Izdatel, Moscow (1959).
18. P. Ehrlich, *Z. Anorg. Chem.* **259**, 1 (1949).
19. G. Gnutzmann and W. Klemm, *Z. Anorg. Allgem. Chem.* **309**, 181 (1961).
20. K. Nassau, L. V. Cherry and W. E. Wallace, *Phys. Chem. Solids* **16**, 123 (1960).
21. R. M. Bozorth, B. T. Matthias, H. Suhl, E. Corenzwit and D. D. Davis, *Phys. Rev.* **115**, 1595 (1959).
22. A. Iandelli, *Natl. Phys. Lab. Gt. Brit. Proc. Symp.* No. 9, I, Paper 3F (1959).
23. A. E. Dwight, *Trans. ASM* **53**, 479 (1961).
24. V. B. Compton and B. T. Matthias, *Acta Cryst.* **12**, 651 (1959).
25. N. C. Baenziger and J. L. Moriarty, Jr., *Acta Cryst.* **14**, 948 (1961).
26. T. H. Geballe, B. T. Matthias, V. B. Compton, E. Corenzwit, G. W. Hull, Jr. and L. D. Longinotti, *Phys. Rev.* **137**, 119 (1965).
27. H. J. Wallbaum, *Arch. Eisenhüttenw.* **14**, 521 (1940–41).
28. S. E. Haszko, *Trans. AIME* **218**, 958 (1960).
29. O. J. C. Runnalls, *Can. J. Chem.* **34**, 133 (1956).
30. D. M. Poole, M. G. Bale and P. G. Mardon (Eds.), *Plutonium, 1960*, p. 267. Cleaver-Hume Press, Ltd., London (1961).
31. M. Korchynsky and R. W. Fountain, *Trans. AIME* **215**, 1033 (1959).
32. H. J. Wallbaum and H. Witte, *Z. Metallk.* **31**, 185 (1939).
33. N. C. Baenziger, R. E. Rundle, A. I. Snow and A. S. Wilson, *Acta Cryst.* **3**, 34 (1950).
34. B. J. Beaudry, J. F. Heafling and A. H. Daane, *Acta Cryst.* **13**, 743 (1960).
35. P. I. Kripyakevich, M. Yu Teslyuk and D. P. Frankevich, *Kristallografiya* **10**, 422 (1965).
36. R. P. Elliott, *Trans. ASM* **53**, 321 (1961).
37. P. Duwez and H. Martens, *Trans. AIME* **194**, 72 (1952).
38. P. Duwez and J. L. Taylor, *Trans. ASM* **44**, 495 (1952).
39. R. P. Elliott, *Constitution of Binary Alloys*, M. Hansen, McGraw-Hill, New York (1958).
40. V. G. Sederman, *Phil. Mag.* **18**, 343 (1934).
41. S. Valentiner, *Z. Metallk.* **44**, 49 (1953).
42. K. Kuo, *Acta Met.* **1**, 720 (1953).
43. R. M. Bozorth, B. T. Matthias, H. Suhl, E. Corenzwit and D. D. Davis, *Phys. Rev.* **115**, 1595 (1959).
44. W. E. Wallace, *Rare Earth Alloys*, p. 212, D. Van Nostrand Co., Inc., Princeton, N.Y. (1961).
45. M. Copeland and H. Kato, in J. F. Nachmand and C. E. Lundin (Eds.), *Rare Earth Research*, p. 133, Gordon and Breach, Science Publishers, Inc., N.Y. (1962).

46. E. M. Savitskii, V. F. Terekhova, I. V. Burov and O. D. Christyakov, *Russ. J. Inorg. Chem.* **6**, 883 (1961).
47. V. F. Novy, R. C. Vickery and E. V. Kleber, *Trans. AIME* **221**, 580 (1961).
48. A. E. Dwight, U.S. At. Energy Comm. ANL-6516, 259 (1961).
49. P. G. Mardon, H. R. Haines, J. H. Pearce and M. B. Waldron, *J. Inst. Metals* **86**, 166 (1957–8).
50. F. W. Schonfeld *et al. Metallurgy and Fuels,* Progress in Nuclear Energy Ser, V, Vol. 2, Pergamon Press, N.Y. (1959).
51. F. H. Ellinger, in A. S. Coffinberry and W. N. Miner (Eds.), *The Metal Plutonium,* p. 281, Univ. of Chicago Press, Chicago (1961).
52. P. Gordon and A. R. Kaufmann, *Trans. AIME* **188**, 182 (1950).
53. C. J. Clews, *J. Inst. Metals* **77**, 577 (1950).
54. R. F. Domagala, J. J. Rausch and D. W. Levinson, *Trans. ASM* **53**, 137 (1961).
55. E. T. Hayes, A. H. Roberson and W. L. O'Brian, *Trans. ASM* **43**, 888 (1951).
56. W. M. Hubbard, E. Adams and J. V. Gilfrich, *J. Appl. Phys.* **31**, 3685 (1960).
57. V. F. Novy, R. C. Vickery and E. V. Kleber, *Trans. AIME* **221**, 585 (1961).
58. A. E. Dwight and M. F. Nevitt, U.S. At. Energy Comm. ANL-6099, 76 (1959).
59. A. Taylor, N. J. Doyle and B. J. Kagle, *J. Less-Common Metals* **3**, 265 (1961).
60. B. C. Giessen, I. Rump and N. J. Grant, *Trans. AIME* **224**, 60 (1962).
61. H. Braun and E. Rudy, *Z. Metallk.* **51**, 360 (1960).
62. T. Heumann and M. Kniepmeyer, *Z. Anorg. Allgem. Chem.* **290**, 191 (1957).
63. A. E. Wood and V. B. Compton, *Acta Cryst.* **11**, 429 (1958).
64. A. E. Dwight, J. W. Downey and R. A. Conner, Jr., *Trans. AIME* **212**, 337 (1958).
65. B. T. Matthias, V. B. Compton and E. Corenzwit, *Phys. Chem. Solids* **19**, 130 (1961).
66. T. J. Heal and G. I. Williams, *Acta Cryst.* **8**, 494 (1955).
67. K. H. Lieser and H. White, *Z. Metallk.* **43**, 396 (1952).
68. D. T. Peterson, P. F. Diljak and C. J. Vold, *Acta Cryst.* **9**, 1036 (1956).
69. A. S. Coffinberry and F. H. Ellinger, *Proc. U.N. Intern. Conf. Peaceful Uses At. Energy Geneva,* **9**, 138 (1955).
70. R. F. Domagala, D. J. McPherson and M. Hansen, *Trans. AIME* **197**, 73 (1953).
71. P. Duwez and C. B. Jordan, *J. Am. Chem. Soc.* **73**, 5509 (1951).
72. R. P. Elliott, Armour Research Foundation, Chicago, Ill., Tech. Rept. 1, OSR Tech. Note OSR-TN-247, 38 (August 1954).
73. C. P. Nash, F. M. Boyden and L. D. Whittig, *J. Am. Chem. Soc.* **82**, 6203 (1960).
74. G. W. Wensch and D. D. Whyte, U.S. Atomic Energy Comm. LA-1304, 28 (1951).
75. E. M. Benson and M. V. Nevitt, U.S. At. Energy Comm. ANL-6099, 75 (1959).
76. F. H. Ellinger, E. M. Cramer and C. C. Land, in W. D. Wilkinson (Ed.) *Extractive and Physical Metallurgy of Plutonium and its Alloys,* Interscience Publishers, Inc., New York (1960).
77. A. Classen and W. G. Burgers, *Z. Krist.* **A86**, 100 (1933).
78. P. Pietrokowsky, *Trans. AIME* **200**, 219 (1954).

Tables 8.1b References

1. K. Strnat, G. Hoffer, W. Ostertag and J. C. Olsen, *11th Ann. Conf. on Mag. & Mag. Mats, San Francisco* (November 1965).
2. K. Strnat and G. Hoffer, *Intermag. Conf. Stuttgart, Germany* (April 20–22, 1966).
3. J. Farrell and W. E. Wallace, *Inorg. Chem.* **5**, 105 (1966).
4. E. M. Savitskii, V. F. Terekhova, I. V. Brov, I. A. Markova and O. P. Naumkin, *Alloys of Rare Earth Metals,* Acad. Sci. USSR Press (1962).
5. E. A. Shrabek and W. E. Wallace, *J. Appl. Phys.* **34**, 1356 (1963).
6. K. Nassau, L. V. Cherry and W. E. Wallace, *J. Phys. Chem. Solids* **16**, 131 (1960).

Table 8.2a References

1. J. Théry and R. Collongues, *C.R. Acad. Sci. Paris* **254**, 685 (1962).

2. E. J. W. Verwey and E. L. Heilmann, *J. Chem. Phys.* **15**, 174 (1947).
3. P. B. Braun, *Nature* **170**, 1123 (1952).
4. L. M. Foster and H. C. Stumph, *J. Am. Chem. Soc.* **73**, 1590 (1951).
5. G. Blasse, *Philips Res. Rpts. Supplements* (1964).
6. G. H. Jonker, *Trabajos de la Tercera Reunion Internacional Sobre Reactividad De Los Solidos, Madrid* **1**, 413 (1957).
7. J. Donohue and W. Shand, *J. Am. Chem. Soc.* **69**, 222 (1947).
8. V. M. Goldschmidt, *Skr. Akad. Oslo* **81**, 107 (1926).
9. I. Lindqvist, *Acta Chem. Scand.* **4**, 1066 (1950).
10. G. E. Bacon and F. F. Roberts, *Acta Cryst.* **6**, 57 (1953).
11. W. Rüdorff and B. Reuter, *Z. Anorg. Chem.* **253**, 194 (1947).
12. A. P. B. Sinha, W. R. Sinjana and A. B. Biswas, *Acta Cryst.* **10**, 439 (1957).
13. F. Bertaut, F. Forrat and J. Dulae, *C.R. Acad. Sci. Paris* **249**, 726 (1959).
14. B. Reuter and G. Marx, *Naturwiss.* **47**, 539 (1960).
15. P. Cossee, Thesis, Leiden (1956).
16. E. Prince, *Phys. Rev.* **102**, 674 (1956).
17. M. Lensen, *Ann. Chim.* **4**, 891 (1960).
18. D. G. Wickham and W. J. Croft, *Phys. Chem. Solids* **7**, 351 (1958).
19. D. B. Rogers, R. J. Arnott, A. Wold and J. B. Goodenough, *Phys. Chem. Solids* **24**, 347 (1963).
20. F. Bertaut and C. Delorme, *C.R. Acad. Sci. Paris* **239**, 504 (1954).
21. S. Holgesson and A. Karlsson, *Z. Anorg. Chem.* **183**, 384 (1924).
22. E. Whipple and A. Wold, *J. Inorg. Nucl. Chem.* **24**, 23 (1962).
23. E. Prince and R. G. Treuting, *Acta Cryst.* **9**, 1025 (1956).
24. C. Delorme, *Bull. Soc. Franc. Minér.* **81**, 79 (1958).
25. M. Robbins and L. Darcy, *J. Phys. Chem. Solids* **27**, 741 (1966).
26. S. C. Abrahams and B. A. Calhoun, *Acta Cryst.* **6**, 105 (1953).
27. *Structure Reports* **11**, 498 (1947).
28. F. K. Lotgering, *Philips Res. Repts.* **11**, 190 (1956).
29. A. Oles, *Compt. Rend.* **260**, 6075 (1965).
30. G. E. Bacon, *Acta Cryst.* **5**, 684 (1952).
31. E. W. Gorter, *Nature* **165**, 798 (1950).
32. G. E. Bacon and F. F. Roberts, *Acta Cryst.* **6**, 57 (1953).
33. A. Boxer, J. F. Ollom and R. F. Rauchmiller, Technical Documentary Report No. ML-TDR-64-224 (July 1964).
34. M. Huber, *C.R. Acad. Sci. Paris* **244**, 2524 (1957).
35. J. E. Weidenborner, N. R. Stemple and Y. Okaya, *Acta Cryst.* **20**, 761 (1966).
36. T. F. W. Barth and E. Posnjak, *Z. Krist.* **82**, 325 (1932).
37. K. Muramori and S. Miyahara, *J. Phys. Soc. Japan* **15**, 1906 (1960).
38. A. Lecerf, Thesis, Bordeaux (1962).
39. R. Plumier and A. Tardieu, *C.R. Acad. Sci. Paris* **257**, 3858 (1963).
40. S. Greenwald, S. J. Pickart and F. H. Grannis, *J. Chem. Phys.* **22**, 1597 (1954).
41. E. W. Gorter, *Philips Res. Repts.* **9**, 403 (1954)
42. J. M. Hastings and L. M. Corliss, *Phys. Rev.* **104**, 328 (1956).
43. G. Villers, A. Lecerf and M. Rault, *Compt. Rend.* **260**, 3017 (1965).
44. J. M. Hastings and L. M. Corliss, *Rev. Mod. Phys.* **25**, 114 (1953).
45. G. Blasse, *Philips Res. Repts.* **18**, 383 (1963).
46. P. K. Baltzer and J. G. White, *J. Appl. Phys.* **29**, 445 (1958).
47. E. G. Larson, R. J. Arnott and D. G. Wickham, *Phys. Chem. Solids* **23**, 1771 (1962).
48. H. Spandau and T. Ullrich, *Z. Anorg. Chem.* **274**, 271 (1953).
49. T. Moeller and G. L. King, *J. Amer. Chem. Soc.* **75**, 6060 (1953).
50. M. Huber, *C.R. Acad. Sci., Paris* **252**, 2744 (1961).
51. H. Hayashi, W. Nakayama, M. Yosida, T. Kozuka, M. Mizuno, K. Yamamoto, T. Yamamoto and T. Noguchi, *Nagoya Kogyo Gijutsu Shikensku Hokqku* **13**, 285 (1964).
52. A. E. Ringwood, *Nature* **198**, 79 (1963).
53. P. Poix and A. Michel, *C.R. Acad. Sci. Paris* **255**, 2446 (1962).

54. A. Durif-Varambon, E. F. Bertaut and R. Pauthenet, *Ann. Chim.* **1**, 525 (1956).
55. A. E. Ringwood, *Amer. Min.* **44**, 659 (1959).
56. R. H. Forster and E. O. Hall, *Acta Cryst.* **18**, 859 (1965).
57. R. Roy and D. M. Roy, *Amer. Min.* **39**, 340 (1954).
58. D. G. Wickham, W. Menyuk and K. Dwight, *Phys. Chem. Solids* **20**, 316 (1961).
59. J. C. Bernier, P. Poix and A. Michel, *C.R. Acad. Sci. Paris* **256**, 5583 (1963).
60. H. Hayaski, N. Nakayama, M. Yosida, T. Kozuka, M. Mizuno, K. Yamamoto, T. Yamamoto and T. Noguchi, *Nagoya Kogyo Gijutsu Shikensho Hokaku* **13**, 291 (1964).
61. P. Poix and A. Michel, *C.R. Acad. Sci. Paris* **255**, 2446 (1962).
62. J. Dulac and A. Durif, *C.R. Acad. Sci. Paris* **251**, 747 (1960).
63. R. W. Harrison and E. J. Delgrosso, *J. Electrochem. Soc.* **110**, 205 (1963).
64. H. Hahn and K. F. Schroder, *Z. Anorg. Chem.* **269**, 135 (1952).
65. J. Flahaut, L. Domange, M. Guittard, M. Ourmitchi and J. Kam. Su Kam, *Bull. Soc. Chim. Fr.* 2382 (1961).
66. H. Hahn and W. Klingler, *Z. Anorg. Chem.* **263**, 177 (1950).
67. H. Hahn, *Z. Anorg. Chem.* **264**, 184 (1951).
68. J. Flahaut, L. Domange, M. Guittard and S. Fahrat, *C.R. Acad. Sci. Paris* **253**, 1956 (1961).
69. H. Hahn, C. deLorent and B. Harder, *Z. Anorg. Chem.* **283**, 138 (1956).
70. H. Hahn and B. Harder, *Z. Anorg. Chem.* **288**, 257 (1956).
71. W. F. deJong and M. W. V. Willems, *Z. Anorg. Chem.* **161**, 311 (1927).
72. H. Hahn, G. Franck, W. Klingler, A. Störger and G. Störger, *Z. Anorg. Chem.* **279**, 241 (1955).
73. G. Blasse, *J. Inorg. Nucl. Chem.* **25**, 230 (1963).
74. J. C. Joubert and A. Durif, *Compt. Rend.* **256**, 4403 (1963).
75. J. C. Joubert and A. Durif-Varambon, *Bull. Soc. Franc. Mineral Crist.* **86**, 92 (1963).

Table 8.2b References

1. E. W. Gorter, *Philips Res. Rpts.* **9**, 403 (1954).
2. G. Blasse, *Philips Res. Rpts.* **18**, 383 (1963).

Table 8.2c References

1. I. E. Campbell, *High Temperature Technology*, Wiley, New York (1959).
2. S. J. Schneider, NBS Monograph 68 (October 1963).
3. J. C. Joubert and A. Durif, *Compt. Rend.* **256**, 4403 (1963).
4. O. H. Krikorian, *Thermal Expansion of High Temperature Materials*, UCRL-6132, (September 1960).

Table 8.3a References

1. F. Bertaut, A. Deschamps, R. Pauthenet and S. Pickart, *J. Phys. Radium* **20**, 404 (1959).
2. V. Adelskold, *Arkiv. Kemi Mineral Geol.* **12A**, No. 29 (1938).
3. A. Wittman, K. Seifert and N. Nowotny, *Montash. Chem.* **89**, 225 (1958).
4. A. B. Chase and G. M. Wolten, *J. Am. Ceram. Soc.* **48**, 276 (1965).

Table 8.3b References

1. V. Adelskold, *Arkiv. Kemi Mineral Geol.* **12A**, No. 29 (1938).
2. A. Aharoni and M. Schieber, *Phys. Rev.* **123**, 807 (1961).
3. R. Panthenet and G. Rimet, *C.R. Acad. Sci.* **249**, 1875 (1959).
4. G. H. Jonker, H. P. J. Wijn and P. B. Braun, *Philips Tech. Rev.* **18**, 145 (1956/7).
5. J. Smit and H. P. J. Wijn, *Ferrites*, John Wiley, New York (1959).

CHAPTER 9

CORUNDUM AND RELATED STRUCTURES

THE corundum structure and an ordered corundum structure (ilmenite) are covered in this chapter.

9.1. Al_2O_3 Structure, Corundum $D5_1$, *R3c*, Rhombohedral, Hexagonal

The corundum structure can be visualized as layers of hexagonal close-packed oxygen atoms with small metal atoms in two-thirds of the octahedrally coordinated holes between the oxygen atoms. As can be seen from Fig. 9.1a, the *a*-axis of the unit cell is not selected so that the oxygen atoms fall in the center of the basal plane and on the center of the *a*-axis as they were in the perovskite related layered structures. The small metal atoms are on $0, 0, z; \frac{1}{3}, \frac{2}{3}, z$ and $\frac{2}{3}, \frac{1}{3}, z$ lines. The idealized arrangement of these metal atoms is shown in Fig. 9.1b. Atoms placed at the actual *z* values also are shown. The layer sequence is also given in Fig. 9.1b.

The atomic positions are twelve aluminum atoms at $(0, 0, 0; \frac{1}{3}, \frac{2}{3}, \frac{2}{3}; \frac{2}{3}, \frac{1}{3}, \frac{1}{3}) + 0, 0, z; 0, 0, \bar{z}; 0, 0, \frac{1}{2} + z; 0, 0, \frac{1}{2} - z$ and eighteen oxygen atoms at $(0, 0, 0; \frac{1}{3}, \frac{2}{3}, \frac{2}{3}; \frac{2}{3}, \frac{1}{3}, \frac{1}{3}) + x, 0, \frac{1}{4}; 0, x, \frac{1}{4}; \bar{x}, \bar{x}, \frac{1}{4}; \bar{x}, 0, \frac{3}{4}; 0, \bar{x}, \frac{3}{4}; x, x, \frac{3}{4}$ with $x = 0.31$ and $z = 0.35$.

Oxides with the corundum structure are listed in Table 9.1a.

TABLE 9.1a. *Compounds with the Corundum Structure*

Compound	Cell size (Å)		Refs.
	a_0	c_0	
α-Al_2O_3	4.763	13.003	1, 2
Cr_2O_3	4.954	13.584	1
α-Fe_2O_3	5.035	13.75	3
α-Ga_2O_3	4.979	13.429	4
Rh_2O_3	5.11	13.82	5
Ti_2O_3	5.148	13.636	1, 6
V_2O_3	5.105	14.449	1

Properties

The oxides, V_2O_3, Ti_2O_3 and Fe_2O_3 are antiferromagnetic with Néel temperatures of 173°, 660° and 950°K, respectively.

Aluminum oxide is the most widely used compound listed in Table 9.1a.

I

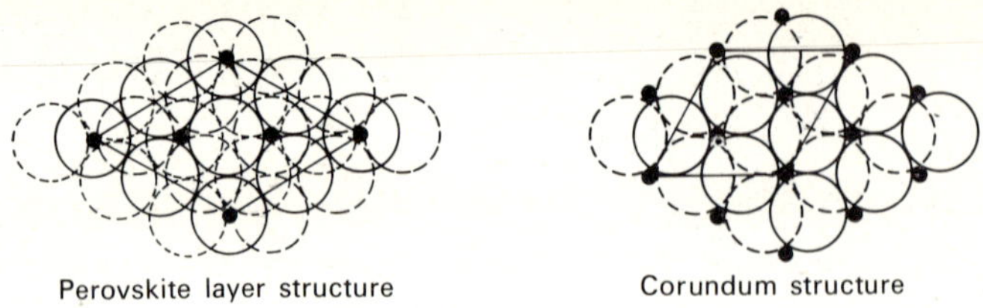

FIG. 9.1a Selecting unit cells from close packed structure for perovskite layer structure and corundum structure

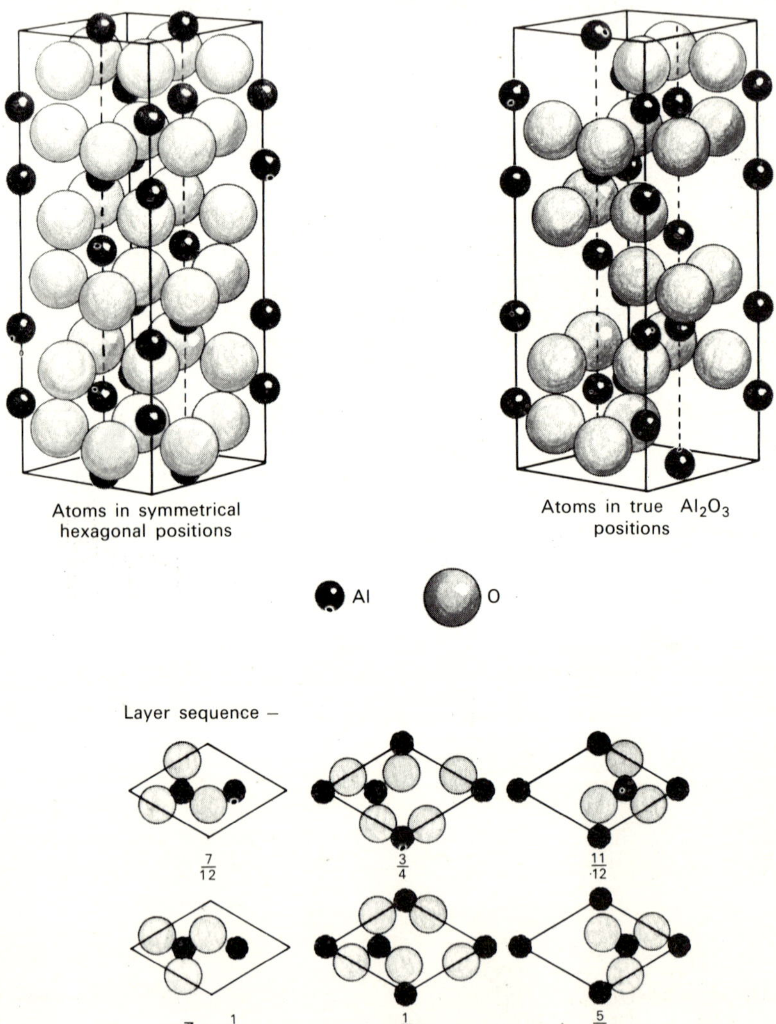

FIG. 9.1b The corundum structure

It is used in watch bearings, phonograph needles and pressure-resistant windows. Hot-pressed powders also are employed as electrical insulators, windows or radomes transparent to microwaves, envelopes for lamps and electronic devices. When it is bonded together by silica glass, aluminum oxide is the basis of refractory bricks, crucibles and spark-plug insulators. Al_2O_3 has a dielectric constant of 9.3, hardness of 2600 kg/mm^3 and a modulus of 57×10^6 psi. Because it has the same structure as Cr_2O_3, it is an excellent host for the Cr^{3+} ion. When a small amount of Cr^{3+} is substituted for Al^{3+} in Al_2O_3, ruby is formed. In addition to its use in jewelry, ruby is employed as a laser material, which fluoresces at 0·6943 μ at 300°K.

Other properties of Al_2O_3, along with those of other oxides, are listed in Table 9.1b.

TABLE 9.1b. *Properties of Compounds with the* Al_2O_3 *Structure*

Material	Melting point (°C)	Coefficient of thermal expansion (10^{-6}/°C)
Al_2O_3	2050(1)	8.3(4)
Cr_2O_3	2440(2)	8.6(5)
Fe_2O_3		12.3(5)
Ti_2O_3	2130(3)	

9.2. $FeTiO_3$ Structure, Ilmenite, *R*3

The ilmenite structure can be visualized as the corundum structure with Fe and Ti ions alternating in layers along the *c*-axis of the unit cell (see Fig. 9.2). The layer sequence is also shown.

9.3. Discussion

Structures, Oxides

The corundum structure contains oxygen atoms in a close-packed arrangement with metal cations in two-thirds of the octahedral holes, while the ilmenite structure is an ordered corundum structure in which cations of one element are situated between every other layer. This octahedral arrangement of oxygen atoms about the cations makes these sites attractive for small metal ions. The larger metal cations which form sesquioxides seem to prefer adopting the A-rare earth and C-rare earth structure. This is reasonable since the coordination about the cation in the A-rare earth structure is seven, and oxygen vacancies in a CaF_2 structure where the coordination about the cation is eight produce the C-rare earth structure. Thus, oxides such as La_2O_3 and Nd_2O_3 form the A-rare earth structure, Y_2O_3 and Sc_2O_3 the C-rare earth structure and Al_2O_3, Cr_2O_3, Fe_2O_3, Ti_2O_3, etc., the corundum structure. When two small ions of appropriate size and charge are substituted for aluminum in Al_2O_3, the ilmenite structure is formed. The

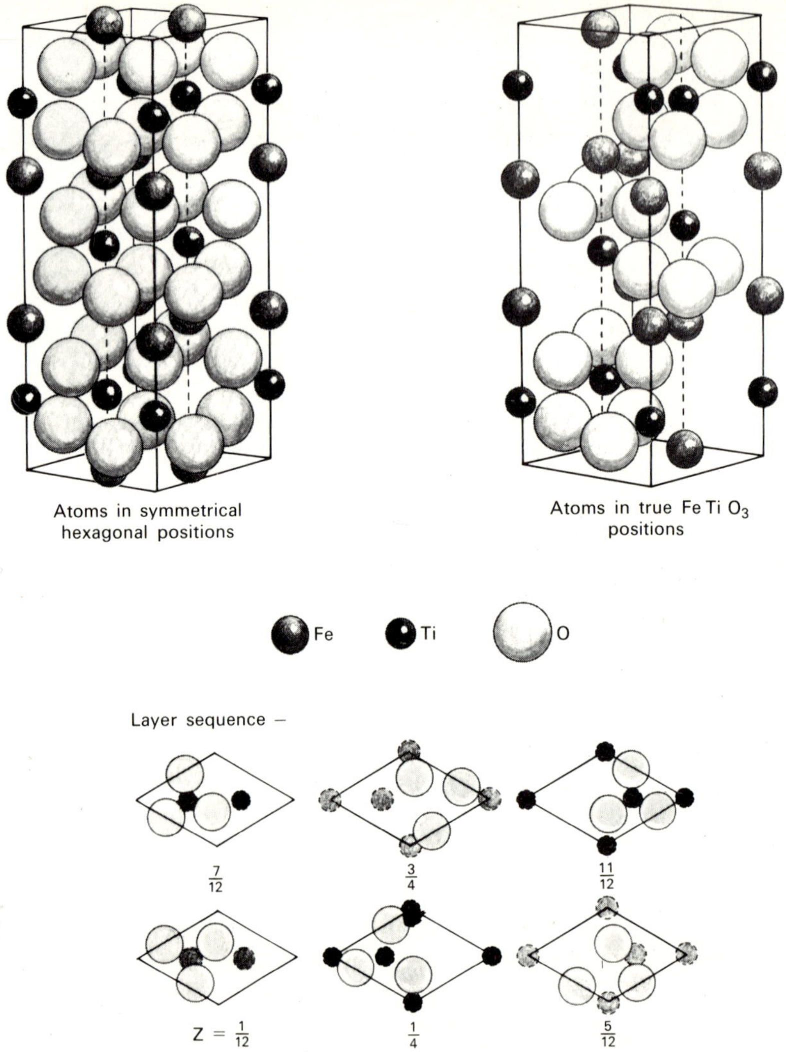

FIG. 9.2 The ilmenite structure

ordering is probably due to the fact that the charges of the two ions are different. When the size of the A ion in ABO_3 type compounds gets large enough in relation to the B ion, such as is found in $CdTiO_3$, the perovskite structure is formed. The compound $CdTiO_3$ only has the perovskite structure at high temperatures but when A is still larger, as in $SrTiO_3$, the compound forms the perovskite structure at room temperature.

The oxides V_2O_3 and Ti_2O_3 are also of interest because of their semiconducting properties. Neither one is stable in air at elevated temperatures and will oxidize to V_2O_5 and TiO_2, respectively. However, the most useful sesquioxide is Al_2O_3 because of its high melting point (2050°C), hardness, and resistance to reduction. Aluminum oxide has been employed as a crucible material and for ceramic structural applications. Its high electrical resistivity also makes it useful as an insulating substrate for electronic components. It falls into the same category of ceramics as MgO, HfO_2 and ZrO_2.

Since Cr_2O_3 has the same structure as Al_2O_3, aluminum oxide has been used as a host for Cr^{3+} to form ruby. Aside from its value in jewelry, it also has been used as a laser. Ruby is probably the most popular crystalline laser material.

Table 9.1*a References*

1. R. H. Newnham and Y. M. deHaan, *Z. Krist.* **117**, 235 (1962).
2. H. Saalfeld, *Naturwiss.* **48**, 24 (1961).
3. R. Brill, *Z. Krist.* **88**, 177 (1934).
4. H. E. Swanson, R. K. FuYat and G. M. Ugrinic, *NBS Circular 539* **4**, 25 (1955).
5. W. H. Zachariasen, *Skrifter Norske Bidenskaps-- Akad. Oslo I. Mat.-- Naturv. Kl.* No. 4 (1928).
6. S. C. Abrahams, *Phys. Rev.* **130**, 2230 (1963).

Table 9.1*b References*

1. P. B. Kantor, E. N. Fomichev and V. V. Kandiјba, *Izmeritel'n Tekhn* **5**, 27 (1966).
2. W. G. Bradshaw and C. O. Mathews, *Properties of Refractory Materials, Collected Data and References,* LMSD-2466 (June 24, 1958).
3. L. J. Cronin, *Am. Ceram. Soc. Bull.* **30**, 234 (1951).
4. J. A. Kohn, P. G. Colter and R. A. Potter, *Am. Min.* **41**, 355 (1956).
5. O. H. Krikorian, *Thermal Expansion of High Temperature Materials,* UCRL-6132 (September 1960).

CHAPTER 10

BETA-TUNGSTEN TYPE AND RELATED STRUCTURES

THE β-tungsten type structures have a body-centered arrangement of atoms of one element with two atoms of other elements in each of the faces of a cubic unit cell. Related structures have this basic arrangement of cations with anions added (see Table 10.0).

TABLE 10.0. *Body Centered Cubic Structure*

|
10.1
β-Tungsten Structure
|
10.2
Garnet Structure

10.1. *β*-Tungsten Structure, Nb_3Sn Type, A15, *P3m*, Cubic

The β-tungsten structure (Fig. 10.1a) was thought to be the structure of a modification of tungsten, but was later determined to be the structure of W_3O. One of the best known compounds with this structure is the superconductor Nb_3Sn. The unit cell of Nb_3Sn contains tin atoms at the corners and center of a cube and two niobium atoms in each of the faces lying on a line which is parallel to the cell edges and whose center is at the center of the face. The lines joining the niobium atoms in each face are perpendicular to each other in adjacent faces.

The tin atoms are at 0, 0, 0 and $\frac{1}{2}, \frac{1}{2}, \frac{1}{2}$ and the niobium atoms are at $\frac{1}{4}, 0, \frac{1}{2}$; $\frac{1}{2}, \frac{1}{4}, 0$; $0, \frac{1}{2}, \frac{1}{4}$; $\frac{3}{4}, 0, \frac{1}{2}$; $\frac{1}{2}, \frac{3}{4}, 0$; $0, \frac{1}{2}, \frac{3}{4}$. Phases with this structure are listed in Table 10.1a.

Properties

Some of the compounds with this structure have been found to have high superconducting transition temperatures (see Table 10.1b and Fig. 10.1b).

Recently a metal alloy $Nb_3(Al, Ge)$ has been found to be a superconductor at 20°K, the highest transition temperature yet measured. The previous front runner, Nb_3Sn, with a transition temperature of 18°K maintained that position for many years. Other phases with the β-tungsten structure such as Nb_3Al, V_3Si and Nb_3Ga also have high transition temperatures. A number

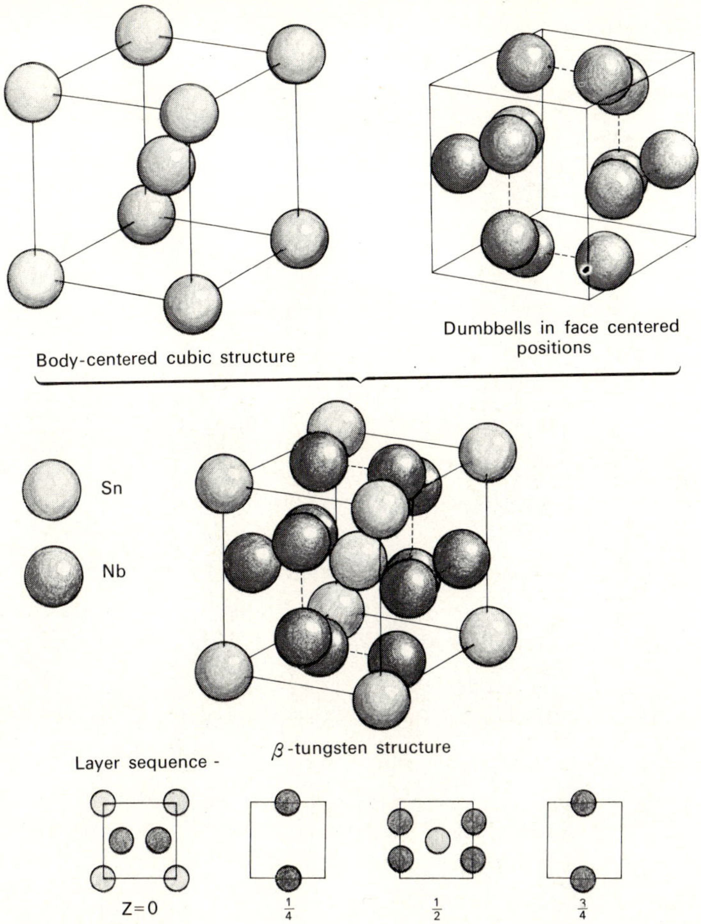

FIG. 10.1a The β tungsten structure

of researchers feel that the chain of atoms in the faces of the cubic unit cells may be the cause of the high transition temperatures found for these compounds.

TABLE 10.1a. *Phases with the β-Tungsten Structure*

Phase	Cell size, a_0(Å)	Refs.
$AlMo_3$	4.950	1
$AlNb_3$	5.187	2
AsV_3	4.75	3
$AuNb_3$	5.21	4
$AuTi_3$	5.094	5
AuV_3	4.88	4
$AuZr_3$	5.482	1
$BiNb_3$	5.320	6
CoV_3	4.675	7
Cr_3Ge	4.623	8
Cr_3Ir	4.668	9
Cr_3Pt	4.706	10
Cr_3Rh	4.656	10
Cr_3Ru	4.673	11
β-Cr_3Si (18 at.% Si)	4.576	12
$GaMo_3$	4.943	13
GaV_3	4.816	13
$GeMo_3$	4.933	14
$GeNb_3$	5.1743	15
GeV_3	4.769	8
γ-$HgTi_3$ (540–760°C)	5.189	16
$HgZr_3$	5.558	16
$InNb_3$ (H.P.)	5.303	6
$IrMo_3$	4.959	17
$IrNb_3$	5.139	18
$IrTi_3$	5.009	19
IrV_3	4.79	15, 19
Mo_3Os	4.973	17
Mo_3Sn	5.094	20
Nb_3Os	5.121	21
Nb_3Pt	5.147–5.166	22
Nb_3Rh	5.115	10
Nb_3Sb	5.262	13, 15
Nb_3Sn	5.289	21
NiV_3	4.71	23
PbV_3	5.76	1
$PtTi_3$	5.024	24
PtV_3	4.814	19
RhV_3	4.767	10
$SbTi_3$	5.217	19, 25
SbV_3	4.932	13
SiV_3	4.721	26
SnV_3	4.96	27

TABLE 10.1b. *Superconducting Transition Temperatures for Compounds with the β-Tungsten Structure*

Phase	T_c (°K)	Refs.
$AlMo_3$	0.58	1
$AlNb_3$	17.5	2
$AuNb_3$	11.5	3
AuV_3	0.74	1
$AuZr_3$	0.92	1
$BiNb_3$ (H.P.)	2.05	4
Cr_3Ir	0.45	5
Cr_3Ru	2.15	5
$GaMo_3$	0.76	1
$GaNb_3$	14.5	6
GaV_3	16.8	7
$GeNb_3$	6	8
$GeMo_3$	1.4	9
GeV_3	6.0	9
$InNb_3$ (H.P.)	9.2	4
$IrMo_3$	8.8	1
$IrNb_3$	1.7	10
$IrTi_3$	5.4	11
Mo_3Os	7.2	10
Mo_3Si	1.3	9
Nb_3Os	1	1
Nb_3Pt	10.9	3
Nb_3Rh	2.6	3
Nb_3Sn	18	10
$PbZr_3$	0.76	1
$PtTi_3$	0.58	1
PtV_3	2.83	11
$SbTi_3$	5.8	11
SiV_3	16.8	1
$SnTa_3$	6.4	12
SnV_3	3.8	12

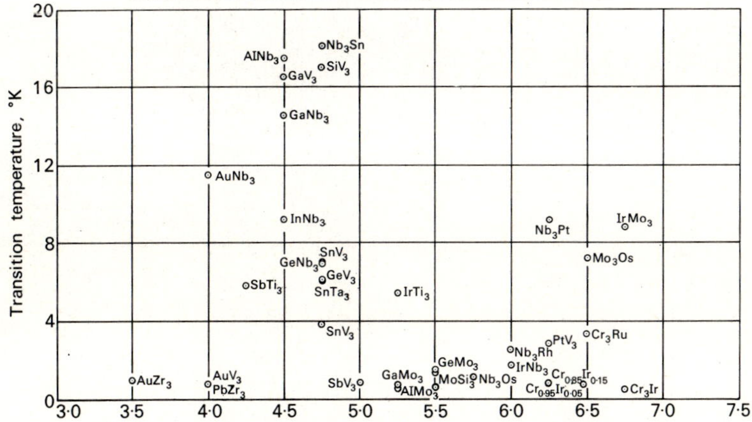

FIG. 10.1b T_c vs. $\bar{e}/A$ for superconductors with the β-tungsten structure

10.2. Garnets, $A_3B'_2B''_3O_{12}$, *Ia3d*, Cubic

The cubic unit cell of a garnet contains eight units of $A_3B'_2B''_3O_{12}$. The phases with this structure are listed in Table 10.2a (p. 247). In the ninety-six oxygen ions, there are three types of sites which are filled with metal ions, sixteen tetrahedrally coordinated sites filled with B′ ions, twenty-four octahedrally coordinated sites for the B″ ions, and twenty-four dodecahedrally coordinated sites for the A ions.

The arrangement of the metal atoms in each octant of the unit cell is similar to that in the A15 structure. The B″ ions are located at the corners and center of a cube and one B′ and one A ion are in each face lying on a line parallel to the edges. These A–B′ dumb-bells are situated so they are perpendicular to each other in adjacent faces. Figure 10.2a shows how an octahedron of oxygen ions at the corner of an octant and dodecahedron are joined to a tetrahedron.

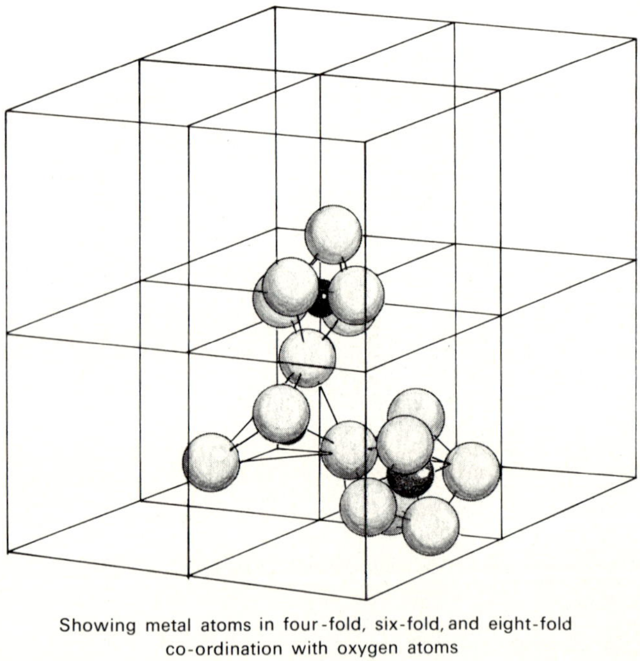

Showing metal atoms in four-fold, six-fold, and eight-fold co-ordination with oxygen atoms

FIG. 10.2a The garnet structure

There are four types of octants. Figure 10.2b shows these octants and how they are placed together in the unit cell. The layer sequence showing metal ions in the unit cell are shown in Fig. 10.2c.

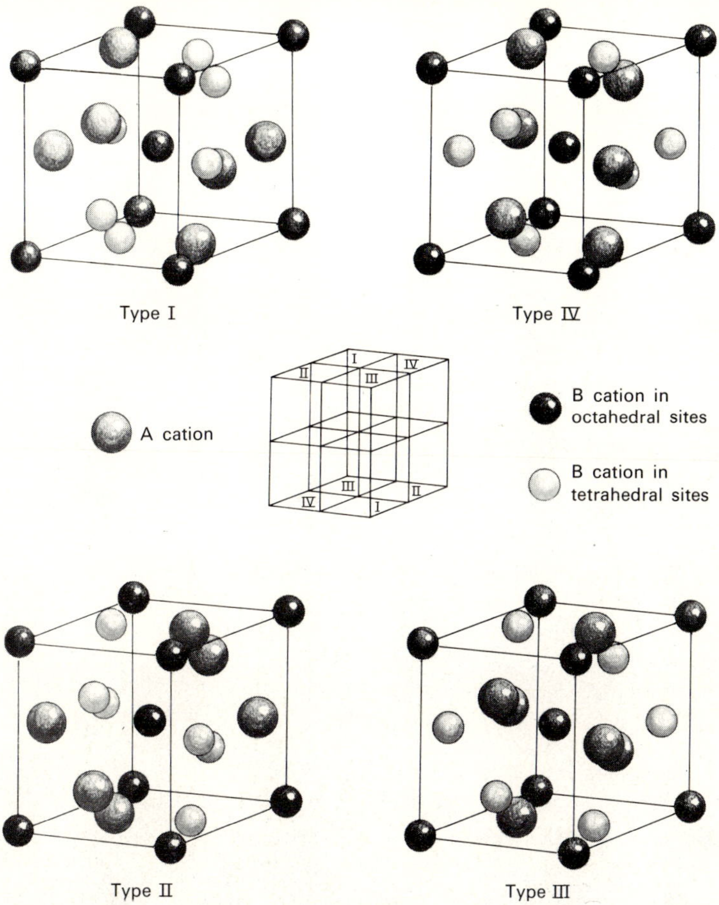

FIG. 10.2b The octants of the garnet unit structure

Properties

The minerals $Fe_3Al_2Si_3O_{12}$, $Ca_3(Fe,Ti)_2Si_3O_{12}$, $Ca_3Al_2Si_3O_{12}$, $Mg_3Al_2Si_3O_{12}$, $Mn_3Al_2Si_3O_{12}$ and $Ca_3Cr_2Si_3O_{12}$ are garnets. Probably the best known compounds with this structure are $Y_3Fe_5O_{12}$ (YIG) and $Y_3Al_5O_{12}$ (YAG). Because of its magnetic properties, YIG has found many uses in microwave applications. Its narrow line width and small anisotropic

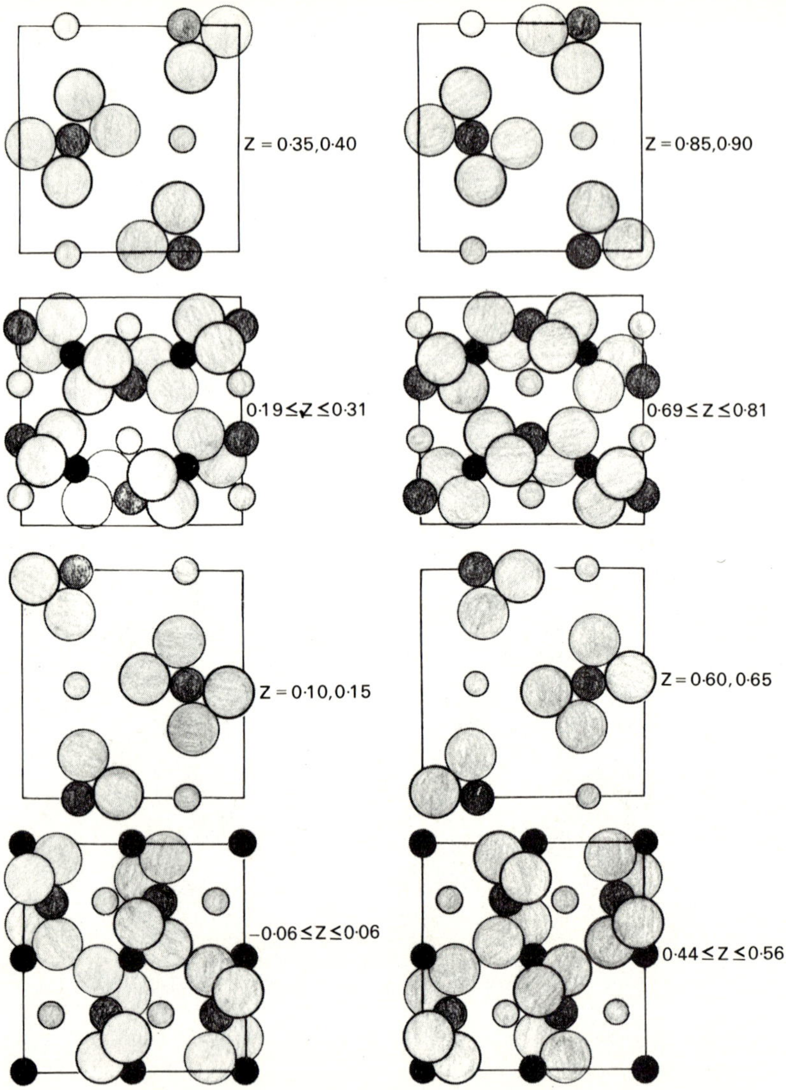

FIG. 10.2c The garnet structure: layer sequence

field make it especially suitable for low-frequency applications. It has been used in ferromagnetic amplifiers, resonance filters and harmonic generators.

A number of garnet phases have been produced by substituting Si^{3+}, Cr^{3+}, Al^{3+}, and Ga^{3+} for Fe^{3+} and many rare earth ions for Y^{3+}. Substitution of Al^{3+} and Ga^{2+} in the tetrahedrally coordinated sites lowers the Curie

TABLE 10.2a. *Phases with the Garnet Structure*

Phase	a_0 (Å)	Refs.
$Dy_3Al_5O_{12}$	12.042	1
$Dy_3Fe_5O_{12}$	12.414	2
$Er_3Al_5O_{12}$	11.981	1
$Er_3Fe_5O_{12}$	12.349	2
$Eu_3Fe_5O_{12}$	12.498	2
$Gd_3Al_5O_{12}$	12.12	3
$Gd_3Fe_5O_{12}$	12.44	3
$Ho_3Al_5O_{12}$	11.981	1
$Ho_3Fe_5O_{12}$	12.380	2
$Lu_3Fe_5O_{12}$	12.277	2
$Nd_3Fe_5O_{12}$	12.60	2
$Sm_3Fe_5O_{12}$	12.530	2
$Sm_3Ga_5O_{12}$	12.432	4
$Tb_3Al_5O_{12}$	12.074	1
$Tb_3Fe_5O_{12}$	12.447	2
$Tm_3Al_5O_{12}$	11.957	1
$Tm_3Fe_5O_{12}$	12.325	2
$Y_3Al_5O_{12}$	12.01	4
$Ca_3Al_2Ge_3O_{12}$	12.1117	5
$Ca_3Cr_2Ge_3O_{12}$	12.262	5
$Ca_3Fe_2Ge_3O_{12}$	12.320	6
$Ca_3Fe_3GeVO_{12}$	12.418	7
$Ca_3Fe_2Sn_3O_{12}$	12.728	8
$Ca_3Fe_2Si_3O_{12}$	12.048	9
$Ca_3TiCoGe_3O_{12}$	12.356	5
$Ca_3TiMgGe_3O_{12}$	12.35	5
$Ca_3TiNiGe_3O_{12}$	12.341	5
$Ca_3SnCoGe_3O_{12}$	12.47	6
$Ca_3V_2Ge_3O_{12}$	12.32	10
$Ca_3V_2Si_3O_{12}$	12.070	11
$Ca_3ZrCoGe_3O_{12}$	12.54	6
$Ca_3ZrMgGe_3O_{12}$	12.51	6
$Ca_3ZrNiGe_3O_{12}$	12.50	6
$Cd_3Al_2Ge_3O_{12}$	12.08	12
$Cd_3Al_2Si_3O_{12}$	11.82	10
$Cd_3Cr_2Ge_3O_{12}$	12.70	12
$Cd_3Fe_2Ge_3O_{12}$	12.26	12
$Cd_3Ga_2Ge_3O_{12}$	12.19	12
$CdGd_2Mn_2Ge_3O_{12}$	12.473	6
$CoGd_2Co_2Ge_3O_{12}$	12.402	6
$CoGd_2Mn_2Ge_3O_{12}$	12.473	6
$CoY_2Co_2Ge_3O_{12}$	12.300	6
$CuGd_2Mn_2Ge_3O_{12}$	12.475	6
$Gd_2CaFe_2Si_3O_{12}$	12.366	13
$GdCa_2Fe_2FeSi_2O_{12}$	12.222	13
$GdCa_2Fe_3Ge_2O_{12}$	12.389	13
$Gd_3Co_2GaGe_2O_{12}$	12.45	6
$Gd_3Fe_2Al_3O_{12}$	12.267	14
$Gd_3Fe_4AlO_{12}$	12.408	13
$Gd_3Fe_3Al_2O_{12}$	12.338	13
$Gd_3Fe_2Al_3O_{12}$	12.265	13
$Gd_3MgFeFe_2SiO_{12}$	12.385	13
$Gd_3Mg_2GaGe_2O_{12}$	12.425	6

TABLE 10.2a *(cont.)*

Phase	a_0 (Å)	Refs.
$Gd_3Mn_2GaGe_2O_{12}$	12.550	6
$Gd_3Ni_2GaGe_2O_{12}$	12.401	6
$Gd_3Zn_2GaGe_2O_{12}$	12.464	6
$MgGd_2Mn_2Ge_3O_{12}$	12.31	6
$Mn_3Al_2Ge_3O_{12}$	11.902	13
$Mn_3Al_2Si_3O_{12}$	11.64	15
$Mn_3Cr_2Ge_3O_{12}$	12.027	12
$Mn_3Fe_2Ge_3O_{12}$	12.087	12
$Mn_3Ga_2Ge_3O_{12}$	12.00	12
$MnGd_2Mn_2Ge_3O_{12}$	12.555	6
$Mn_3NbZnFeGe_2O_{12}$	12.49	6
$MnY_2Mn_2Ge_3O_{12}$	12.392	6
$NaCa_2Co_2V_3O_{12}$	12.431	5
$NaCa_2Cu_2V_3O_{12}$	12.423	5
$NaCa_2Mg_2V_3O_{12}$	12.446	5
$NaCa_2Ni_2V_3O_{12}$	12.373	5
$NaCa_2Zn_2V_3O_{12}$	12.439	5
$YCa_2Zr_2Fe_3O_{12}$	12.684	15
$Y_3Fe_2Al_3O_{12}$	12.161	15

point of YIG from 560°K linearly with concentration. Other ferrimagnetic phases with the garnet structure are listed in Table 10.2b. Total substitution of Al^{3+} in the octahedral and tetrahedral sites results in the formation of $Y_3Al_5O_{12}$, one of the most interesting laser host materials found to date. Neodymium has been substituted in the Y^{3+} position and Cr^{3+} in the Al^{3+} position, alone and together. When both are substituted energy transfer has been observed between energy levels of the two ions.

TABLE 10.2b. *Ferromagnetic Curie Points of Garnets*

Compound	Curie point (°C)	Refs.
$Eu_3Fe_5O_{12}$	290	1
$Gd_3Fe_5O_{12}$	292	2
$Sm_3Fe_5O_{12}$	292	3
$Y_3Fe_5O_{12}$	280	2

10.3. Discussion

Structure (Relations)

The β-tungsten structure has atoms of one element in a body-centered cubic arrangement and two atoms of the other element in each of the faces. The garnet has the same arrangement for the cations in each octant of its unit cell with these cations in either four, six- or eightfold coordination of oxygen ions around them.

Intermetallics

The compounds with these structures have extremely interesting properties. Some of the intermetallic compounds with the β-tungsten structure have the highest superconducting transition temperatures and carry the largest currents, without becoming normal, of any materials. This has been attributed to the fact that it is an "open" structure with a great deal of empty space, that it contains chains of atoms formed by the combination of the two atoms in each face of the unit cell, and that some of the compounds which form this structure have ideal electron per atom ratios (~ 5) (see Fig. 10.1a). All of these factors are empirical rules but they have provided the direction for researchers trying to find new superconducting compounds. In this search these compounds have been used to form solid solution, and the phases produced have had superconducting transition temperatures between those of the end members, although the relationship between the composition and transition temperature has not been exactly linear.

Note that the ordered Cu_3Au type compounds also have the same atomic ratio of elements. The compounds Nb_3Sn and Nb_3Ge have the β-tungsten structure, while Nb_3Si has the Cu_3Au structure. This has has been attributed to the fact that tin and germanium have larger atomic radii than the silicon atom.

Oxides

Compounds with the garnet structure also have interesting properties. Those phases such as $Y_3Al_5O_{12}$ are good laser host materials for the rare earth ions and Cr^{3+}. The large eight oxygen coordinated holes are ideal for the rare earth ion and the octahedrally-coordinated holes are of appropriate size for Cr^{3+}.

When paramagnetic ions (those with unpaired electrons) are in the aluminum ion sites, their superexchange interaction through oxygen ions leads to ferromagnetic behavior, $Y_3Fe_5O_{12}$ being the most widely studied example. The structure provides the environment for a large number of different substituted ions and thus the properties of these compounds can be modified as can be done in the spinels.

Table 10.1*a References*

1. B. T. Matthias, T. H. Geballe and V. B. Compton, *Revs. Mod. Phys.* **35**, 1 (1963).
2. P. S. Swartz, *Phys. Letters* **9**, 448 (1962).
3. K. Bachmayer and H. Nowotny, *Mh. Chem.* **86**, 741 (1955).
4. E. A. Wood and B. T. Matthias, *Acta Cryst.* **9**, 534 (1956).
5. P. Duwez and C. B. Gordon, *Acta Cryst.* **5**, 213 (1952).
6. D. H. Kilpatrick, *J. Phys. Chem. Solids* **25**, 1213 (1964).
7. P. Duwez, *Trans. AIME* **191**, 564 (1951).
8. H. J. Wallbaum, *Naturwiss.* **32**, 76 (1944).

9. B. T. MATTHIAS, T. H. GEBALLE, V. B. COMPTON, E. CORENZWIT and G. W. HULL, JR., *Phys. Rev.* **128**, 588 (1962).
10. P. GREENFIELD and P. A. BECK, *J. Metals.* **8**, 265 (1956).
11. R. M. WATERSTRAT and J. S. KASPER, *Trans. AIME* **209**, 872 (1957).
12. H. J. GOLDSCHMIDT and J. A. BRAND, *J. Less-Common Metals* **3**, 34 (1961).
13. E. A. WOOD, V. B. COMPTON, B. T. MATTHIAS and E. CORENZWIT, *Acta Cryst.* **11**, 604 (1958).
14. A. W. SEARCY, R. J. PEAVLER and H. J. YEARIAN, *J. Am. Chem. Soc.* **74**, 566 (1952).
15. M. V. NEVITT, *Trans. AIME* **212**, 350 (1958).
16. P. PIETROKOWSKY, *J. Metals* **6**, 219 (1954).
17. E. RAUB, *Z. Metallk.* **45**, 23 (1954).
18. A. G. KNAPTON, *J. Inst. Metals* **87**, 28 (1958–59).
19. B. T. MATTHIAS, V. B. COMPTON and E. CORENZWIT, *Phys. Chem. Solids* **19**, 130 (1961).
20. D. H. KILLPATRICK, *J. Phys. Chem. Solids* **25**, 1499 (1964).
21. S. GELLER, B. T. MATTHIAS and R. GOLDSTEIN, *J. Am. Chem. Soc.* **77**, 1502 (1955).
22. H. KIMURA and A. ITO, *Nippon Kinzoku Gakkaishi* **25**, 88 (1961).
23. W. ROSTOKER, *J. Inst. Metals* **80**, 698 (1961).
24. H. NISHIMURA and T. HIRAMATSU, *Nippon Kinzoku Gakkaishi* **21**, 469 (1957).
25. A. KJEKSHUS, *10th Inter-Scand. Chem. Conf., Stokholm* (1959).
26. H. J. WALLBAUM, *Z. Metallk.* **31**, 362 (1939).
27. G. D. CODY, J. J. HANAK, G. T. MCCONVILLE and F. D. ROSI, *Proc. Intern. Conf. Low Temp. Phys., 7th, Toronto, Ont.*, p. 382 (1960).

Table 10.1*b References*

1. B. T. MATTHIAS, T. H. GEBALLE and V. B. COMPTON, *Rev. Mod. Phys.* **35**, 1 (1963).
2. P. S. SWARTZ, *Phys. Rev. Letters* **9**, 448 (1962).
3. E. A. WOOD and B. T. MATTHIAS, *Acta Cryst.* **9**, 534 (1956).
4. D. H. KILLPATRICK, *J. Phys. Chem. Solids* **25**, 1213 (1964).
5. B. T. MATTHIAS, T. H. GEBALLE, V. B. COMPTON, E. CORENTZWIT and G. W. HULL, JR., *Phys. Rev.* **128**, 588 (1962).
6. E. A. WOOD, V. B. COMPTON, B. T. MATTHIAS and E. CORENZWIT, *Acta Cryst.* **11**, 604 (1958).
7. J. H. WERNICK, F. J. MORIN, S. L. HSU, D. DORSI, J. P. MAITA and J. E. KUNZLER, *J. Appl. Phys.* **32**, 325 (1961).
8. B. T. MATTHIAS, T. H. GEBALLE, R. H. WILLENS, E. CORENZWIT and G. W. HULL, JR., *Phys. Rev.* **139**, A1501 (1965).
9. G. F. HURDY and J. K. HULM, *Phys. Rev.* **89**, 884 (1953).
10. B. T. MATTHIAS, *Phys. Rev.* **97**, 741 (1955).
11. B. T. MATTHIAS, V. B. COMPTON and E. CORENZWIT, *J. Phys. Chem. Solids* **19**, 130 (1961).
G. D. CODY, J. J. HANAK, G. T. MCCONVILLE and F. D. ROSI, *RCA Rev.* **25**, 338 (1964).

Table 10.2*a References*

1. C. B. RUBENSTEIN and R. L. BARNS, *Am. Mineralogist* **49**, 1499 (1964).
2. G. P. ESPINOSA, *J. Chem. Phys.* **37**, 2344 (1962).
3. A. L. GENTILE and R. ROY, *Am. Mineralogist* **45**, 701 (1960).
4. M. L. KEITH and R. ROY, *Am. Mineralogist* **39**, 1 (1954).
5. A. DURIF, *Sol. State Phys. Electron Telecommun. Proc. Intern. Conf., Brussels, 1958*, p. 500 (1960).
6. S. GELLER, C. E. MILLER and R. G. TREUTING, *Acta Cryst.* **13**, 179 (1960).
7. S. GELLER, G. P. ESPINOSA, R. C. SHERWOOD and H. J. WILLIAMS, *J. Appl. Phys.* **36**, 321 (1965).
8. S. GELLER, R. M. BOZORTH, M. GILLEO and C. E. MILLER, *Phys. Chem. Solids* **12**, 111 (1960).
9. B. J. SKINNER, *Am. Min.* **41**, 428 (1956).
10. H. STRUNZ and P. JACOB, *Neues Jahrb. Mineral, Montash.* **1960**, 78 (1960).
11. R. G. J. STRENS, *Am. Mineralogist* **50**, 260 (1964).
12. H. STRUNZ, H. FREIGANY and B. CONTAY, *Neues Jahrb. Mineral. Montash,* **1960**, 47 (1960).

13. S. GELLER, H. J. WILLIAMS, R. C. SHERWOOD and G. P. ESPINOZA, JR., *Appl. Phys.* **36**, 88 (1965).
14. C. E. MILLER, *Phys. Chem. Solids* **17**, 229 (1961).
15. S. GELLER, R. M. BOZORTH, C. E. MILLER and D. D. DAVIS, *Phys. Chem. Solids* **13**, 28 (1960).

Table 10.2*b References*

1. G. VILLERS, J. LORIERS and C. CLERC, *C.R. Acad. Sci.* **255**, 1196 (1962).
2. G. VILLERS, J. LORIERS and C. CLAUDEL, *C.R. Acad. Sci.* **247**, 710 (1958).
3. J. LORIERS and G. VILLERS, *C.R. Acad. Sci.* **252**, 1590 (1961).

CHAPTER 11

GRAPHITE TYPE AND RELATED STRUCTURES

THE graphite type structures are made up of sheets of atoms formed from six-membered rings with wide separations between the sheets. In a related $CuAl_2$ structure there are alternating sheets of Cu and Al atoms. This relationship is shown in Table 11.0.

TABLE 11.0. *Layered Structures*

11.1	11.2	11.3
Graphite structure	Boron nitride structure	$CuAl_2$ structure

11.1. Graphite Structure, *C6mc*

In crystalline graphite, sheets of six-membered rings are situated so that the atoms in alternate (1st and 3rd) layers lie over one another, and the second layer is displaced in such a way that the sides of the rings in the three layers remain parallel but only three atoms in a ring of the first layer lie directly over three atoms in the second layer (see Fig. 11.1).

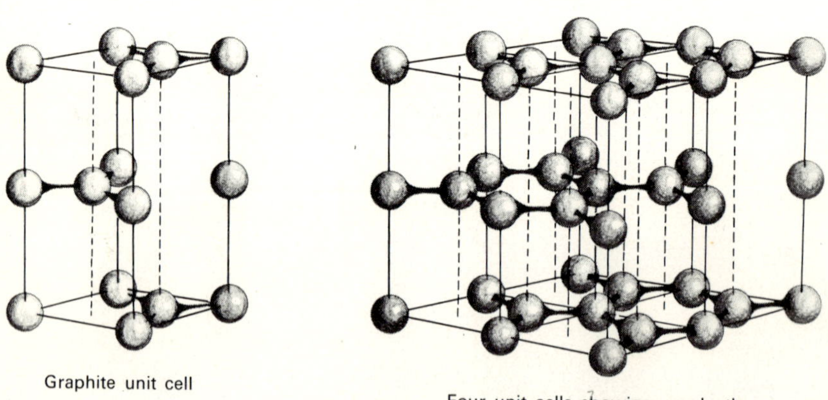

Graphite unit cell

Four unit cells showing covalently bonded layers

FIG. 11.1. The graphite structure.

The unit cell of graphite is a hexagonal unit cell of $a = 2.456$ Å and $c = 6.696$ Å. There is also a rhombohedral form of graphite. The pseudo-cell has the same base as hexagonal graphite and a height 50% longer ($a_0' = 2.456$ Å, $c_0' = 10.044$ Å). The actual c-axis of hexagonal graphite can vary from 6.7 Å to 6.88 Å, depending on the degree of crystallinity or of disorientation of layers. In graphites prepared by decomposing carbon containing gases such as methane on a hot substrate the degree of crystallinity depends on the temperature of preparation. At very low temperatures amorphous

but the sheets appear to be rotated with respect to one another. At temperatures above 2200°C crystalline graphite is formed. It is interesting that this "pyrolytic graphite" containing two-dimensional crystallinity can be crystallized in the c-direction by heating at 2400°C or above. This material has become quite important since pyrolytic carbon is highly oriented with the c-axis perpendicular to the substrate surface; thus it has nearly single crystal properties.

A number of complex structures are formed when graphite is treated with molten potassium, oxidized, treated with acids or heated in fluorine. It is presumed that compounds such as $C_{16}K$, C_8K, $C_{11}H_4O_5$, $[C_{24}]^+ (HSO_4)^-$, $2H_2SO_4$, and (CF)m were formed containing these foreign atoms or ions between the layers. The graphite appears to swell and large interlamellar spacings are observed.

Properties

The C—C bond length is 1.42 Å in the sheets, and 3.35 Å between sheets. This long bond distance gives graphite its lubricatory properties. In addition, graphite has highly anisotropic properties. The properties of graphite in the sheets are similar to those of a metal, while the properties perpendicular to the sheets are more like those of semiconductors. Thus, the electrical and thermal conductivity of single crystal or highly oriented graphite are much greater in the plane of the sheets than they are perpendicular to the sheets.

11.2. Boron Nitride Structure, *C6/mnc*

The crystal structure of boron nitride is closely related to that of graphite except that the atoms in one layer lie directly above those in the next and the six-membered rings are made up of alternating B and N atoms. The atoms in the layers are so arranged that the B atom in one layer lies directly above a nitrogen atom in the next and vice versa (Fig. 11.2).

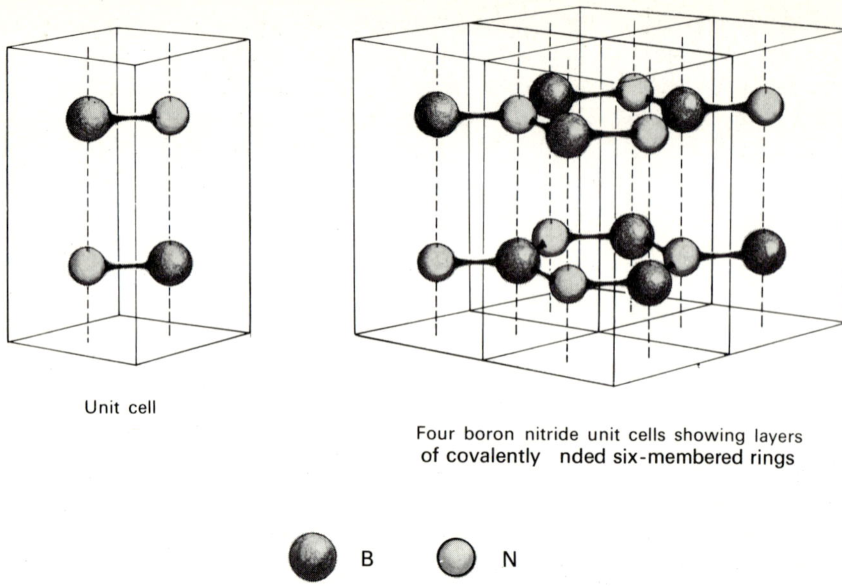

FIG. 11.2. The boron nitride structure.

Properties

The properties of BN are quite different from those of graphite. It is an extremely good insulator, it is very inert and is one of the few materials which will contain molten boron.

Like graphite, it can be produced in a highly oriented state by decomposing gases such as BCl_3 and NH_3 on a hot substrate. This pyrolytic material has found numerous uses in high-temperature technology. At 55 kbar pressure and 1500°C, BN forms a cubic structure which is extremely hard like diamond and is called borazone.

11.3. $CuAl_2$ Type, C16, *I4/mcm*, Tetragonal

The unit cell of $CuAl_2$ consists of Cu–Cu units at the centers of the vertical edges of a tetragonal cell and half-way up a $\frac{1}{2}$, $\frac{1}{2}$, z-line, and a square of Al atoms at the base of the cell and another parallel but rotated 90° at the center. The Cu–Cu units are vertical in the cells and the squares are parallel to the base of the cell. Thus, there are sheets of Cu atoms as shown in Fig. 11.3 with sheets of Al atoms between them.

The Cu atoms are at

$$0, 0, \tfrac{1}{4};\ 0, 0, \tfrac{3}{4};\ \tfrac{1}{2}, \tfrac{1}{2}, \tfrac{1}{4};\ \tfrac{1}{2}, \tfrac{1}{2}, \tfrac{3}{4}$$

and the Al atoms are at

$$x, \tfrac{1}{2}+x, 0;\ \bar{x}, \tfrac{1}{2}-x, 0;\ \tfrac{1}{2}+x, \bar{x}, 0;\ \tfrac{1}{2}-x, x, 0;\ \tfrac{1}{2}+x, x, \tfrac{1}{2};$$
$$\tfrac{1}{2}-x, \bar{x}, \tfrac{1}{2};\ x, \tfrac{1}{2}-x, \tfrac{1}{2};\ \bar{x}, \tfrac{1}{2}+x, \tfrac{1}{2}.\quad x \cong \tfrac{1}{6}$$

Phases with this structure are listed in Table 11.3.

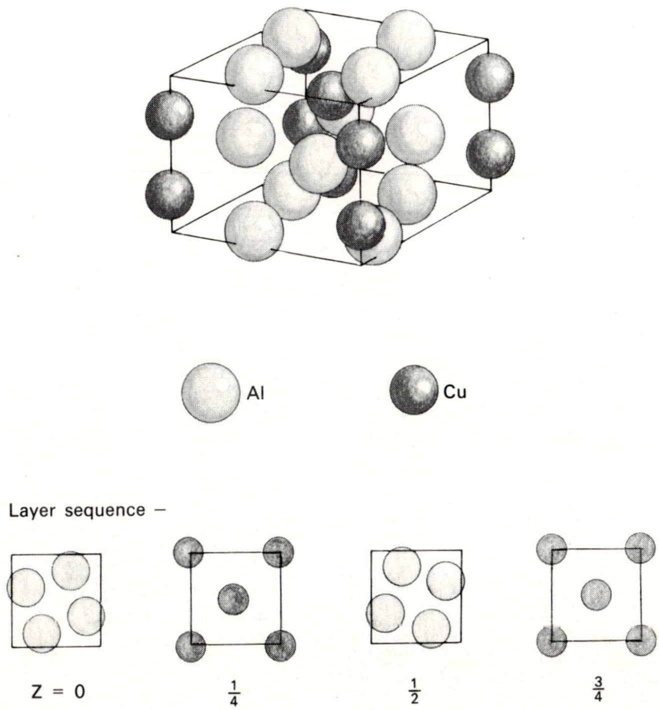

FIG. 11.3. The Al_2Cu structure.

Properties

The two borides, Fe_2B and Co_2B, are ferromagnetic with Curie points of 739° and 510°C, respectively. The melting points of Co_2B and Ta_2Si, Zr_2Si and Ni_2B are 1400°, 2460°, 2220° and 1100°C, respectively, and the phases Cr_2B and Fe_2B decompose at 1890° and 1389°C respectively.

TABLE 11.3. *Phases with the* $CuAl_2$ *Structure*

Phases	Cell size (Å)		Refs.
	a_0	c_0	
Intermetallics			
ϕ-$AgIn_2$	6.883	5.615	1
$AgTh_2$	7.56	5.84	2
θ-Al_2Cu	6.066	4.874	3
$AlTh_2$	7.616	5.861	4
$AuNa_2$	7.417	5.522	5
$AuPb_2$	7.325	5.655	6
$AuTh_2$	7.42	5.95	2
$BeTa_2$	5.99	4.85	7
$CoSc_2$	6.375	5.616	8
$CoSn_2$	6.361	5.452	9
$CuTh_2$	7.28	5.75	10
$FeGe_2$	5.911	4.951	6
$FeSn_2$	6.539	5.311	11
$MnSn_2$	6.660	5.445	12
Pb_2Pd	6.849	5.833	6
Pb_2Rh	6.664	5.865	6
$PdTh_2$	7.33	5.93	8
Sb_2Ti	6.666	5.817	13
Sb_2V	6.555	5.635	14
$SiTa_2$	6.157	5.039	15
$SiZr_2$	6.6120	5.2943	16
Th_2Zn	7.60	5.64	17
Borides			
Co_2B	5.016	4.220	18
Cr_2B	5.180	4.316	19
Fe_2B	5.109	4.249	20
Mn_2B	5.148	4.208	21
Mo_2B	5.543	4.735	22
Ni_2B	4.990	4.245	18
β-Ta_2B	5.778	4.864	23
W_2B	5.564	4.740	22

11.4. Discussion

The graphite, boron nitride and Al_2Cu structures are layer structures. The spacings between layers in the graphite and boron nitride structures

sional. Thus, the properties of these materials are different in the directions perpendicular and parallel to the layers (see Table 11.4). Thermal conductivity, electrical conductivity and tensile strength are vastly different in

TABLE 11.4. *Properties of Pyrolytic Graphite and Boron Nitride*

	Pyrographite	Boron nitride
Thermal conductivity, watts/cm^2/°C/cm	$\parallel$ c: 0.03 $\perp$c: 1.6–3.9	
Thermal expansion, /°C	$0{\cdot}66 \times 10^{-6}$	same as graphite
Electrical resistivity, ohm-cm	$\parallel$ c: 0.25 $\perp$c: 2×10^{-4}	1000°C 1500°C $\parallel$ c: 10^{10} 10^5 $\perp$c: 10^8 10^3
Tensile strength, psi	15,000–20,000	

the two directions. In graphite, the electrical conductivity in the layers is metal-like and perpendicular to these layers it is like a semiconductor. Boron nitride, on the other hand, is an insulator, but a better insulator in one direction than the other. The difference in electron configuration of C and B and N makes graphite a much better conductor than BN, but the structures give both materials their anisotropic properties.

Table 11.3 References

1. M. F. GENDRON and R. E. JONES, *J. Phys. Chem. Solids* **23**, 405 (1962).
2. J. R. MURRAY, *J. Inst. Met.* **84**, 91 (1955).
3. A. J. BRADLEY, J. H. GOLDSCHMIDT and H. LIPSON, *J. Inst. Metals* **63**, 149 (1938).
4. H. N. VANNUCHT, *Philips Res. Repts.* **16**, 1 (1961).
5. W. HAUCKE, *Z. Electrochem.* **43**, 712 (1937).
6. H. J. WALLBAUM, *Z. Metallk.* **35**, 218 (1943).
7. E. GANGLBERGER, E. LAUBE and H. NOWOTNY, *Monatsh. Chem.* **96**, 242 (1965).
8. T. H. GEBALLE, B. T. MATTHIAS, V. B. COMPTON, E. CORENZWIT, G. W. HULL and L. D. LONGINOTTI, *Phys. Rev.* **137**, A119 (1965).
9. O. NIAL, *Z. Anorg. Chem.* **238**, 287 (1938).
10. B. T. MATTHIAS, V. B. COMPTON and E. CORENZWIT, *Phys. Chem. Solids* **19**, 130 (1961).
11. F. LIHL and H. KIRNBAUER, *Monatsh. Chem.* **86**, 1031 (1955).
12. H. NOWOTNY and K. SCHUBERT, *Z. Metallk.* **37**, 17 (1946).
13. H. NOWOTNY and J. PESL, *Monatsh. Chem.* **82**, 336 (1951).
14. H. NOWOTNY, R. FUNK and J. PESL, *Monatsh. Chem.* **82**, 519 (1951).
15. H. NOWOTNY, H. SCHACHNER, R. KIEFFER and F. BENESOVSKY, *Monatsh. Chem.* **84**, 1 (1953).
16. P. PIETROKOWSKY, *Acta Cryst.* **7**, 435 (1954).
17. W. C. BAENZIGER, R. E. RUNDLE and A. I. SNOW, *Acta Cryst.* **9**, 93 (1956).
18. T. BJURSTRÖM, *Arkiv. Kemi. Min. Geol.* **11A**, 1 (1933).
19. F. BERTAUT and P. BLUM, *Compt. Rend.* **236**, 1055 (1953).
20. T. BJURSTRÖM and H. ARNFELT, *Z. Phys. Chem.* **B4**, 469 (1929).
21. R. KIESSLING, *Acta Chem. Scand.* **4**, 146 (1950).
22. R. KIESSLING, *Acta Chem. Scand.* **1**, 893 (1947).
23. R. KIESSLING, *Acta Chem. Scand.* **3**, 603 (1949).

CHAPTER 12

SUMMARY OF THE STRUCTURE OF PHASES

AT THIS point, a brief summary of the structure of various types of compounds is in order. The details can be found in the appropriate chapters.

12.1. Intermetallics, Sulfides, Selenides and Tellurides

Because of the large number of intermetallic phases, only a few of their general features will be discussed (see Chapter 2). The solid solutions can be divided into A_1A_1, A_2A_2, B_1B_1, and B_2B_2 alloys where A_1, A_2, B_1 and B_2 are the elements in different regions of the periodic table.

The A_1B_1 type of phases with the CsCl and NaTl structures are probably the closest to being ionic. When the B_1 atoms form a lattice network and the A_1 atoms are situated in the holes in the structure of ionic intermetallics, the intermetallics are called Zintl phases.

One of the largest groups of intermetallics are the phases with the $MgZn_2$ and $MgCu_2$ structure. Intermetallics of this type, whose formation depends mainly on geometric considerations, are called Laves phases.

Another large number of intermetallics are the electron phases described by Hume-Rothery. These are phases which appear at electron per atom ratios of $\frac{3}{2}$, $\frac{21}{13}$, and $\frac{7}{4}$ when B elements are added to an A element. The ones described in this book are the β-phases which have electron per atom ratios of $\frac{3}{2}$ and which order to form the CsCl structure. The best known phase of this type is β-CuZn.

Some of the most important A_2B_2 phases have the NiAs structure. The magnetic phases MnBi and MnSb are examples.

The BB type phases with the zinc blende structure are probably the most interesting ones of this group. These are the III-V and II-VI semiconductor compounds which have been so widely studied in recent years.

There are a number of A_1B_2 intermetallics such as MgS which have the sodium chloride structure. These atoms are octahedrally coordinated. Many others have the CaF_2 structure in which the coordination number of the Ca ion is eight. Compounds such as Mg_2Ge and Mg_2Pb have the anti-form of this structure. There are also a number of anti-C-rare earth and anti-A-rare earth type intermetallics like Mg_3P_2 and Mg_3Bi_2, respectively.

Of practical importance are the selenides and tellurides for thermoelectric

TABLE 12.2. *Structure of Borides*

STRUCTURES

[1] NaCl	[5] NiAs
[2] Zinc blende	[6] CaB_6
[3] CaF_2	[7] $CuAl_2$
[4] AlB_2, (H.T.)	[8] W_2B_5

Be_2B[3] Mg_3B_2 MgB_2[4]												
					Cr_2B[7] CrB CrB_2[4]	Mn_2B[7]			Co_2B[7]	Ni_2B[7]	Cu_2B_2 . . .	PB[2] AsB[2]
CaB_6[6]	ScB_2[4]		TiB TiB_2[4]	VB VB_2[4]	Cr_3B_2	MnB MnB_2[4]	Fe_2B[7] FeB		CoB CoB_2	Ni_3B_2 NiB		
SrB_6[6]	YB_6[6]		Ti_2B_8[8] ZrB[1] ZrB_2	NbB_2[4]	MoB_2[4] Mo_3B_4 Mo_2B[7]		RuB_2[4] Ru_2B_5[8]			PdB[5]	AgB_2[4]	
BaB_6[6]	LaB_6[6]		HfB HfB_2[4]	Ta_2B_4[7] TaB_2	WB WB_2 W_2B_5[8] WB_4 W_2B[7]		OsB_2[4] Os_2B_5[8]			Pt_2B	AuB_2[4]	
CeB_4 CeB_6[6]		PrB_6[6]	NdB_6[6]	SmB_6[6]	GdB_6[6]	TbB_6[6]	ErB_6[6]	TmB_6[6]	YbB_6[6]	LuB_6[6]		
ThB_2 ThB_6[6]			UB_2[4] UB_4 UB_{12}	PuB_2[4]								

applications, the β-tungsten type intermetallics for use as superconductors, and the III-V and II-IV compounds for semiconductor uses.

12.2. Borides

The borides are high melting point, highly conducting, hard materials and, except for Mg_3B_2, they are unreactive with water and acids. Their structures are presented in Table 12.2.

Beryllium boride has the fluoride structure and many of the other borides with the A_2B formula have the $CuAl_2$ structure in which there are layers of B atoms.

In the FeB, CrB and MoB structures there are zigzag chains of boron atoms. The AB_2 borides with the AlB_2 structure have layers of B atoms with layers of A atoms in between. The UB_4 and CaB_6 type structures contain a lattice framework of B atoms and B_6 units in the former and B_6 units in the latter. The boride UB_{12} has a structure in which B_{12} units form a lattice network.

12.3. Carbides

The carbides can be classified as ionic (see Table 12.3a), metallic or covalent. The elements of groups I, II and III form carbides which are ionic. These compounds are transparent and nonconducting and when hydrolyzed with water or acids yield hydrocarbons.

TABLE 12.3a. *Ionic Carbides*

	Be_2C										
Li_2C_2											
	Mg_2C_3										
	MgC_2							Al_4C_3			
Na_2C_2	Mg_2C_2							Al_2C_6			
K_2C_2	CaC_2				Cu_2C_2	ZnC_2					
Rb_2C_2	SrC_2	YC_2				CdC_2					
Cs_2C_2	BaC_2	LaC_2			Au_2C_2						
CeC_2											
Ce_2C_6	PrC_2	NdC_2	SmC_2	GdC_2	TbC_2	DyC_2	HoC_2	ErC_2	TmC_2	YbC_2	LuC_2
		UC_2									

The compounds Be_2C and Al_4C_3 on hydrolysis give methanides, Mg_2C_3 yields allylene and the rest of the carbides given above, along with ThC_2 and VC_2, yield mostly or all acetylene. The carbide Be_2C has the fluorite structure, and the AC_2 carbides have structures related to the sodium chloride structure where the C_2 units replace the chlorine ions.

The metallic carbides are shown in Table 12.3b. The AC type carbides, except for MoC and WC, have the sodium chloride structure and the A_2C carbides form a hexagonal close-packed structure.

They are hard, good conductors, high melting materials and are chemically unreactive. The other carbides of the A_3C_2 or A_3C type shown below yield hydrocarbons when reacted with water or acids. The carbide W_2C has the CdI_2 structure.

TABLE 12.3b. *Metallic Carbides*

	V_2C					
TiC	VC	Cr_3C_2	Mn_3C	Fe_3C	Co_3C	Ni_3C
		Mo_2C				
ZrC	NbC	MoC				
		W_2C				
HfC	TaC	WC				
ThC	UC	NpC	PuC			

The covalent carbides SiC and B_4C are also chemically unreactive and hard. The most common form of SiC has the zinc blende structure. Boron carbide has a structure related to the tetragonal boron structure.

12.4. Halides

A summary of the structures of the halides is presented in Table 12.4. All of the alkaii and silver halides except for AgI, CsCl, CsBr and CsI have the sodium chloride structure. These latter cesium halides and TlCl and TlBr adopt the CsCl structure. In general, the AX halides with $R_A/R_X > 0.73$ form the CsCl structure with most of the others adopting the NaCl structure. Thallium iodide and AgI have different structures.

The fluorides of the smaller cations with the general formula AX_2 form the rutile structure with a six coordination number for A while most of the larger A atoms prefer the eight-fold coordination in the fluorite structure. Other halides $CaCl_2$, $CaBr_2$ and CuF_2 crystallize in distorted rutile structures. Barium chloride, $BaBr_2$, BaI_2, $PbCl_2$, $PbBr_2$, $SmCl_2$, $EuCl_2$ and sometimes PbI_2 have the $PbCl_2$ structure in which the lead atom is surrounded with nine halide atoms. Strontium bromide, $SmBr_2$ and $EuBr_2$ crystallize in a structure related to that of $PbCl_2$. Cadmium iodide, CaI_2 and PbI_2 have the CdI_2 structure. The very small beryllium atoms, when combined with fluorine atoms, prefer the fourfold coordination in the β-cristobalite structure.

Among the trifluorides, the smaller metals Mo, Ta and Nb have a coordination number of six in the ReO_3 structure. The larger metals form fluorides which have the tysonite structure containing the cation coordinated with five nearest neighbors, and six more nearby.

A large number of these same elements form trichlorides and tribromides with the $Y(OH)_3$ structure in which the coordination number of the Y atoms is nine.

The halides also form complex structures such as the K_2PtCl_6, the perovskite and the ordered perovskite structure.

12.5. Hydrides

In classifying the hydrides (Table 12.5), the alkali metal and alkaline earth hydrides can be considered to be salt-like. They have high melting points, high boiling points, and conduct in the fused state. The alkali metal hydrides

TABLE 12.4. *Structures of Halides*

STRUCTURES

[1] NaCl	[5] CaF_2	[9] BeF_2
[2] CsCl	[6] $PbCl_2$	[10] ReO_3
[3] rutile	[7] CdI_2	[11] $Y(OH)_3$
[4] distorted rutile	[8] $SrBr_2$	[12] Tysonite structure

LiF[1]	BeF_2											
LiCl[1]												
LiBr[1]												
LiI[1]												
NaF[1]	MgF_2[3]											
NaCl[1]												
NaBr[1]												
NaI[1]												
KF[1]	CaF_2[5]					MnF_2[3]	FeF_2[3]	CoF_2[3]	NiF_2[3]	CuF_2[4]	ZnF_2[3]	
KCl[1]	$CaCl_2$[4]									CuCl		
KBr[1]	$CaBr_2$[4]									CuBr		
KI[1]	CaI_2[7]									CuI		
RbF[1]	SrF_2[5]			NbF_3[10]	MoF_3[10]					AgF[1]	CdF_2[5]	
RbCl[1]	$SrCl_2$[5]									AgCl[1]		
RbBr[1]	$SrBr_2$[8]									AgBr[1]		
RbI[1]										AgI	CdI_2[7]	
CsF[1]	BaF_2[5]			TaF_3[10]							HgF_2[5]	PbF_2[5, 6]
CsCl[2]	$BaCl_2$[6]	$LaCl_3$[11]								TlCl[2]		$PbCl_2$[6]
CsBr[2]	$BaBr_2$[6]									TlBr[2]		$PbBr_2$[6]
CsI[2]	BaI_2[6]									TlI		PbI_2[7]
CeF_3[12]	PrF_3[12]	NdF_3[12]		SmF_3[12]	EuF_3[12]							
$CeCl_3$[11]	$PrCl_3$[11]	$NdCl_3$[11]	$PmCl_3$[11]	$SmCl_3$[11]	$EuCl_3$[11]	$GdCl_3$[11]	$CeBr_3$[11]	$PrBr_3$[11]				
				$SmCl_2$[6]	EuF_2[5]							
				$SmBr_2$[8]	$EuCl_2$[6]							
					$EuBr_2$[8]							
AcF_3[12]	UF_3[12]	NpF_3[12]	PuF_3[12]	AmF_3[12]								
$AcCl_3$[11]	UCl_3[11]	$NpCl_3$[11]	$PuCl_3$[11]	$AmCl_3$[11]								
$AcBr_3$[11]	UBr_3[11]	$NpBr_3$[11]										

TABLE 12.5. *Structures of Hydrides*

[1] NaCl structure
[2] rutile structure
[3] CaF_2 structure
[4] wurtzite structure
[5] BiF_3 structure

LiH[1]	BeH_2										B_2H_6	CH_4	NH_3	H_2O	HF
NaH[1]	MgH_2										$(AlH_3)_x$	SiH_4	PH_3	H_2S	HCl
KH[1]	CaH_2[2]		TiH_2	VH	CrH_3	FeH	CoH	NiH							
				VH_3		FeH_2	CoH_2	NiH_2	CuH[4]	ZnH_2	Ga_2H_6	GeH_4	AsH_3	H_2Se	HBr
RbH[1]	SrH_2		ZrH_2	NbH_2[3]	MoH_x			Pd_2H							
								$PdH_{0\ 6}$		AgH		SnH_4	SbH_3	H_2Te	HI
		LaH_3[5]													
CsH[1]	BaH_2	LaH_2[3]	HfH_2	TaH_x	WH_x							PbH_4	BiH_3		
		CeH_2[3]	PrH_2[3]	NdH_2[3]				TbH_2[3]		DyH_2[3]		HoH_2[3]	ErH_2[3]	TmH_2[3]	LuH_2[3]
		AcH_2[3]													

have the sodium chloride structure, and MgH_2 has the rutile structure. Transition, lanthanide and actinide metals form metallic hydrides. Unlike the salt-like hydrides, these hydrides are less dense than the parent metals. A few of the more electropositive lanthanide and actinide elements form hydrides which are partially salt-like and partly metallic. Many of these dihydrides as well as NbH_2 have the calcium fluoride structure. All of these hydrides can be prepared by direct combination of the metal and hydrogen. Many of the transition elements will take up the hydrogen in various amounts and release it upon heating.

The B_2 elements in the periodic table form molecular hydrides. These are volatile, and undergo thermal decomposition.

12.6. Nitrides

The nitrides also can be divided into ionic, metallic and covalent nitrides. The ionic nitrides, Li_3N and A_3N_2 listed in Table 12.6, have high melting points and hydrolyze in water yielding hydroxides and ammonia.

TABLE 12.6. *Nitrides*

Li_3N	$\underline{Be_3N_2}$									BN
	$\underline{Mg_3N_2}$									AlN
	$\underline{Ca_3N_2}$	ScN	TiN	VN	CrN	Mn_4N	Fe_4N	Cu_3N	$\underline{Zm_3N_2}$	GaN
	Sr_3N_2	YN	ZrN	NbN	Mo_2N				$\underline{Cd_3N_2}$	InN
	Ba_3N_2	LaN	HfN	TaN	W_2N					
		ThN	$\underline{U_2N_3}$ / UN_2							

The underlined nitrides form the C-rare earth structure. The nitrides which have the sodium chloride structure are hard, inert and good conductors. These are the intermetallic nitrides. The nitride Cu_3N has the ReO_3 structure and Mn_4N and Fe_4N have a structure related to the perovskite structure.

The covalent nitride BN has a graphite-like structure and AlN, GaN, InN have the wurtzite structure.

12.7. Oxides

The structures of the binary oxides are summarized in Table 12.7. Cu_2O, Ag_2O and Pb_2O have the cuprite structure. The cations are coordinated with four oxygen ions in this structure, with four coplanar oxygen ions in the structure of PdO and PtO, and tetrahedrally in the structure of BeO (wurtzite or zinc blende) and ZnO (wurtzite). However, most of the AO oxides have the sodium chloride structure. Some of the smaller A atoms form distorted or defect variations of this structure.

Many of the transition metals and aluminum and gallium form sesquioxides with the corundum structure. These metal atoms are octahedrally coordinated with oxygen atoms. The larger trivalent metal ions combine to form oxides with the C-rare earth structure. These are Sc^{3+}, Y^{3+} and the small

TABLE 12.7. *Structure of the Oxides*

[1] Cuprite structure
[2] NaCl structure
[2*] Distorted or defect NaCl structure
[3] Corundum structure
[4] C-rare earth structure
[5] A-rare earth structure
[6] Rutile
[6*] Distorted rutile structure
[7] Fluorite or antifluorite structure
[7*] Fluorite at (H.T.)

BeO—'wurtzite', zinc blende structures
ZnO—wurtzite structure
ReO_3—ReO_3 structure
SiO_2—silica structure

$Li_2O^{[7]}$	BeO														
$Na_2O^{[7]}$	$MgO^{[2]}$												$Al_2O_3^{[3]}$	SiO_2	
$K_2O^{[7]}$	$CaO^{[2]}$	$Sc_2O_3^{[4]}$	$TiO^{[2*]}$ $Ti_2O_3^{[3]}$ $TiO_2^{[6]}$		$VO^{[2*]}$ $V_2O_3^{[3]}$ $VO_2^{[5*]}$ V_2O_5	$Cr_2O_3^{[3]}$ $CrO_2^{[6]}$	$MnO^{[2*]}$ $Mn_2O_3^{[4]}$ $MnO_2^{[6]}$ Mn_3O_4	$FeO^{[2*]}$ $Fe_2O_3^{[3]}$ Fe_3O_4	$CoO^{[2*]}$ Co_2O_3 Co_3O_4	$NiO^{[2*]}$	$Cu_2O^{[1]}$ CuO	ZnO	$Ga_2O_3^{[3]}$	$GeO_2^{[6]}$	
$Rb_2O^{[7]}$	$SrO^{[2]}$	$Y_2O_3^{[4]}$	$ZrO^{[2]}$ $ZrO_2^{[7*]}$		$NbO^{[2*]}$ $NbO_2^{[6]}$ Nb_2O_5	$MoO_2^{[6*]}$ MoO_3	$TcO_2^{[6*]}$	$RuO_2^{[6]}$	$Rh_2O_3^{[3]}$	PdO	$Ag_2O^{[1]}$ AgO	$CdO^{[2]}$	$In_2O_3^{[4]}$	$SnO_2^{[6]}$	$TeO_2^{[6]}$
	$BaO^{[2]}$	$La_2O_3^{[4]}$	$HfO_2^{[7]}$		$TaO^{[2]}$ $TaO_2^{[6]}$ Ta_2O_5	$WO_2^{[6*]}$ WO_3	$ReO_2^{[6*]}$ ReO_3	$OsO_2^{[6]}$	$IrO_2^{[6]}$	PtO		HgO	$Tl_2O_3^{[4]}$	PbO $Pb_2O^{[1]}$ $PbO_2^{[2]}$ Pb_3O_4	$PoO_2^{[7]}$
			$Ce_2O_3^{[5]}$ $CeO_2^{[7]}$		$Pr_2O_3^{[5]}$ $PrO_2^{[7*]}$	$Nd_2O_3^{[5]}$		$Sm_2O_3^{[4]}$ $SmO^{[2]}$	$EuO_2^{[4]}$		$YbO^{[2]}$				
			$ThO_2^{[7]}$	$PaO^{[2]}$	$UO_2^{[7]}$ $UO^{[2]}$	$NpO_2^{[7]}$ $NpO^{[2]}$	$PuO_2^{[7]}$ $PuO^{[2]}$	$AmO_2^{[7]}$ $AmO^{[2]}$		$CmO_2^{[7]}$					

TABLE 12.8. *Free Energy Data for Carbides, Nitrides and Oxides*(1)

I	II	III	IV	V	VI	VII	VIII			I	II	III	IV
Li_3N −37 Li_2O −134	Be_3N_2 −121 BeO −136 Mg_3C_2 −18 Mg_3N_2 −96 MgO −136											B_4C −13 BN −56 B_2O_3 −286 Al_4C_3 −38 AlN −56 Al_2O_3 −378	SiC −12 SiO_2 −197
	CaC_2 −15		TiC −43	VC −28	Cr_3C_2 −21 Cr_2C_3 −44 Cr_4C −17	Mn_3C −4	Fe_3C 5	Co_3C 9	Ni_3C 9				
	Ca_3N_2 −98	ScN −61	TiN −74	VN −35	CrN −24	Mn_3N_2 −35 Mn_5N_2 −37 Mn_4N −30	Fe_2N 3 Fe_4N 1				Zn_3N_2 −5		
K_2O −76	CaO −144	Sc_2O_3 −389	TiO_2 −212	V_2O_5 −341	CrO_3 −121	MnO_2 −111	Fe_3O_4 −244	Co_3O_4 −182		NiO −51	ZnO −76		
KO_2 −57			Ti_2O_3 −344 TiO −117	V_2O_4 −316 V_2O_3 −277 VO −91	CrO_2 −130 Cr_2O_3 −253	Mn_3O_4 −308 Mn_2O_3 −211 MnO −87	Fe_2O_3 −177 FeO −59	CoO −52					

Sr_3N_2 −77 SrO −134	Y_2O_3 −453	ZrC −43 ZrN −87 ZrO_2 −248	NbC −38 NbN −51 Nb_2O_5 −423 NbO_2 −177	Mo_2C −2 Mo_2N −12 MoO_3 −162 MoO_2 −118		Ag_2O_2 7 Ag_2O −3	CdO −54	SnO_2 −124 SnO – −61
Ba_3N_2 −73 BaO −126	LaN −65 La_2O_3 −403	HfN 81 HfO_2 −253	TaC −38 TaN −54 Ta_2O_5 −457	WC −9 W_2N −11 WO_3 −182 WO_2 −125	ReO_3 −128 Re_2O_7 −252		Hg_2O −13 HgO −14	PbO_2 −52 Pb_3O_4 −148 PbO −45
		CeO_2 −245 Ce_2O_3 −411	Nd_2O_3 −413	PrO_2 −218 Pr_2O_3 −421				
	ThC_2 −50 Th_3N_4 −282 ThO_2 −279	UC_2 −38 U_2C_3 −78 UC −44 U_2N_3 −194 UN −75						

rare earth metal ions. While the coordination number of the metal atoms is also six in this structure, it is derived from the CaF_2 structure in which the metal atoms are coordinated with eight anions. The still larger trivalent ions are coordinated with seven oxygen atoms in the A-rare earth structure.

Among the AO_2 oxides, the smaller A atoms form oxides with the rutile structure, while the larger A atoms prefer the eightfold coordination in the CaF_2 structure. Note that of the group IV metal oxides TiO_2 has the rutile structure, ZrO_2 has the CaF_2 structure at high temperatures and HfO_2 has the undistorted structure. This is expected because the ionic radii increase going down in a group in the periodic table.

The oxide ReO_3 is made up of octahedra of oxygens surrounding the Re atoms and WO_3 has a related structure.

Among the other oxides, PbO forms a layer type structure, Pb_3O_4 has a structure which have $Pb^{4+}O_6$ octahedra joined by Pb^{2+} surrounded with their oxygen ions.

In the complex oxides, some of the structures are related to the NaCl structure. The trirutile structure contains atoms of two different metallic elements such as $ZnSb_2O_6$, ordered so that unit cells of about the same size used to describe the rutile structure are needed to describe this structure. The ilmenite structure also has positions for small atoms of two elements. These atoms are ordered in the aluminum positions of the Al_2O_3 structure in which two-thirds of the octahedral holes between the oxygen atoms are filled.

In the spinel structure, small A ions occupy one-eighth of the tetrahedral holes and small B ions one-half of the octahedral holes between the oxygen ions. The importance of the interaction between the A and B ions in determining the ferrimagnetic properties has already been noted. This is true of the $BaFe_{12}O_{19}$ structure which is related to the spinel structure but contains BaO_3 layers as well.

Since the pyrochlore and scheelite structures are derived from the CaF_2 structure, it might be expected that the cations in these structures would be appropriate for eightfold coordination. However, in the $A_2B_2O_7$ type compounds an oxygen atom is missing from each CaF_2 type unit cell and in $CaWO_4$ the oxygen atoms are moved toward the tungsten atoms. Only the A atom has to be large in these structures.

In the garnet structure there are positions which are eight, six and fourfold coordinated. Thus there are large atoms like Y^{3+} and smaller atoms such as Fe^{2+} and Fe^{3+} in $Y_3Fe_5O_{12}$. These materials are also important ferromagnetic materials.

The perovskite structure has A cations in twelvefold coordination, so A ions should be large and B ions should be small since they are octahedrally coordinated. The ferroelectric properties of the perovskites have been attributed to distortions in the BO_6 octahedra. Other related structures such as

K_2NiF_4, $Sr_3Ti_2O_7$, $Bi_4Ti_3O_{12}$, tungsten bronzes and layer structures also have octahedrally coordinated B ions, while the A ion coordination varies.

12.8. Free Energies of Formation of Carbides, Nitrides and Oxides

The free energies of formation for a number of carbides, nitrides and oxides are presented in Table 12.8 so that their relative stabilities can be compared. The values are given in kilocalories at a temperature of 25°C.

Table 12.8 Reference

1. C. E. WICKS and F. E. BLOCK, Bulletin 605, Bureau of Mines (1963).

APPENDIX

TABLE A-1. *Properties of Miscellaneous Phases*

	Sym.	m.p. (°C)	Resistivity (μ ohm-cm)	Coeff. of thermal exp. (10^{-6}/°C)
Borides				
AlB_{10}	ortho.	d 2100[1]		
AlB_{12}	mono.	d 2150[2]	2×10^{12}[38]	
Be_5B		1160[3]	15×10^{3}[39]	
BeB_2	hex.	1700[3]	2×10^{4}[40]	
BeB_4		>2000[3]	18×10^{3}[39]	
Cr_4B	ortho.	d 1750[4]	176[41]	
Cr_5B_3	tetr.	d 2000[4]		
CrB	ortho.	2050[4]	69[41]	9.5[52]
Cr_3B_4	ortho.			
FeB	ortho.	1540[5]		
LaB_4	tetr.	1800[6]	24[6]	
Mo_2B_5	rhomb.	d 2100[7]	18[42]	
Nb_3B_2	tetr.	d 1950[8]		
NbB	ortho.	2280[8]	65[43]	
Ni_3B	ortho.	1155[9]		
NiB	ortho.	1020[10]		
Ta_3B_2	tetr.	2120[11]		
TaB	tetr.	2430[11]	100[44]	
Ta_3B_4	ortho.	d 2650[11]		
ThB_4	tetr.	2210[12]		
UB_4	tetr.	2495[13]		
UB_{12}	cubic	2235[14]		
VB	ortho.	2250[4]	4[45]	
V_3B_2	tetr.	2070[14]	19[46]	
V_3B_4	ortho.	d 2350[14]		
α-WB	tetr.	2400[15]		
W_2B_5	hex.	2300[5]	43[46]	
YB_2	hex.	2100[16]		
YB_4	tetr.	2800[16]	29[47]	
ZrB_{12}	cubic	2680[17]	60[48]	
Carbides				
Al_4C_3	rhomb.	d 2100[18]		
Co_3C	ortho.	d 2300[18]		
Fe_3C	ortho.	1650[15]		4.6[53]
Mo_2C	ortho.	2410[19]	71[46]	8[54]
Mn_3C	ortho.	1520[5]		7[55]
Nb_2C	ortho.			
Ni_3C	hex.	d 2100[5]		
Ta_2C	ortho.	3400[20]		
V_2C	ortho.	2165[21]		
W_2C	ortho.	2730[22]	76[46]	5.8[54]

TABLE A-1 *(cont.)*

	Sym.	m.p.(°C)	Resistivity (μ ohm-cm)	Coeff. of thermal exp. (10^{-6}/°C)
Nitrides				
Nb_2N	hex.	2420[23]	142[33]	3.3[56]
NbN	hex.	2300[24]	78[39]	10.1[57]
Ta_2N	ortho.	2050[23]	263[49]	5.2[23]
TaN	hex.	3087[25]	128[49]	3.6[23]
V_3N	hex.		123[50]	8.1[43]
Oxides				
B_2O_3		315[26]		
Bi_2O_3		824[26]		
Cs_2O		490[26]		
GeO_2		1115[26]		
MoO_3		820[27]		
Nb_2O_5		1486[26]		
P_2O_5		569[26]		
Re_2O_7		290[26]		
SeO_2		240[26]		
SiO_2		1723[26]		
Ta_2O_5		1872[26]		
V_2O_5		674[26]		
WO_3		1473[26]		
Others (Silicides)				
CaSi	ortho.	1245[28]		
Ca_2Si	ortho.	1000[28]		
CrSi	cubic	1545[29]	129[51]	11.3[58]
$CrSi_2$	hex.	1500[29]	914[51]	
Cr_3Si	cubic	1710[30]	35[51]	
Cr_5Si_3	tetr.	d 1600[30]		14.2[58]
$HfSi_2$	rhomb.	1750[31]		
$LaSi_2$	tetr.	1525[32]		
$MoSi_2$	tetr.	2030[33]	22[51]	8.3[58]
Mo_5Si_3	tetr.	2100[34]		6.7[58]
$NbSi_2$	hex.	2150[29]	50[51]	11.7[58]
Nb_4Si	hex.	d 2580[29]		
Nb_5Si_3	hex.	2400[35]		4.6[59]
$SiTa_2$	tetr.	d 2460[30]	124[51]	
Si_2Ta	hex.	2200[30]	46[51]	8.9[59]
Si_3Ta_5	hex.	2120[30]	55[51]	
SiTi	rhomb.	1920[36]	63[51]	
Si_2Ti	rhomb.	1460[35]	27[51]	
Si_2V	hex.	1660[33]	67[51]	11.2[58]
Si_2W	tetr.	2165[37]	13[51]	7.9[58]
SiZr	rhomb.	d 2150[28]		
Si_2Zr	rhomb.	1700[32]		8.6[58]

Table A-1 References

1. J. COHN, J. KORTZ and A. GIARDINI, *Z. Krist.* **111**, 53 (1948).
2. G. LONG and L. FOSTER, *J. Electrochem. Soc.* **109**, 1176 (1962).

3. G. S. Markevich, Invest. of the Be-B System, Dissertation Abstract, Leningrad (1961).
4. R. Kieffer and F. Benesovsky, *Powder Metallurgy*, **2**, 145 (1958).
5. G. V. Samsonov and Ya. S. Umasnkii, *Hard Compounds of Refractory Metals*, Metallurgizdat, Moscow (1957).
6. R. Johnson and A. Daane, *J. Phys. Chem.* **65**, 909 (1961).
7. R. Steinitz, J. Binder and D. Moskowitz, *J. Metals* **4**, 148 (1952).
8. H. Nowotny, F. Benesovsky and R. Kieffer, *Z. Metallk.* **50**, 417 (1959).
9. P. T. Kolomytsev, *Izv. Akad. Nauk SSSR, Ser. Metallurg. i Toplivo*, **3**, 83 (1960).
10. H. Giebelhausen, *Z. Anorg. Chem.* **91**, 251 (1915).
11. C. Agte and K. Moers, *Z. Anorg. Chem.* **198**, 233 (1931).
12. K. Matterson, H. Jones and N. Moore, *Pulvermet in der Atomkerntechnik*, 4, Plansee-seminar Springer-Verlag, 279 (1962).
13. B. Howlett, *J. Inst. Metals* **88**, 467 (1959-60).
14. H. Nowotny, F. Benesovsky and R. Kieffer, *Z. Metallk.* **50**, 258 (1959).
15. P. Schwarzkopf and F. Glaser, *Z. Metallk.* **44**, 353(1953).
16. C. Lundin, *Rare Earth Metal Phase Diagrams*, Chicago (1959).
17. F. Glaser and B. Post, *J. Metals* **5**, 1119 (1953).
18. M. P. Slavinskii, *Physico-Chemical Properties of the Elements*, Metallurizdat, Moscow (1952).
19. M. Nadler and C. Kempter, *J. Phys. Chem.* **64**, 1471 (1960).
20. F. Ellinger, *Trans. ASM* **31**, 89 (1943).
21. E. Storms and R. McNeal, *J. Phys. Chem.* **66**, 1401 (1962).
22. R. Kieffer and P. Schwarzkopf, *Hard Alloys*, Metallurgizdat, Moscow (1957).
23. T. S. Verkhoglyadova, Dissertation Abstract, Kiev Polytech. Inst., Kiev (1962).
24. H. Giebelhausen, *Z. Anorg. Chem.* **91**, 251 (1915).
25. T. Vasilos and W. Kingery, *J. Am. Ceram. Soc.* **37**, 409 (1954).
26. S. J. Schneider, NBS Monograph 68 (Oct. 1963).
27. L. M. Vitney, *Vestn. Mosk. Univ. Ser. II, Khim.* **21**, 60 (1966).
28. G. V. Smirnov and B. F. Ormont, *Doklady Akad. Nauk SSSR* **96**, 557 (1954).
29. G. V. Samsonov, *Silicides and their Application in Technology*, Izd. An. Ukr. SSR, Kiev (1959).
30. R. Kieffer, F. Benesovsky, H. Nowotny and H. Schacher, *Z. Metallk.* **44**, 242 (1953).
31. G. V. Samsonov and K. I. Portnoi, *Alloys of the Basis of Refractory Compounds*, Oborongiz, Moscow (1961).
32. I. Binder, *J. Am. Ceram. Soc.* **43**, 287 (1960).
33. *Welding Engr.* No. 4 (1958).
34. R. Kieffer and E. Cerivenka, *Z. Metallk.* **43**, 101 (1952).
35. R. Kieffer, F. Benesovsky and H. Schmid, *Monatsh. Chem.* **86**, 413 (1955).
36. M. Hansen, H. Kessler and D. McPherson, ASM Reprint No. 4 (1951).
37. R. Kieffer, F. Benesovsky and E. Galistl, *Z. Metallk.* **43**, 284 (1952).
38. C. Dauben, *U.S. Atomic Energy Comm. Publ.*, USRL-**2888**, 30 (1955).
39. G. S. Markevich, Yu. D. Kondrashev and L. Ya. Markovskii, *Zhur. Neorg. Khim.* **5**, 1783 (1960).
40. L. Y. Markovskii and G. S. Markevich, *Zhur. Neorg. Khim.* **33**, 1667 (1960).
41. P. S. Koslyi, S. N. L'vov, V. F. Nemchenko and G. V. Samsonov, *Izv. Akad. Nauk SSSR, Otdel Tekh. Nauk, Ser. Metalluguzai Toplivo*, **6** (1962).
42. H. Juretschke and R. Steinitz, *Phys. Rev. Solids* **4**, 118 (1958).
43. G. V. Samsonov, *Zhur. Tekh. Fiz.* **26**, 716 (1956).
44. F. Glaser, *J. Metals* **4**, 391 (1952).
45. K. Molers, *Z. Anorg. Chem.* **198**, 243 (1931).
46. S. N. L'vov, V. F. Nemchenko, G. V. Samsonov, *Dokl. Akad. Nauk SSSR* **135**, 577 (1960).
47. R. Johnson and A. Daane, *J. Chem. Phys.* **38**, 425 (1963).
48. B. Bost and F. Glaser, *J. Metals* **4**, 631 (1952).
49. O. Kubaschewski and E. Evans, *Metallurgical Thermochemistry*, London (1958).
50. G. V. Samsonov and T. S. Verkhozlyadova, *Dokl. Akad. Nauk SSSR*, **142**, 612 (1962).
51. V. S. Neshpor and G. V. Samsonov, *Fiz. Tverd. Tela.* **2**, 2101 (1960).

52. V. S. Neshpor and P. S. Kislyi, *Ogneupori* **23**, 231 (1959).
53. A. M. Belikov and A. A. Savinskaya, *Fiz. Metal i Metalloved* **14**, 299 (1962).
54. A. M. Belikov and Ya. S. Umanskii, *Kristallografiya* **4**, 684 (1959).
55. P. Elliott and C. Kempter, *J. Phys. Chem.* **62**, 630 (1958).
56. G. V. Samsonov and T. S. Verkhoglyadova, *Zhur. Neorg. Khim.* **5**, 1231 (1961).
57. I. Gangler, *J. Am. Ceram. Soc.* **33**, 367 (1950).
58. V. S. Neshpor and M. I. Reznichenko, *Fiz. Metal i Metalloved* **15**, 631 (1963).
59. H. Nowotny and E. Laube, *Planseeber Pulvermetal* **9**, 54 (1961).

INDEX

In this index are listed the sections in which structures, properties of phases, and crystallographic data for the phases are found (A-1 indicates the Appendix).

Structures

Properties

Densities

Free Energies of Formation

Ionic Radii

Magnetic Properties

Mechanical Properties
(Hardness, Modulus and Ultimate Tensile Strength Values)

Melting Points

Optical Properties, Lasers

Superconducting Properties

Thermal Conductivities

Thermal Expansion Coefficients

Thermoelectric Properties

Crystallography